Penguin Education

The Economics of Technological Change

Edited by Nathan Rosenberg

Penguin Modern Economics Readings

General Editor

B. J. McCormick

Advisory Board

K. J. W. Alexander
R. W. Clower
G. Fisher
P. Robson
J. Spraos
H. Townsend

The Economics of Technological Change

Selected Readings

Edited by Nathan Rosenberg

Penguin Books

Penguin Books Ltd, Harmondsworth,
Middlesex, England
Penguin Books Inc., 7110 Ambassador Road,
Baltimore, Md 21207, U.S.A.
Penguin Books Australia Ltd,
Ringwood, Victoria, Australia

First published 1971
This selection copyright © Penguin Books Ltd, 1971
Introduction and notes copyright © Penguin Books Ltd, 1971

Made and printed in Great Britain by
Richard Clay (The Chaucer Press) Ltd
Bungay, Suffolk
Set in Monotype Times

Contents

Introduction

Technological change has occupied a curious sort of underworld in economics for a long time. In the work of the classical economists – Malthus, Ricardo, the Mills – it was not totally ignored. It played, however, the role of a kind of afterthought which modified somewhat the dimensions of an analysis which was undertaken without it, and in which other variables – population growth, capital formation, diminishing returns in agriculture – were regarded as more important. In the middle of the nineteenth century Marx focused attention forcefully upon technological change as the prime mover in capitalist development, but his work had little immediate impact upon the central tradition in economics. In the last decades of the nineteenth century, attention was focused upon the principles of optimal resource allocation, usually within a static framework from which technological change had been deliberately excluded.

Early in the twentieth century Joseph Schumpeter began an influential academic career which played an important role in bringing technological change to the attention of economists. Yet, although technological change played a central role in Schumpeter's analysis of capitalism – both in accounting for its short-run instability and for its dynamic long-run behaviour – Schumpeter in a sense kept the phenomenon at arm's length. That is to say, he treated inventive activity as an essentially exogenous force, which had important economic consequences, but no primary economic causes or antecedents. Although Schumpeter explained the timing of innovation in economic terms, he had relatively little to say about economic factors shaping inventive activity, which seemed to live a life of its own.

By the middle of the 1950s evidence had accumulated which strongly suggested two things: that technological change is a major – many economists would argue *the* major – determinant of the economic growth of rapidly growing economies, and that the forces shaping technological change are, at least to a very large extent, economic, and therefore, far from being an exogenous variable, it can be examined and understood directly in terms of economic analysis. Although this is rather flattering to the economist, who finds that the analytical tools which constitute his stock-in-trade are indispensable to the understanding of such an important collection of problems, it is hardly cause for excessive self-congratulation. Seen in historical perspective, it took an unconscionably long time for economists to turn their analytical techniques upon this range of problems. Nevertheless, the articles reprinted in this volume convincingly demonstrate, among other things, that economic reasoning provides a powerful apparatus for understanding both the causes and the consequences of technological change.

The perspective offered here is additionally valuable because the public has in the recent past been treated – if that is the right word – to a range of extreme views on the subject. At one extreme is a train of thought, going back to the Point Four of President Truman's 1949 inaugural address, suggesting that there exist some small number of 'technological fixes' with which the poverty and misery afflicting so much of the human race can be easily overcome. At the other extreme are those who argue that technological change is the carrier of forces which are, inevitably, destructive of the most highly cherished human values. In this volume we will not be concerned directly with either of these perspectives: with technology-as-panacea or technology-as-Armageddon. Rather, the focus will be upon technology as a complex phenomenon which can be studied empirically, and whose causes and effects are amenable to treatment with the analytical tools of the economist. It will be readily conceded that this approach does not touch, much less exhaust, all that is of interest in this vast subject. On the other hand there is little question that someone who has not acquired some minimum competence in this approach simply does not possess the intellectual 'table-stakes' for a serious examination of the problems – and the promise – posed by technological change.

Part One
The Nature and the Process of Technological Change

The Readings reprinted in this first section offer some
historical and analytical perspectives on the phenomenon of
technological change. Schumpeter, who has exerted a
powerful influence in shaping our way of thinking about the
subject, strongly emphasized the discontinuous and
disruptive nature of technological change. Indeed, Schumpeter's
whole conception of the historical performance of capitalism
centres upon the manner in which new technologies are
introduced and their subsequent short-run destabilizing and
disequilibrating effects upon the system into which they are
introduced. Since the long-run growth in productivity has been
essentially the result of the introduction of new production
functions, Schumpeter sees capitalism's short-run instability
and long-run growth as inseparably interrelated.

Usher, one of the most acute and discerning
twentieth-century scholars of the history of technology, places
great stress upon the elements of continuity and the
cumulative significance, to the inventive process, of large
numbers of changes, each one of small magnitude.
Furthermore, Usher was too much of an historian to be
willing to treat an invention as some sort of *deus ex machina*.
In his general concern with the emergence of novelty in history
he paid careful attention to the specific factors which
conditioned or *set the stage for* the successful inventor.

The paper by Ruttan is a valuable attempt to achieve clarity
among the three related but very distinct concepts of
'invention', 'innovation', and 'technological change'. In so
doing he also manages deftly to suggest how Usher's theory
may be used to complement Schumpeter's where the latter's
theory is weak and perhaps defective.

Finally, in spite of the fact that technological change involving significant alterations in the final product remains highly intractable from the viewpoint of theoretical analysis, Blaug's paper convincingly demonstrates that conventional economics can shed a flood of light upon cost-reducing process innovations. Blaug presents a cogent, critical survey of a large body of literature and argues that the neo-classical conception of technological change still provides the best available framework for organizing and arranging our knowledge of this subject.

1 J. Schumpeter

The Instability of Capitalism

J. Schumpeter, 'The instability of capitalism', *Economic Journal*, 1928, pp. 361–86.

Economic stability under static conditions

The many 'instabilities' created by the War `and by post-war vicissitudes, whilst very properly engaging the attention of economists in all countries both as to diagnosis and as to remedial policy, do not, in themselves, present to science any new or startling problems. There is nothing strange in the fact that events such as the breakdown of Russia or, generally, disturbances arising from without the sphere of economic life, should affect its structure, its data and its working. In this paper I shall disregard them entirely, and deal merely with the question whether or not the capitalistic system is stable in itself – that is to say, whether or not it would, in the absence of such disturbances, show any tendency towards self-destruction from inherent economic causes, or towards out-growing its own frame. The interest of such an investigation is primarily scientific; still, an answer to that question is not without some diagnostic value, and, therefore, not without some, if remote, bearing upon policy; especially as there is, it seems to me, a marked tendency to reason upon post-war figures and about post-war problems, exactly as if they reflected something like the normal working of our economic system, and to proceed, on this basis, to conclusions about the system as such.

By way of clearing the ground, it may be well, firstly, to distinguish the kind of stability or instability we propose to discuss, from other phenomena covered by the same terms. Looking, for instance, at France, with her stationary population and enterprise and her vast colonial empire, and at the opposite state of things in Italy, the observer may well have an impression of instability – let us call it 'political' instability – which, however, has nothing to do with economic instability in our sense; for in the economic systems of these countries there might still be perfect stability. Or if we assume a state of things in which the whole of the industry of a country is monopolized by one single firm, we should probably agree in calling such a system unstable in a very obvious sense – let us label the case as one of 'social instability' – whilst it could be highly stable economically. Instability in still another sense would obtain in a system, for which equilibrium wages were at a point below what workers will put up with – although there need not be any tendency in the economic conditions themselves to produce any change at all *by the mere working of the system*. Finally, special cases of instability may arise from particular influences from without, which cannot properly be charged to the economic system at all. England's return to the gold standard is a case in point. 'Stabilizing' the pound at what was, viewed from the standpoint of existing conditions, an artificial value, naturally meant dislocating business, putting a premium on imports and a tax on exports, intensifying losses and unemployment, thereby creating a situation eminently unstable. But this instability is evidently due to the act of politicians, and not to the working of the system which, on the contrary, would have evolved a value of the pound exactly fitting the circumstances. In short, the economic stability we mean, although it *contributes* to stability in other senses, is not *synonymous* with them, nor does it *imply* them. This view must, of course, seem highly superficial to anyone who assumes the existence of as close a relation between the economic and other spheres of social life as, for instance, Marx did. As, however, it would be waste of time to prove to English readers the necessity of separating these several spheres, I may confine myself to these remarks.

Secondly, we have to define what we mean by 'our economic system': we mean an economic system characterized by private property (private initiative), by production for a market and by the phenomenon of credit, this phenomenon being the *differentia specifica* distinguishing the 'capitalist' system from other species, historical or possible, of the larger genus defined by the two first characteristics. Although few things seem to me to be more firmly established by historical research than the fact that economic history cannot be divided into epochs corresponding to different systems, it is still permissible to date the *prevalence* of capitalistic methods from about the middle of the eighteenth century (for England), and to call the nineteenth century κατ' ἐξοχήν the time of *competitive*, and what has so far followed, the time of increasingly '*trustified*', or otherwise 'organized', 'regulated', or 'managed', capitalism.

Thirdly, capitalism may be stable or not, simply in the sense that it may be expected to last or not. Its history might be full of the most violent fluctuations or even catastrophes – as it undoubtedly has been so far – and these fluctuations or catastrophes might even be inherent in its working – which precisely is what we want to form an opinion about – and we might still, in a real sense, have to call it 'stable' if we have reason to expect it to last. Whenever we mean no more than this – that is to say, when we merely mean to speak of the question of what may be termed the institutional survival of capitalism, we will henceforth speak of the capitalist *order* instead of the capitalist *system*. When speaking of the stability or instability of the capitalist *system*, we shall mean something akin to what businessmen call stability or instability of business conditions. Of course, mere instability of the 'system' would, if severe enough, threaten the stability of the 'order', or the 'system' may have an inherent tendency to destroy the 'order' by undermining the social positions on which the 'order' rests.

The businessman's meaning of stability we have now to translate into the language of theory. It will shorten matters and facilitate exposition if I state at the outset that, barring differences on a number of particular points, the following remarks run entirely on Marshallian lines. But I could equally well call them Walrasian lines. For within serious economic theory there are no such things as 'schools' or differences of principle, and the only fundamental cleavage in modern economics is between good work and bad. The basic lines are the same in all lands and in all hands: there are differences in exposition, in the manner – and mannerism – of putting things, for example, according to the relative importance different authors attach, respectively, to rigour and generality or to vicinity to 'real life'. Then there are differences in technique, the very greatness of Menger, Böhm-Bawerk and Wieser, for example, consisting in their having achieved so much with such shockingly clumsy and primitive tools, the use of which was an insurmountable bar to correctness. There are, furthermore, differences in individual pieces of the analytic machine – as, for example, between the Walrasian and the Marshallian demand curves, or between the role assigned to coefficients of production respectively by Marshall and Walras, Pareto and Barone. Finally, there are differences as to particular problems, the most important of which are the theories of interest and of the business cycle. But this is all. There is no difference in fundamentals – Clark's productivity or Walras' equilibrium or the Austrian imputation or Marshall's substitution or Wicksell's compound of Walras and Böhm-Bawerk being all of them in the last analysis the same thing, and all, in spite of appearances to the contrary, equally far removed from, and at the same time and in the same sense descendants of, Ricardo's patchwork.

The economic system in the sense of conditions and processes reduces itself for the purposes of Theory to a system in the scientific sense of the word – a system, that is, of interdependent quantities – variables and parameters – consisting of quantities of commodities, rates of commodities and prices, mutually determining each other. This system has been found to be stable, and its stability to be amenable to rational proof, under static conditions. Not as stable, it is true, as economists would have held sixty years ago, when most of them – nearly all, in fact, except the Marxists – would have most confidently asserted absolute stability both of the capitalist *order* and the capitalist *system*: stability has fared very much as the theory of maximum satisfaction did. Just as newer methods, whilst yielding correct proof of what they left of the competitive maximum, have considerably taken away from its importance, so similarly, whilst showing that we have, generally, as many equations as we have 'unknown' quantities, and therefore a determined state of equilibrium corresponding to a given set of certain data which turns out to be stable under appropriate conditions, they have also shown that the exceptions to this general 'determinateness' are considerable. Even apart from cases such as the possibility of the offer curve of labour[1] curling back or such as the case of the

1. This, of course, does not make equilibrium entirely indeterminate, but only makes the system have several, mostly two, different solutions.

value of money in a system of bimetallism without legal ratio,[2] we have many instances where equilibrium cannot be said to be determinate. The case where both supply and demand are inelastic, is an example.[3] It may be said, for example, that the home demand for wheat in the United States is highly inelastic within a considerable interval of price. Supply, again, though very variable, is equally inelastic – if it be permitted to apply this term to supply for shortness sake – within intervals of time too short to allow for extension or contraction of acreage; and this may, perhaps, partially explain the instability of American farming.

But although illustrations of this and other cases abound, the determinateness of static equilibrium under competitive conditions is yet a broad basic fact, and this equilibrium is stable, provided that supply price[4] – the price of 'willingness to sell' – is an increasing function of quantity of product. This condition rests on the fundamental fact that the extending of production by

2. It is worth while emphasizing, however, that there is no indeterminateness when two or more commodities circulate as money and every transaction is concluded specifically in one of them. The instability only arises if contracts are in terms of 'money' generally, so that payment can be made in any of those commodities.

3. Another has been pointed out by Wicksell, *Geldwert und Güterpreise:* if coefficients of production be constant and if there be no alternative use for the factors of production – their quantity being, moreover, fixed – then there would be indeterminateness of their shares in the product. Still others have been discussed by Taussig (1921), and Divisia (1928), p. 410. This case of indeterminateness arises only from the absence of any true marginal utility of money. It has been pointed out before by Professor Cassel, and is, of course, easily remedied.

4. The supply price schedule meant here is the series of supply prices at which, given the methods of production actually in use and embodied in given plants and under given general conditions and trade practices, the respective quantities of product would be forthcoming. The schedule, therefore, refers, in an obvious sense, to a point of time. It does not, however, take account of chance occurrences, such as momentary market situations on the one hand; and it does not, on the other hand, take account of any but marginal adjustments, *capable of being decomposed into infinitesimal steps*: so it might be called a short period, normal. But the objections to this would be the implication of the existence of some long-period normal and, besides, the emphasis which this manner of expression lays on the element of time, whilst the important thing is not the lapse of time as such, but what happens during it.

any given industry means withdrawing quantities of factors of production from increasingly 'important' other uses, which, of course, does not show within single firms – any more than the influence on demand price of increasing output shows within the field of action of single firms in a state of pure competition – but is yet the force the balancing of which against decreasing marginal utilities of product determines the distribution of resources between industries. There is, it is true, an interval for practically every industry in which this condition is not satisfied, owing to the tendency which it embodies being overcompensated by fixed costs distributing themselves over an increasing number of units of product. As long as this is the case, there cannot be a point of stable equilibrium.[5] But the effect of this spends itself necessarily and, therefore, stable equilibrium will nevertheless eventually emerge, although there may, and often will, be a prior instability – instability of the kind which is one of the sources of what is called 'overproduction'.

Any other cause of 'increasing cost' is excluded by the static hypothesis, the justification for accepting such an arrangement being that it separates clearly different sets of phenomena, which stand in need of different treatment. Innovations in productive and commercial methods, in the widest sense of the term – including specialization and the introduction of production on a scale different from the one which ruled before – obviously alter the *data* of the static system and constitute, whether or not they have to do with 'invention', another body of facts and problems. And so does that part of 'external economies', which is represented by such instances as the trade journal, the bureau of standards, the 'pooling' of reserve stocks of materials incident to the presence of a large market in them and so on. The reader

5. Not even if, in the familiar illustration, the demand curve cut the supply curve negatively. For even then it must be to the interest of every single producer, who *ex hypothesi* neglects the influence of his own action on price, to go on producing in this case. Whilst this lasts, there is *movement* towards equilibrium (and this distinguishes *this* case of 'increasing returns' fundamentally from others), but not equilibrium itself. Whilst other cases of the compound called 'increasing returns' *vires acquirunt eundo,* and thereby may lead up to a monopoly, this one can hardly do so. It may offer, however, instances of increasing cost for an industry as a whole in the face of the presence of decreasing unit cost in every single firm.

is asked to stay judgement about the exclusion of these things until later. Here it is only necessary to point out that we should have to emphasize the heterogeneous nature of all these phenomena the very moment we included them. In any case we should have to recognize that there is no 'law of decreasing cost' of the same kind as, and symmetrical to, the law of increasing cost.[6] The relation of the two can, perhaps, be best seen by means of the analogy with the 'demand side' of the problem. Empirically we evidently could arrive in very many cases at demand curves which would slope upwards instead of down (see, for example, Professor Moore's demand curve for pig iron). And there are, of course, very many similar cases, the special point of interest about the pig-iron curve being the fact that its periodicity is indicative of the business cycle. Nobody, however, thinks less on that account of what is universally considered to be the 'true' slope of the theoretic demand curve. Everybody, on the contrary, recognizes that what happens in such cases is a shifting – by which term we mean to cover inexactly not only displacement but also distortion – of the theoretic curves, every one of which retains its fundamental characteristic in obedience to the 'law' it

6. By law of increasing cost we may mean four things entirely independent of one another: firstly, we may, as above, mean what is of the very essence of the economic process and, indeed, only another way of stating the law of satiable wants, that the significance of successive doses of means of production must always increase as they are drawn into any one industry for the reason that they are actually or virtually taken away from others. Secondly, we may, as pointed out before, mean that successive doses of any one factor of production applied to a constant quantity of the others yield a decreasing physical increment of product, everything, especially method, remaining the same. The most 'practical' way of making use of this proposition is to consider a given plant, embodying both a given method of production and an inelastic set of supplementary costs, and to vary elements of prime cost one at a time. This is perhaps the best tool we have to deal with the routine work of the management of a single firm. It has, however, nothing whatever to do, thirdly, with a community being driven in the process of expansion of production to exploit less and less fertile productive opportunities. This has been well stated in Professor Sraffa's (1925) acute study, and commented on by Professor Pigou in the issue for June 1927. And, fourthly, there is the prophecy to which Ricardo owes the epithet of pessimist, that improvements (in agriculture) of productive methods will in the long run fail to counterbalance increasing costs in the second and third sense, in case population should keep on increasing.

has been constructed to represent, and that any curve displaying a positive slope is merely a statistical[7] or historical curve fitted through a family of successive theoretic ones. The same applies to – if I be permitted to waive for the sake of shortness the objections to speaking of so doubtful a thing – supply curves. There is only one theoretic supply curve; and it slopes upwards in all cases. Changes of data do not make it slope down, but shift it, or, more correctly, break it off[8] and start a new one. And through these changing positions – in all of which these curves retain their slope and meaning – we may, if we so choose, fit historical curves, which will certainly often slope down. They will, in fact, display no regularity at all. It may not even be quite easy, in some cases, to guard against the supreme misfortune of total cost being actually smaller for a greater output than for a lesser one, for changes of *data*, once admitted, would sometimes produce this result, which could not, in competitive circumstances, be handled

7. The theoretic curve can, of course, be determined statistically without ceasing to be a theoretic curve, the above distinction not turning on the fact, or possibility, of statistical determination, but on whether or not the curve expresses or illustrates a *theorem*, thereby acquiring logical unity as distinguished from what could be termed 'descriptional' unity. Now I am far from overrating the importance of this distinction: on the one hand, theory itself is only a way of describing facts; on the other hand, any descriptional unity may, by some progress of analysis, turn into a logica unity any moment – in fact, the frontier between the two continually shifts in the progress of science. But this is no reason for simply ignoring it and for co-ordinating things, which do not stand on the same plane.

8. This links up with another distinction, the importance of which is best seen by means of an example: Böhm-Bawerk's theory of interest stresses the importance of the 'roundabout' process of production. But it is not the *running* of production of a given degree of roundaboutness which matters, but *the act of introducing* greater 'roundaboutness'. There is a drop – in its nature discontinuous, irregular, 'unpredictable' and 'historically' unique – in costs the moment production starts on the new plan (on *any* successful new plan, no matter whether it involves roundaboutness or not), but there is no further and continuous saving of costs per unit of product in the running of it. Generalizing: changes of *data* may be represented by lines connecting the displaced and distorted theoretic curves. If they are small and frequent, these lines may themselves *look* like our curves. But they never *are* theoretic curves and have not, in this sense, any theoretic meaning.

by assuming that the larger quantity would be produced but partially destroyed.[9]

There is nothing new or startling in thus limiting the scope of this part of our analytic engine. In fact, we are doing no more than to sum up what has been an unmistakable doctrinal tendency ever since it came to be recognized, firstly, that increasing cost in the sense of decreasing physical response to productive effort applied to a constant quantity of one of the factors is no peculiarity of agriculture, but a general phenomenon – a phenomenon which, given the same conditions, applies to all kinds of production and, given other conditions, does not apply even to agriculture; secondly, that there is a more fundamental tendency at work to make the second derivative of total cost with respect to output positive, and one which has nothing to do with the physical 'law of decreasing returns', whence the difficulty of filling certain empty boxes. We are merely clinching, on the one hand, what seems to us to be the true real-cost-phenomenon, and, on the other hand, what seems to us to be both the meaning of economic 'statics' and the nature of static equilibrium. That this is perfectly in keeping with the fundamental drift of Marshallian analysis, I will try to show in a footnote.[10]

9. See Schultz (1927), p. 441. Therefore the assumption $dy/dx > 0$ remains arbitrary, unless reinforced by Cunynghame's criterion $dy/dx > y/x$.

10. Marshall, indeed, repeatedly protests against the limitations of the static apparatus (see especially a letter of his to Professor John B. Clark). Now if it were true that reasoning by means of it is 'too far removed from life to be useful', then the greater part of the analysis of the *Principles* would be useless – as would be the greater part of any exact science: for Marshallian analysis rests just as much on static assumptions as Professor Clark's structure. But it is not true. There is nothing unduly abstract in considering the phenomena incident to the running of economic life under given conditions taken by themselves. On the contrary, it means giving this class of problems the treatment they require. And Marshall himself has contributed substantially to the perfection of this treatment by forging such invaluable tools as his consumer's surplus and his quasi-rent. He has, furthermore, made use of static assumptions both in his theory of distribution and in the fundamentals of his catallactics; in fact, in one decisive point, when dealing with refinements calling for rigour of analysis, he has confined his argument to increasing cost. And he has, finally, himself insisted on the irreversibility of, and on the difficulties peculiar to, a declining supply curve, and come, in doing so, very near to saying much

There seem to be, however, two other sources of instability due to indeterminateness within the precincts of the 'static' system. By universal consent, single monopoly yields determined and stable equilibrium, but dual and multiple monopoly, or, generally, the case in which firms can and do take account of their own influence on price, is held, by very high authorities, to fail to do so. Cournot's treatment and the objections raised against it, first by Bertrand and then by Edgeworth, are well known. As this case is not only more important practically than either of the cases of 'free, pure or simple' competition on the one hand, and of single monopoly on the other, but also the more general one in a theoretic sense – for the competitive hypothesis is, after all, an additional condition and very much in the nature of a crutch – the breach in our wall seemed a rather serious one. To clear up

---

the same as what has been said above. Loyalty to tradition, aversion to appearing too 'theoretical' – which carried so much weight with him – and that tendency of his, to which we owe so much in other respects, to take short cuts to the problems of practical life, may account for his not taking the final step and for what I cannot but agree with Mr Keynes in considering the least satisfactory part of his analysis, successfully assailed by Professor Sraffa. This entailed a string of consequences, but fundamentally what we have said is but a development of a trend overlaid indeed by other things, but yet present in the *Principles*.

We may add the weight of Professor Pigou's authority. For in the article quoted in a previous note, he excludes, for the sake of 'logical coherence' of the cost function, the bulk of those phenomena which we ourselves propose to exclude for the same reason. He, indeed, even rules out what we have called the fundamental law of cost $\{\varphi''(x) > 0\}$. But this he does merely on the technical ground that it is 'impossible to construct a cost function' in the event of changes in the relative values of factors of production being liable to occur in consequence of changes in the scale of production of an industry. On the other hand, he does not entirely rule out external economies. But what he retains of them are merely 'variations in aggregate costs associated with, and due to, variations in the scale of output' (Pigou, 1927, p. 189); and if we insert, as we must, the word 'automatically' in this sentence, very few, if any, cases will be found to answer the criterion, as has been pointed out by Professor Young (1913), p. 678. Of course, expansion and improvement are closely allied in real life. But, as we shall try to explain in the text, the main causation is the one from improvement to expansion and cannot adequately be dealt with by static analysis at all. If this be correct, Professor Pigou's position will be seen to approach closely the one taken up in the text, if the reader take hold of the fact that econ-

the matter has been one of the last of the many services Knut
Wicksell has rendered to science.[11]

The simplest form of the second case of what I call 'correspec-
tive prices' is presented by exchange between two monopolists.
It is again Professor Edgeworth's authority which accounts for
well-nigh universal acceptance of the view – first expressed by
him in his *Mathematical Psychics* – that there is indeterminateness of price within an interval (on the contract curve) which must
in general be considerable. He even went so far as to describe the
state of things in a trustified economic world as a 'chaos'. Here,
therefore, is a rich source of instability opened up. Naturally, any
theorist might well be tempted to link up what instabilities he
sees with this possible explanation of them. Nor can we reply by
pointing to the fact that prices fixed by trusts display in many
and important instances much less fluctuation than could be expected under competitive conditions; for non-economic forces,
pressure of public opinion or fear of government action, for
instance, might account for that. And the authority of Professor
Edgeworth has been reinforced by the not less weighty authority
of Professor Pigou.

Now it is perfectly true that there is, in this case, just as in the

---

omies, before becoming 'external', must generally be internal ones in
some firm or firms of the same *or some other* industry.

I do not mean, furthermore, to raise by what I have said objections to the
attempts to determine cost functions statistically. On the contrary, I am a
humble admirer of the pioneer work done by Professor H. L. Moore and
his followers, even though I beg leave to point out that to speak of 'moving
equilibria' may prove misleading, in the face of the fact that what really
happens is *destruction* of equilibria in the received meaning of this term.

11. It is with reluctance that I contradict the great shade of Edgeworth.
But there seems to be no warrant to assume indeterminateness in the case
of what Professor Pigou calls monopolistic competition. Taking into consideration the limiting instance only, that of duopoly, which can be easily
generalized, and assuming both competitors to be in exactly the same
position, we are, firstly, faced by the fact that they cannot very well fail to
realize their situation. But then it follows that they will hit upon, and
adhere to, the price which maximizes monopoly revenue for both taken
together (as, whatever the price is, they would, in the absence of any preference of consumers for either of them, have to share equally what monopoly
revenue there is). The case will not differ from the case of conscious combination – in principle – and be just as determinate. The only other alternative
which presents itself in the absence of any hope of driving the competitor
out of the market, is best 'visualized' by starting from one monopolist con-

case of one-sided monopoly, much less *guarantee* of a tendency towards equilibrium prices actually asserting itself. We have much less reason to expect that monopolists will, in either case, charge an equilibrium price, than we have in the case of perfect competition; for competing producers *must* charge it as a rule under penalty of economic death, whilst monopolists, although having a *motive* to charge the monopolistic equilibrium price, are not forced to do so, but may be prevented from doing so by other motives. Furthermore, it is quite true also that such things as bluffing, the use of non-economic force, a will to force the other party to their knees, have much more scope in the case of two-sided monopoly – just as cut-throat methods have in the case of limited competition – than in a state of perfect competition.

But there is yet more than academic interest in stating that our theory does not break down at this point. Equilibrium is determinate even in this case – even if we take so extreme an instance as a trade union comprising all the workmen of a country, quite sure of the allegiance of its members, capable of preventing immigration from abroad or from other strata of society, and an employers' union similarly constructed. If we assume that each

---

trolling the market and then introducing a second one (Cournot's procedure). It is perhaps more 'realistic' to assume that the first monopolist will not, as would be to his ultimate advantage, readily surrender half of his market to the newcomer, but that the latter will have to force his way in. And this case is equally determinate, as has been shown by Wicksell in his review article on Professor Bowley's (1925) 'Groundwork'. Taking, as the unit of the price p, that price at which the output would be zero, and, similarly, as the unit of the quantity sold x, that quantity which could be disposed of at the price zero (Edgeworth), we have $p = 1 - x$. A single monopolist would, if there are no costs, maximize px and charge a price of $1/2$, selling $1/2$. The second man, having to face this situation, will obviously maximize *his* output, x, multiplied by price – that is, $x_2 p = x_2(1/2 - x_2)$, and, therefore, sell $1/4$. Whereupon the first will have to readjust *his* output, x_1, and to offer $3/8$ and so on. This finally leads to a limit at the price of $1/3$, when each of the two sell $1/3$, the price being higher and the quantity sold smaller than under competition. There is nothing absurd in this. It cannot be objected that neither of the two competitors is justified to assume, in deciding on how to adjust his output, that the other will stick to *his*. For no such assumption is really involved, the above argument aiming only at describing the process of *tâtonnement*, out of which the equilibrium price is finally bound to emerge, and things would remain substantially the same if some of the steps were to drop out – just as the equilibrium of perfect competition

party has a definite monopoly-demand-curve and knows the curve of the other; that each party wants to get the best terms it can – the workmen's union offering varying amounts of labour and providing for those of its members who may have to be kept unemployed – without attempting to attain victories or to inflict defeats; and that the contract is to cover the whole period of account (the *uno actu* condition), then the barter point between the parties is perfectly determined, and *not* only the range within which there will be barter. It could be indeterminate only for reasons which would make the case indeterminate also in competition. Nor can it be held that the assumptions alluded to are so very far from reality. They are, if anything, nearer to reality than the assumptions implied in the idea of theoretically perfect competition: it is, for instance, much more common than observers believe whose attention is naturally focused on abnormal cases, for employers and workmen to meet in precisely the frame of mind assumed, and to view with misgivings all the economic, political and social risks of holding out or of a struggle, which may turn out bad business even in the case of success. By proceeding by way of Walras' *prix crié par hazard* or simply by inspecting the two schedules plotted against one another, our statement will too readily be found to hold good to make it necessary to give formal proof.[12]

---

does not necessarily come about by every one of the theoretical steps of bidding actually taking place in practice. Nor can it be said that the two monopolists would, on reaching what we have called the equilibrium price, try to retrace their steps. For neither of them could do so singly without losing his customers. They could do so only together – the case would become one of single monopoly.

12. The well-known Edgeworthian apparatus commonly used to prove the contrary merely shows that the *elements described by it* do not suffice to determine more than a range. Professor Bowley in his 'Groundwork' reaches, in dealing with the case of one employer and one workman, the result of incompatibility of the respective maxima only by implying that the workman could produce the product by himself. The 'Groundwork' contains, however, two most suggestive approaches to the problem of universal monopoly, the one embodied in a note carrying that title, the other leading to the theorem that there is determinateness in the case of *either* the products *or* the factors – but not both of them – being monopolized. Arguments analogous to those of our text seem to show that at least the same sort of determinateness obtains in these cases too.

So there is rather more of stability[13] about the economic system than we should expect on most of the authoritative statements. But how much this amounts to depends entirely on the nature of that other restriction, which we have introduced alongside of the competitive assumption just discarded: the 'static state', which we define both by a distinguishable set of facts and by an analytic apparatus or theoretical point of view. The set of facts consists in the sum of operations which form the essence of the ever-recurring circular process of production and consumption and which make up a self-contained whole. It is no valid objection to say that this process cannot be thought of independently of growth or, generally, change. For it can. Just as a child's blood circulation, although going on concurrently with its growth or, say, pathological change in its organs, is yet capable of being singled out and dealt with as a distinct real phenomenon, so that fundamental circular process can be singled out and dealt with as a distinct real phenomenon, and *every analyst*[14] *and every businessman does so deal with it* – the latter

13. This stability is of the same nature, and its exact proof of the same value, as the stability of any other exact system. Of course, it is compatible with a large amount of instability in the actual phenomenon. Part of this instability is unimportant, both for theoretical and for practical purposes; another part, whilst practically important, is yet uninteresting in a discussion of principles; still another, however, has, as we shall see, both practical and theoretical importance. None of these groups of cases affects the fundamental importance of exact proof of stability in the sense meant, as would be obvious everywhere except in economics, where the sterility incident to the prevalence of interest in the 'practical problem' has yet to be overcome and where scientific refinement is still an opprobrium. But it must be borne in mind that our arrangement excludes all important cases of determined but unstable equilibrium. For the above argument, therefore, and within our meaning of terms, determinateness spells economic stability under static conditions, although, of course, these two things do not coincide logically and always require separate proof. The shortest way to satisfy oneself on this point is by verifying the statement that, of all cases of equilibrium known to Marshallian analysis, only the stable ones remain – apart from chance equilibria which occur during the process of Walrasian *tâtonnement* – for a static theory as above defined. Correct proof of this stability has not been given so far, but does not seem to meet with any great difficulty.

14. Of course, only a minority of economists are aware of the fact. And some of those who are, spoil the edge of the tool by speaking of a 'stationary' state. Some of these, again, construct a state of harmonious progress to

realizing that it is one thing to figure out the outlay on, and the income from, a building in given circumstances and another thing to form an idea about the future prospects of the neighbourhood, or that it is one thing to manage an existing building and another to pull it down and replace it by another of a different kind. Nor is our analogy with the circulation of the blood idle. For the first complete analysis of the static economic process, Quesnay's, was directly inspired by Harvey's discovery. The analytic apparatus or theoretic point of view of statics is presented by the concept of a determined equilibrium, the use of which, however, is not absolutely confined to the explanation of the circular process, as temporary equilibria occur outside of this process.

Because a set of facts, which form a coherent whole and are, in many cases, capable of statistical separation from the rest, corresponds to static theory, the static state is not merely a methodological device, still less a pedagogical one. And its range is much widened by the fact that it is not a state of rest. It is firstly, of course, no state of absence of motion, as it implies the everchanging flow of productive services and consumers' goods, although this flow is looked upon as going on under substantially unchanging conditions. But, secondly, conditions need not be entirely constant. We can allow seasonal oscillations. We can also allow, without leaving the precincts of statics, chance variations, provided reaction to them is merely adaptive, in the sense of an adaptation *capable of being brought about by infinitesimal steps*. And we can, finally, deal with the phenomenon of mere growth of population, of capital and, consequent thereupon, of the National Dividend. For these changes occur continuously, and adaptation to them is essentially continuous. They may *condition* discontinuous changes; but they do not, directly and by their mere presence, bring them about. What they do bring about automatically are only variations at the margins.[15] Increase of

occupy the ground between 'statics' and what too obviously lies outside of it. There is no objection to such a construction. But it is not always recognized that, owing to the fact that it implies consideration of long periods, the 'normal', which pertains to it, is much bolder and much more dangerous an abstraction than the static one.

15. Although, therefore, even these influences do not work within a given state of equilibrium and do not tend towards a given centre of gravitation,

population, for instance, will, by itself, merely tend to make labour cheaper, and diagnosis of the state of any particular nation in any particular point of time will have to recognize this as a real and distinct element of the situation, however much it may be compensated by other factors. From this it follows that mere growth is not in itself a source of instability of either the system or the order of capitalism, within the meaning given to 'stability' in this paper. This disposes of some, if not most, theories of 'disproportionality', past and present, and gives further help towards 'localizing' causes of instability.

Stability and progress

This might very well be all: economic life, or the economic element in, or aspect of, social life might well be essentially passive and adaptive and *therefore, in itself, essentially stable*. The fact that reality is full of discontinuous change would be no disproof of this. For such change could without absurdity be explained by influences from without, upsetting equilibria that would, in the absence of such influences, obtain or only shift by small and determined steps along with what we have called continuous growth. We could, of course, even then fit trend lines through the facts succeeding one another historically; but they would merely be expressions of whatever has happened, not of distinct forces or mechanisms; they would be statistical, not theoretical; they would have to be interpreted in terms of particular historic events, such as the opening up of new countries in the nineteenth century, acting on a given rate of growth – and not in terms of the working of an economic mechanism *sui generis*. And if analysis could not detect any purely economic forces with-

but displace this centre and propel the economic organism away from the old position, the static apparatus is admirably competent to deal with them. Treatment of such questions has been called 'dynamics' by some authorities, foremost among whom was E. Barone. It would, perhaps, be best to drop the terms statics and dynamics altogether. Certainly they are misnomers, when used in the sense given to them in the text, and care should be taken not to think of them by way of analogy with their meanings in mechanics and not to confuse the different meanings attached to them by different writers. All the different meanings, I suppose, lead back to John Stuart Mill, who owes the suggestion to Comte, who, in his turn, expressed indebtedness to the zoologist de Blainville.

in the system making for qualitative and discontinuous change, we should evidently be driven to this conclusion,[16] which can never lack verification, as there are always outside influences to point to, and as a great part of the facts of non-equilibrium must in any case be explained largely on such lines, whether there be a definite piece of non-static mechanism in them or not.

Now it is always unsafe, and it may often be unfair, to attribute to any given author or group of authors clear-cut views of comprehensive social processes, the diagnosis of which must always rest largely on social vision as distinguished from provable argument. For no author or group of authors can help recognizing many heterogeneous elements, and it is always easy to quote passages in proof of this. The treatment of the history of the analysis of value, cost and interest affords examples in point,[17]

16. As a matter of fact, this is what the position of our highest authorities comes to. It is certainly the position of Ricardo and John Stuart Mill, whose discussion of 'progress' mainly turns on the question of relative growth of population and capital, occasionally affected by improvement of methods of production, which they glance at in passing as a disturber of the normal course of things. Such is the position, too, of Walras or, for that matter, of Böhm-Bawerk, who both of them seem convinced that everything of a purely economic nature must needs fit into one homogeneous body of doctrine, which is frankly 'static' with Walras, whilst Böhm-Bawerk always rejected the static conception precisely because it excludes some things which yet are undoubtedly 'purely economic'. John B. Clark is the one outstanding exception, but Marshall, although embracing within his wide horizons every one of the elements essential to a distinct theory of 'dynamics', still forced all of them into a frame substantially 'static'. The present writer believes that some of the difficulties and consequent controversies about Professor Pigou's argument in his *Economics of Welfare* are traceable to the same source, and his work on *Industrial Fluctuations* is a monument to the view that economic life, in itself essentially passive, is being continually disturbed and propelled by 'initial impulses' coming from outside.

17. Even within the narrower precincts of problems such as these, it has become a fashion – a justified reaction, perhaps, from the opposite vice – to interpret older authors so very broadly as to make them '*see*' everything and *definitely say* nothing, and to frown on another way of stating their views as ungenerous. I submit, however, firstly that whilst this attitude is the correct one in evaluating individual theorists – provided that the same generous broadness be vouchsafed to all – it is not useful in bringing out characteristics; secondly, that mere 'recognition' of a fact means nothing unless the fact be welded into the rest of the argument and made to do theoretic work.

and it must be left to the reader to form his own opinion about the correctness or otherwise of our thus formulating what seems to us to be received doctrine: industrial expansion, automatically incident to, and moulded by, general social growth – of which the most important purely economic forces are growth of population and of savings – is the basic fact about economic change or evolution or 'progress'; wants and possibilities develop, industry expands in response, and this expansion, carrying automatically in its wake increasing specialization and environmental facilities, accounts for the rest, changing continuously and organically its own data.

Grounds for dissent from this view present themselves on several points, but I am anxious to waive objections in order to make stand out *the* objection. Without being untrue, when taken as a proposition summing up economic history over, say, a thousand years,[18] it is inadequate, or even misleading, when meant to be a description of that mechanism of economic life which it is the task of economic theory to explain, and it is no help towards, but a bar to, the understanding of the problems and phenomena incident to that mechanism. For expansion is *no* basic fact, capable of serving in the role of a cause, but is itself the result of a more fundamental 'economic force', which accounts both for expansion and the string of consequences emanating from it. This is best seen by splitting up the comprehensive phenomenon of general industrial growth into the expansion of the single industries it consists of. If we do this for the period of predominantly competitive capitalism, we meet indeed at any given time with a class of cases in which both entire industries and single firms are drawn on by demand coming to them from outside and so expanding them automatically; but this additional

18. Different sets of problems require different distances from the objects of our interest; and different propositions are true from different distances and on different planes of argument. So, e.g., for a certain way of describing historic processes, the presence of a military commander of Napoleonic ability may truly be said to be of causal importance, whilst, for a survey farther removed from details, it may have hardly any importance at all. Our analytic apparatus consists of heterogeneous pieces, every one of which works well on some of the possible 'planes' of argument and not at all on others, the overlooking of which is an important, and sometimes the only, source of our controversies.

demand practically always proceeds, as a secondary phenomenon,[19] from a primary change in some other industry – from
textiles first, from iron and steam later, from electricity and chemical industry still later – which does not *follow*, but *creates*
expansion. It *first* – and by its initiative – expands its own production, thereby creates an expansion of demand for its own and,
contingent thereon, other products, and the general expansion of
the environment we observe – increase of population included –
is the *result* of it, as may be visualized by taking any one of the
outstanding instances of the process, such as the rise of railway
transportation. The way by which every one of these changes is
brought about lends itself easily to general statement: it is by
means of new combinations of existing factors of production,
embodied in new plants and, typically, new firms producing
either new commodities, or by a new, i.e. as yet untried, method,
or for a new market, or by buying means of production in a new
market. What we, unscientifically, call economic progress means
essentially putting productive resources to uses *hitherto untried in
practice*, and withdrawing them from the uses they have served
so far. This is what we call 'innovation'.

What matters for the subject of this study is merely the

19. We may conveniently enumerate, partly anticipating and partly
repeating, the more important types of those secondary phenomena, which
we hold received opinion, neglecting the primary phenomenon, exclusively
deals with, and which would not entirely, but almost entirely, be absent
without the primary one:

(a) Expansion of some industries called forth by primary expansion in
others, as stated above: if a new concern establishes itself, grocers' businesses
will expand in the neighbourhood and so will producers of subsidiary
articles. *The expansion of all industries, which do not themselves display any
break in their practice during the time under consideration* is to be accounted
for thus.

(b) If the primary change results in turning out better tools of production, naturally this will expand the industries which use them. This must be
taken account of in judging the comparative success of some State-managed
railways surrounded by private industries, which force on them improved
engines, fittings, and so on.

(c) Every given change starts from a given environment, and would be
impossible without its facilities. But every given environment embodies the
results of previous primary change, and, therefore, cannot be taken, except
within static theory, as an ultimate datum, acting autonomously, but is
itself, in great part, a secondary phenomenon.

essentially discontinuous character of this process, which does not lend itself to description in terms of a theory of equilibrium. But we may conveniently lead up to this by insisting for the moment on the importance of the difference between this view and what I have called the received one. Innovation, unless it consists in producing, and forcing upon the public, a new commodity, means producing at smaller cost per unit, breaking off the old 'supply schedule' and starting on a new one. It is quite immaterial whether this is done by making use of a new invention or not; for, on the one hand, there never has been any time when the store of scientific knowledge had yielded all it could in the way of industrial improvement, and, on the other hand, it is not the knowledge that matters, but the successful solution of the task *sui generis* of putting an untried method into practice – there may be, and often is, no scientific novelty involved at all, and even if it be involved, this does not make any difference to the nature of the process. And we should not only, by insisting on invention, emphasize an irrelevant point – irrelevant to our set of problems, although otherwise, of course, just as relevant as, say, climate – and be thereby led away from the relevant one, but we should also be forced to consider in-

---

(d) So is, in great part, what we have called growth. This is specially clear in the case of saving, the amount of which would be very much smaller in the absence of its most important source, the entrepreneurs' profits. It is also true as to increase of population. And expansion, incident to what would be left of growth in the absence of primary change, would soon be quenched by a (physical) law of decreasing returns acting sharply. *This, then, is the main reason why we think so little of the autonomous – as distinguished from secondary – importance of external economies incident to mere expansion and of what is left of increasing returns*, if we exclude all that is either primarily or secondarily due to the cause we are about to consider.

(e) Industrial evolution inspires collective action in order to force improvement on lethargic strata. Of this kind was, and is, government action on the Continent for improving agricultural methods of peasants. This is not 'secondary' in the sense we mean it, but if it comes to creating external economies by non-economic influence, it has nevertheless been due so far mainly to some previous achievement in some private industry.

(f) Successful primary change is followed by general reorganization within the same industry, more and more other firms following the lead of some, both because of the profits to be gained and the losses to be feared. During this process, what have at first been the internal economies of the

ventions as a case of external economies.[20] Now this hides part of the very essence of the capitalist process. This kind of external economies – and, in fact, nearly every kind, even the trade journal must, unless the product of collective action, be somebody's business – characteristically comes about by first being taken up by one firm or a few – by acting, that is, as an internal economy. This firm begins to undersell the others, part of which are thereby definitely pushed into the background to linger there on accumulated reserves and quasi-rents, whilst another part copies the methods of the disturber of the peace. *That* this is so we can see every day by looking at industrial life; it is precisely what goes on, what is missing in the static apparatus and what accounts both for dissatisfaction with it and for the attempts to force such phenomena into its cracking frame – instead of, as we think it natural to do, recognizing and explaining this as a distinct process going on along with the one handled by the static theory. *Why* this is so is a question which it would lead very far to answer satisfactorily. Successful innovation is, as said before, a task *sui generis*. It is a feat not of intellect, but of will. It

leaders soon become external economies for the rest of the firms, whose behaviour need be no other than one of passive adaptation (and expansion) to what *for them* is environmental advantage. But for us, the observers, to look upon the process as one of adaptation to expanding environment is to miss the salient point.

(g) Incident to all the phenomena glanced at, are, among other things, secondary gains going to all kinds of agents, who do not display any initiative. There is, however, another, a secondary, initiative, stimulated by the possibility of such gains becoming possible – extensions of businesses, speculative transactions and so on, calculated to secure them. The periodic rise and fall of the level of prices – an essential piece, as we shall see, of the mechanism of change in competitive capitalism – carries in its wake extensions and, to finance them, applications for credit merely due to the fact of prices rising, which greatly intensify the phenomenon. And this secondary phenomenon is being as a rule realized much more clearly by observers than the primary phenomenon which gives rise to it.

Our analysis neither overlooks nor denies the importance of these things. On the contrary, it aims at showing their cause and nature. But in a statement of fundamental principles within so short a compass they cannot loom large in the picture.

20. There is another point which arises out of the usual treatment of these things: nobody can possibly deny the occurrence or relevance of those great breaks in industrial practice which change the data of economic life from time to time. Marshall, therefore, distinguishes these, which he calls 'sub-

is a special case of the social phenomenon of leadership.[21] Its difficulty consisting in the resistances and uncertainties incident to doing what has not been done before, it is accessible for, and appeals to, only a distinct type which is rare. Whilst differences in aptitude for the routine work of 'static' management only result in differences of success in doing what everyone does, differences in this particular aptitude result in only some being able to do this particular thing at all. To overcome these difficulties incident to change of practice is the function characteristic of the entrepreneur.

Now if this process meant no more than one of many classes of 'friction', it certainly would not be worth our while to dissent

---

stantive' inventions and which he deals with as chance events acting from outside on the analogy, say, of earthquakes, from inventions which, being of the nature of more obvious applications of known principles, may be expected to arise in consequence of expansion itself. This distinction is insisted upon by Professor Pigou in the paper quoted above. This view, however, cuts up a homogeneous phenomenon, the elements of which do not differ from one another except by degree, and is readily seen to create a difficulty similar to that of filling the empty boxes. Exactly as the failure to distinguish different processes leads, in the case of the boxes, to a difficulty in distinguishing between groups of facts – and leads, also, to that state of discussion in which some authors hold that most industries display *increasing*, others that most industries display *decreasing*, still others, that normally any industry shows *constant*, returns – so it is obviously impossible to draw any line between those classes of innovations, or, for that matter, inventions; and the difficulty is not one of judging particular cases, but one of principle. For *no* invention is independent of existing data; and *no* invention is *so* dependent on them as to be automatically produced by them. In the case of important invention, change in data is great; in the case of unimportant invention it is small. But this is all, and the *nature* of the process and of the special mechanism set in motion is always the same.

21. This does not imply any glorification. Leadership itself does not mean only such aptitudes as would generally command admiration, implying, as it does, narrowness of outlook in any but one direction and a kind of force which sometimes it may be hardly possible to distinguish from callousness. But economic leadership has, besides, nothing of the glamour some other kinds of leadership have. Its intellectual implications may be trivial; wide sympathies, personal appeal, rhetorical sublimation of motives and acts count for little in it; and although not without its romance, it is in the main highly unromantic, so that any craving for personal hero-worship can hardly hope for satisfaction where, among, to be sure, other types, we meet with slave-trading and brandy-producing puritans at the historic threshold of the subject.

from the usual exposition on that account, however many facts might come under this heading. But it means more than this: its analysis yields the explanation of phenomena which cannot be accounted for without it. There is, first, the 'entrepreneurial' function as distinct from the mere 'managerial' function – although they may, and mostly must, meet one another in the same individual – the nature of which only shows up within the process of innovation. There is, secondly, the explanation of entrepreneurs' gain, which emerges in this process and otherwise gets lost in the compound of 'earnings of management',[22] the treating of which as a homogeneous whole is unsatisfactory for precisely the same reason which, by universal consent, makes it unsatisfactory so to treat, say, the income of a peasant tilling his own soil, instead of treating it as a sum of wages, rent, quasi-rent and, possibly, interest. Furthermore, it is *this* entrepreneurs' profit which is the primary source of industrial fortunes, the

Apart from this source of possible objections, there is a much more serious one in the mind of every well-trained economist, whom experience has taught to think little of such intrusions into theory of views savouring of sociology, and who is prone to associate any such things with a certain class of objections to received doctrine, which continually turn up however often they may have been refuted – sublimely ignorant of the fact – such as objections to the economic man, to marginal analysis, to the use of the barter hypothesis and so on. The reader may, I think, satisfy himself that no want of theoretic training is responsible for statements which I believe to tally fundamentally with Marshallian analysis.

No difficulty whatever arises as to verification. That new commodities or new qualities *or new quantities* of commodities are forced upon the public by the initiative of entrepreneurs – which, of course, does not affect the role of demand within the static process – is a fact of common experience; that one firm or a small group of firms leads in the sense meant above, in the process of innovation, thereby creating its own market and giving impulse to the environment generally, is equally patent (and we do not deny facts of other complexion – the secondary or 'consequential' ones); and all we are trying to do is to fit the analytic apparatus to take account of such facts without putting its other parts out of gear.

22. The function in question being a distinct one, it does not matter that it appears in practice rarely, if ever, by itself. And whoever cares to observe the behaviour of businessmen at close quarters will not raise the objection that new things and routine work are done, as a rule, indiscriminately by the same manager. He will find that routine work is done with a smoothness wholly absent as soon as a new step is to be taken, and that there is a sharp cleavage between the two, insuperable for a very worthy type of manager.

history of every one of which consists of, or leads back to, successful acts of innovation.[23] And as the rise and decay of industrial fortunes is *the* essential fact about the social structure of capitalist society, both the emergence of what is, in any single instance, an essentially temporary gain, and the elimination of it by the working of the competitive mechanism, obviously are more than 'frictional' phenomena, as is that process of under-selling by which industrial progress comes about in capitalist society and by which its achievements result in higher real incomes all round.

Nor is this all. This process of innovation in industry by the agency of entrepreneurs supplies the key to all the phenomena of capital and credit. The role of credit would be a technical and a subordinate one in the sense that everything fundamental about the economic process could be explained in terms of goods, if industry grew by small steps along coherent curves. For in that case financing could and would be done substantially by means of the current gross revenue, and only small discrepancies would

---

This extends far into the realm of what we are wont to consider as automatic change, bringing about external economies and increasing returns. Take the instance of a business letting out motor cars on the principle 'drive yourself'. A mere growth of the neighbourhood, sufficient to make such a business profitable, does not produce it. Someone has to realize the possibility and to found the firm, to get people to appreciate its services, to get the right type of cars and so on. This implies solution of a legion of small problems. Even if such a firm already exists and further environmental growth make discontinuous extension feasible, the thing to be done is not so easy as it looks. It would be easy for the trained mind of a leading industrialist, but it is not so for a typical member of the stratum which does such business.

23. It is, as has been said in a previous note, not the *running* of a business according to new plan, but the act of *getting it* to run on a new plan, which accounts for entrepreneurs' profits, and makes it so undesirable to try to express them by 'static' curves, which describe precisely the phenomena of the 'running' of it. The theoretical reason for our proposition is, that either competition or the process of imputation must put a stop to any 'surplus' gain, even in a case of monopoly, in which the value of the patent, the natural agent or of whatever else the monopoly position is contingent on, will absorb the return in the sense that it will no longer be profit. But there is also a 'practical' observation to support this view. No firm ever yields returns indefinitely, if only run according to unchanged plan. For everyone comes the day when it will cease to do so. And we all of us know that type of industrial family firm of the third generation which is on the road to that state, however conscientiously it may be 'managed'.

need to be smoothed. If we simplify by assuming that the whole circular process of production and consumption takes exactly one period of account, no instruments or consumers' goods surviving into the next, capital – defined as a monetary concept – and income would be exactly equal, and only different phases of one and the same monetary stream. As, however, innovation, being discontinuous and involving considerable change and being, in competitive capitalism, typically embodied in new firms, requires large expenditure previous to the emergence of any revenue, credit becomes an essential element of the process. And we cannot turn to savings in order to account for the existence of a fund from which these credits are to flow. For this would imply the existence of previous profits, without which there would not be anything like the required amount – even as it is, savings usually lag behind requirements – and assuming previous profits would mean, in an explanation of principles, circular reasoning. 'Credit-creation', therefore, becomes an essential part both of the mechanism of the process and of the theory explaining it. Hence, saving, properly so called, turns out to be of less importance than the received doctrine implies, for which the continuous growth of saving – accumulation – is a mainstay of explanation. Credit-creation is the method by which the putting to new uses of existing means of production is brought about through a rise in price enforcing the 'saving' of the necessary amount of them out of the uses they hitherto served ('enforced savings' – see Mr Robertson's 'imposed lacking').

Finally, it cannot be said that whilst all this applies to individual firms, the development of whole industries might still be looked at as a continuous process, a comprehensive view 'ironing out' the discontinuities which occur in every single case. Even then individual discontinuities would be the carriers of essential phenomena. But, besides, for a definite reason that is not so. As shown both by the typical rise of general prices and the equally typical activity of the constructional trades in the prosperity phase of the business cycle, innovations cluster densely together. So densely, in fact, that the resultant disturbance produces a distinct period of adjustment – which precisely is what the depression phase of the business cycle consists in. *Why* this should be so, the present writer has attempted to show elsewhere

(Schumpeter, 1911; see also 1927).[24] *That* it is so, is the best single verification and justification of the view submitted, whether we apply the criterion of its being 'true to life' or the criterion of its yielding explanation of a phenomenon *not itself implied in its fundamental principle*.

If, then, the putting to new uses of existing resources is what 'progress' fundamentally consists in; if it is the nature of the entrepreneur's function to act as the propelling force of the process; if entrepreneur's profits, credit, and the cycle prove to be essential parts of its mechanism – the writer even believes this to be true of interest – then industrial expansion *per se* is better described as a consequence than as a cause; and we should be inclined to turn the other way round what we have termed the received chain of causation. In this case, and as those phenomena link up so as to form a coherent and self-contained logical whole, it is obviously conducive to clearness to bring them out boldly; to relegate to one distinct body of doctrine the concept of equilibrium, the continuous curves and small marginal variations, all of which, in their turn, link up with the circuit flow of economic routine under constant data; and to build, alongside of this, and *before* taking account of the full complexity of the 'real' phenomenon – secondary waves, chance occurrences, 'growth' and so on – a theory of capitalist change, assuming, in so doing, that non-economic conditions or data are constant and automatic and gradual change in economic conditions is absent. But there is no difficulty in inserting all this. And it would seem to follow that

24. The failure of the price level to rise in the United States during the period 1923–6 will be seen to be no objection but a further verification of this theory. It has, however, been pointed out to the writer, by a very high authority, that prices did also fail to rise in the United States in the prosperity immediately preceding the War. It could be replied that the factors which account for the stability 1923–6 had been active already before the War. But the US Bureau of Labour figures for 1908–13 are 91, 97, 99, 95, 101, 100. See also Professor Persons' chart in *Review of Economic Statistics*, January 1927. It may be well to mention that constructional trades and their materials need not necessarily show their activity fully by *every* index. Iron, e.g., being an international commodity, need not rise in price if the phases of the cycle do not quite coincide in different countries. As a matter of fact, they generally do. But the right way to deal with iron and steel is to use the Spiethoff index (production + imports – exports), and this has, so far, always worked satisfactorily.

the organic analogy is less adapted to express faithfully the nature of the process than many of us think; although, of course, being a mere analogy, it may be so interpreted as not to imply anything positively wrong and as to avoid the idea of an equilibrium growth *ad instar* of the growth of a tree, which it may, but need not necessarily, suggest.

Summing up the argument and applying it to the subject in hand, we see that there is, indeed, one element in the capitalist process, embodied in the type and function of the entrepreneur, which will, *by its mere working and from within* – in the absence of all outside impulses or disturbances and even of 'growth' – destroy any equilibrium that may have established itself or been in process of being established; that the action of that element is not amenable to description by means of infinitesimal steps; and that it produces the cyclical 'waves' which are essentially the form 'progress' takes in competitive capitalism and could be discovered by the theory of it, if we did not know of them by experience. But by a mechanism at work in, and explaining the features of, periods of depression, a new equilibrium always emerges, or tends to emerge, which absorbs the results of innovation carried out in the preceding periods of prosperity. The new elements find their equilibrium proportions; the old ones adapt themselves or drop out; incomes are rearranged; prosperity inflation is corrected by automatic self-deflation through the repayment of credits out of profits, through the new consumers' goods entering the markets and through saving stepping into the place of 'created' credits. So the instabilities, which arise from the process of innovation, tend to right themselves, and do not go on accumulating. And we may phrase the result we reach in our terminology by saying that there is, though instability of the *system*, no economic instability of the *order*.

The instability due to what we conceive to be the basic factor of purely economic change is, however, of very different importance in the two historic types of capitalism, which we have distinguished.

Innovation in competitive capitalism is typically embodied in the foundation of new firms – the main lever, in fact, of the rise of industrial families; improvement is forced on the whole branch by the processes of underselling and of withdrawing from them their means of production, workmen and so on shifting to the new firms; all of which not only means a large amount of disturbance as an incident, but is also effective in bringing about the result, and to change 'internal' economies into 'external' ones, only *as far as* it means disturbance. The new processes do not, and generally cannot, evolve out of the old firms, but place themselves side by side with them and attack them. Furthermore, for a firm of comparatively small size, which is no power on the money market and cannot afford scientific departments or experimental production and so on, innovation in commercial or technical practice is an extremely risky and difficult thing, requiring supernormal energy and courage to embark upon. But as soon as the success is before everyone's eyes, everything is made very much easier by this very fact. It can now, with much-diminished difficulty, be copied, even improved upon, and a whole crowd invariably does copy it – which accounts for the leaps and bounds of progress as well as for setbacks, carrying in their wake not only the primary disturbance, inherent to the process, but a whole string of secondary ones and *possibilities* although no more than possibilities, of recurrent catastrophes or crises.

All this is different in 'trustified' capitalism. Innovation is, in this case, not any more embodied *typically* in new firms, but goes on, within the big units now existing, largely independently of individual persons. It meets with much less friction, as failure in any particular case loses its dangers, and tends to be carried out as a matter of course on the advice of specialists. Conscious policy towards demand and taking a long-time view towards investment becomes possible. Although credit creation still plays a role, both the power to accumulate reserves and the direct access to the money market tend to reduce the importance of this element in the life of a trust – which, incidentally, accounts for the phenomenon of prosperity coexisting with stable, or nearly stable, prices which we have had the opportunity of witnessing in the United States 1923–6. It is easy to see that the three causes alluded to, whilst they accentuated the waves in competitive, must tend to soften them down in trustified, capitalism. Progress becomes 'automatized', increasingly impersonal and decreasingly a matter of leadership and individual initiative.

This amounts to a fundamental change in many respects, some of which reach far out of the sphere of things economic. It means the passing out of existence of a system of selection of leaders which had the unique characteristic that success in *rising* to a position and success in *filling* it were essentially the same thing – as were success of the firm and success of the man in charge – and its being replaced by another more akin to the principles of appointment or election, which characteristically divorce success of the concern from success of the man, and call, just as political elections do, for aptitudes in a candidate for, say, the presidency of a combine, which have little to do with the aptitudes of a good president. There is an Italian saying, 'Who enters the conclave as prospective pope, will leave it as a cardinal', which well expresses what we mean. The types which rise, and the types which are kept under, in a trustified society are different from what they are in a competitive society, and the change is spreading rapidly to motives, stimuli and styles of life. For our purpose, however, it is sufficient to recognize that the only fundamental cause of instability inherent to the capitalist system is losing in importance as time goes on, and may even be expected to disappear.

Instead of summing up a very fragmentary argument, I wish to emphasize once more, in concluding, that no account whatsoever has been taken of any but purely economic facts and problems. Our diagnosis is, therefore, no more sufficient as a basis for prediction than a doctor's diagnosis to the effect that a man has no cancer is a sufficient basis for the prediction that he will go on living indefinitely. Capitalism is, on the contrary, in so obvious a process of transformation into something else, that it is not the fact, but only the interpretation of this fact, about which it is possible to disagree. Towards this interpretation I have wished to contribute a negative result. But it may be well, in order to avoid misunderstanding, to state expressly what I believe would be the positive result of a more ambitious diagnostic venture, if I may presume to do so in one short and imperfect sentence: capitalism, whilst economically stable, and even gaining in stability, creates, by rationalizing the human mind, a mentality and a style of life incompatible with its own fundamental conditions, motives and social institutions, and will be changed, although not by economic necessity and probably even at some sacrifice of economic welfare, into an order of things which it will be merely matter of taste and terminology to call socialism or not.

References

BOWLEY, A. L. (1925), 'Groundwork', *Ekonomisk Tidskrift*, reprinted in 1927 in *Archiv für Sozialwissenschaft*.

DIVISIA, F. (1928), *Economique Rationnelle*, Gaston Dion.

PIGOU, A. C. (1927), 'The laws of diminishing and increasing cost,' *Econ. J.*, June.

SCHULTZ, H. (1927), 'Theoretical considerations relating to supply', *J. polit. Econ.*, vol. 35, no. 4.

SCHUMPETER, J. (1911), *Theorie der Wirtschaftlichen Entwicklung*, 2nd edn, 1926.

SCHUMPETER, J. (1927), 'The explanation of the business cycle', *Economica*, no. 21.

SRAFFA, P. (1925), 'Relazioni fra costo e quantita prodotta', *Annali di Economia*, vol. 2, Reprinted in *Econ. J.*, December, 1926.

TAUSSIG, F. W. (1921), 'Is market price determinate?', vol. 35, *Quart. J. Econ.*

YOUNG, A. A. (1913), 'Pigou's *Wealth and Welfare*', *Quart. J. Econ.*, August.

2 A. P. Usher

Technical Change and Capital Formation

A. P. Usher, 'Technical change and capital formation', *Capital Formation and Economic Growth*, National Bureau of Economic Research, 1955, pp. 523–50.

Acts of skill and insight

The study of technical change in the economy has been hindered by the failure of students to treat effectively the kinds of novelty that are a normal and continuous consequence of the skilled activities of engineers and technicians and those that are related to acts of insight and the process of invention. In the history of the sciences and technology, primary emphasis has fallen upon the process of invention and, all too frequently, upon selected items in the process as a whole. In the administration of the economy, however, acts of skill have played a commanding part in the diffusion of new technical processes both by fields and by geographic areas.

Many presume that the diffusion of technical knowledge and the applications of known techniques are imitative acts devoid of novelty. Tarde's sharp distinction between imitation and invention is hardly more than a broad generalization of attitudes that have been widely held over long periods of time (see Davis, 1906, pp. 52 and 56–61). But these interpretations rest upon concepts of knowledge, skill and invention that fail to recognize the pervasiveness of novelty in our behavior. They are inconsistent with the concept of emergent novelty that is rapidly developing in the biological and psychological fields.

The distinction between acts of skill and inventions is suggestively drawn by Gestalt psychology. Novelty is to be found in the more complex acts of skill, but it is of a lower order than at the level of invention. As long as action remains within the limits of an act of skill, the insight required is within the capacity of a trained individual and can be performed at will at any time. At the level of invention, however, the act of insight can be achieved only by superior persons under special constellations of circumstance. Such acts of insight frequently emerge in the course of performing acts of skill, though characteristically the act of insight is induced by the conscious perception of an unsatisfactory gap in knowledge or mode of action (Koffka, 1935, pp. 382, 628–31; Köhler, 1929, pp. 371–94; Welford, 1951, pp. 11–27). The principles underlying the distinction are clear, but application to particular cases is difficult, because it is not easy to know whether even a specific individual actually performed the act as and when he chose. Common usage reflects the difficulty of making rigorous classifications of activities that involve any elements of novelty.

This problem of boundaries between acts of skill and inventions has a long history and is an underlying feature of the laws granting monopoly privileges for the introduction of new industries and new processes. Special privileges were at first granted to favorites of the rulers in nearly all European states, but opposition developed and it became customary to base the privileges on some element of novelty. Despite emphasis upon the idea of invention, privileges were given for introducing new trades and processes that were admittedly not inventions. Although this development followed similar lines in several major jurisdictions, it will be enough for our present purpose to confine our attention to English law. A principle first stated by counsel in 1598 was embodied in the important section of the Statute of Monopolies (1624) which became the basis of the patent system in England. This authorized the issue of 'letters patents and grants of privilege for the term of fourteen years, or under, to be made of the sole working or making of any manner of new manufactures within this realm, to the true and first inventor or inventors of such manufactures' (Fox 1947, p. 219). Until the middle of the nineteenth century, judicial decisions emphasized the phrase 'new manufactures within this realm'. The device or process needed only to be new in England; and it did not have to be an invention. As late as 1803 Lord Ellenborough stated the issue sharply. 'There are common elementary materials to work with in machinery, but it is the adoption of those materials to the execution of any particular purpose, that constitutes the invention; and if the application of them be new, if the combination in its nature be essentially new, if it be productive of a new end, and beneficial to the public, it is that species of invention which, protected by the King's Patent, ought to continue to the person the sole right of vending' (Fox, 1947, p. 229).

The Constitution of the United States did not permit the issue of a patent for the introduction of a device or process developed abroad by a person other than the person introducing the process in the United States. But improvements that should be classed as the work of a skilled mechanic were not excluded either by the statute or by the earlier case decisions. This trend was reversed in 1850 by the decision in *Hotchkiss* v. *Greenwood*. The doctrine was laid down that 'unless more ingenuity and skill in applying the old method . . . were required . . . than were possessed by an ordinary mechanic acquainted with the business, there was an absence of that degree of skill and ingenuity which constitute essential elements of every invention' (Fox, 1947, p. 245). This doctrine was adopted both in the United States and in England. The British statute of 1932 explicitly excluded any obvious development of existing knowledge; and the Supreme Court of the United States in 1941 took the position that the new device 'must reveal the flash of creative genius, not merely the skill of the calling' (pp. 239–40 and 247).

It is tempting to fill in the literary background for this identification of invention with the work of genius. As early as the seventeenth century many scientists and inventors were becoming self-conscious about the recognition of the priority of their achievement, for they felt that this constituted their claim to fame. Huygens took pains to state very carefully the relation of his work on the pendulum to the work of Galileo. Newton was offended by Leibnitz's publications on the calculus without the acknowledgements that he felt to be due him (Hirsch, 1931, n.b. pp. 277–317; and Lang-Eichbaum, 1931, pp. 6–7; Zilsel, 1926).

We cannot give space to this problem, but it is important to recognize the danger of this identification of invention with an act of genius. It leads toward an undue emphasis upon a relatively small number of acts which are presented without due regard to the conditions which made them possible, and to a concept of change at infrequent intervals in units of great magnitude, although the simplest effort of analysis makes it clear that acts of insight are numerous, pervasive, and of very small magnitudes. The analysis of behavior, and most particularly of social action, becomes confused and misleading if the transcendental point of view is not carefully and consistently distinguished from the empirical.

All modes of action can be reduced to three categories: innate activities, acts of skill, and inventive acts of insight. The broader descriptions of these categories are clearly formulated in recent research in biology and Gestalt psychology, but so much work remains to be done that definitions of boundaries must be treated tentatively and the interweaving of different types of activity sketched with caution.

Innate activities are unlearned modes of action that develop as responses to the structure of the organism or the biochemical processes that control its functions. Current research and analysis present interpretations of these activities that differ in many important respects from the older concepts of instinct (Morgan, 1930; Stone, 1951; Tinbergen, 1951; Wheeler, 1928). Acts of skill include all learned activities whether the process of learning is an achievement of an isolated adult individual or a response to instruction by other individuals of the same or different species of organisms. Inventive acts of insight are unlearned activities that result in new organizations of prior knowledge and experience. In this meaning the concept was introduced by Köhler (1917) in his study of apes. It has been further developed by Koffka, but it is still incompletely generalized. It is not recognized by those who do not accept the general positions of Gestalt psychology.

Biologists now find grounds for presuming the existence of learning and acts of skill among wide arrays of subhuman organisms, especially among the social insects and many higher mammals. Psychologists recognize the presence of some acts of insight among subhuman organisms, but the basis of inference presents many special problems and results are uncertain. The importance of this new analysis for the social sciences lies in the superiority of these techniques for the study of the boundary between acts of skill and acts of insight at the higher levels in technology and economic administration.

Generalization of the concept of an act of skill leads to its extension to fields of conceptual activity involving interpretations of codes, rules for group behavior, and the execution of policies for individual or group activity. Inventive acts of insight occur in all these conceptual fields. Social activity involves an interweaving of acts of skill with interspersed acts of insight. No particular field should be presumed to involve a single type of action. The term 'innovator', as applied by Schumpeter to the entrepreneur, suggests a kind of differentiation from inventors in the technical fields that is likely to be misleading. Executives, like technicians, must be presumed to perform many acts of skill, and likewise to achieve many acts of insight and invention.

Legal doctrines in the field of patent law suggest the more important gradations of novelty that may be found in administration and engineering. The concept of the act of skill presents three choices for testing the relation of existing knowledge to an improvement. We may take as a measure the achievement of any person trained in the field, that of a superior person with general interests and training, or that of a superior person with special interests and training. The inventive act of insight emerges only under special conditions of a different kind. These conditions cannot be controlled at will. The new perception or novel synthesis is produced by some special constellation of circumstances that invites or requires a special response. Although in a particular instance the act of insight may not involve an essentially different response to circumstance from what we find in an act of skill, the conditions which precipitate the act of insight are not regularly recurrent. In the act of skill there are forms of recurring action which fall within a definable range of variation; and, over time, it is presumed that action will be necessary throughout the whole range of variables. Novelties emerge, but within limits which are defined by the possible changes in conditions. From time to time the performance of an act of skill may result in a new observation of the properties of materials or a new perception of relationships or a new mode of action. Acts of insight thus occur in the normal course of the exercise of skills.

Since acts of skill are so directly related to an established technique of action and to organized systems of knowledge, the individual act of novelty is often ignored. Acts of insight, especially at high levels, are likely to become the focus of all our attention, so that we are prone to ignore individual acts of novelty in long sequences of action which can best be described as processes of invention. Uncritical observation of behavior is likely to give too little stress to elements of novelty in acts of skill, and to present inventive acts of insight as completely unconditioned, isolated actions.

In a formal theory of invention, according to the general principles of Gestalt analysis, it is possible to recognize four distinctive steps in the process of invention: the perception of an unsatisfactory pattern, the setting of the stage, the primary act of insight, and critical revision and development. Acts of insight occur at each step if major elements of novelty are involved (Usher, 1929, ch. 2). The first and last steps in the process commonly involve close relationships with acts of skill. New problems emerge, because some inadequacy of existing knowledge or of current modes of action is perceived. Existing skills are seen to be inadequate. Some measure of failure in the performance of an act of skill touches off a sequence of invention. If stage-setting is deliberately undertaken by systematic experimentation, acts of skill enter at this stage also, but not too clearly. After the major act of insight has occurred, critical revision and development involve a very intimate interweaving of minor acts of insight and acts of skill performed at high levels by persons of special training.

The work of Frank Julian Sprague upon electric traction affords two striking illustrations of the development of new techniques which at the final stages were performed under contracts with sharply restricted terminal dates. Both contracts were designed to demonstrate the possibility of extending to allied fields equipment that was already in use in narrower fields. At that time they were incidental to the activities of an established business, and costs were treated as promotional expenses (Passer, 1953, pp. 241–3 and 271–3). The trolley system at Richmond, Virginia had to submit to rigorous tests before a fixed date, as did the multiple-unit system for rapid transit service, first installed at Chicago. Neither dateline was actually met. By making financial concessions, an extension of time was obtained for the Richmond contract. A preliminary test for the system of multiple control was carried out one day after the date specified, and a full test made ten days later. The delays were due partly to illness and partly to mechanical problems and not to new acts of insight or inventions which occurred after the contracts were signed. The substantive inventions preceded the making of the contracts. In terms of the process of invention, the work that held up the contract was work of critical revision. In professional terminology, it was engineering work; the psychologist would classify it as the exercise of a series of acts of skill. They involved a combination of activities in the fields of technology and entrepreneurship.

These instances also serve to demonstrate the need of much invention after the decisive demonstration of a new technique. Supplementary inventions of tertiary rank were necessary to secure full efficiency, and further secondary inventions were ultimately made. The quality of this work is usually ignored in descriptions of the application of the technique, except in accounts dominated by professional interest in technical detail. Practical application of a new technique does not mean that the process of invention has come to an end. But, in general, it is true that, as development proceeds, acts of skill become increasingly important.

There is no difference in the general character of the behavior of entrepreneurs and technologists. Entrepreneurs and executive directors invent new concepts of social ends and new procedures in social action. They discover new meanings in motivations and new modes of reconciling authority with individual freedom. But these acts of insight are dispersed through a highly diversified array of acts of skill which are often incorrectly classified because they are generalized to a greater degree than acts of skill in the fields commonly regarded as skilled and professional.

In considering the history of science and technology, the acts of insight and the processes of cumulative synthesis that can best be identified with invention are obviously more important than the acts of skill which are performed in the current applications of knowledge to individual and social needs. It is dangerous, however, to presume that the performance of acts of skill does not require abilities of a high order. At lower levels of action mere competence may suffice for much of the activity in the field. At higher levels – in the fields of art and the professions, and in leadership of large groups – mere competence has a narrowly limited value. Virtuosity in performance becomes an essential requirement. Accomplishment with such distinction is possible only to small numbers of individuals, and, even if no major acts of insight are involved in their activity, their achievements are no less important to the social life of the group than the achievements of inventors. At the higher levels, however, acts of skill and insight are so completely interwoven that we cannot easily distinguish them.

Primary inventions and discoveries

The concept of a process of invention requires a notion of a sequence of acts of insight which leads to a cumulative synthesis of many items which were originally independent. Strictly speaking, each act of insight is an achievement of novelty of the highest order. Practically, we characterize as an invention only some concept or device that represents a substantial synthesis of old knowledge with new acts of insight. Common usage, however, does not require us to assume that an invention is practical. In fact, there has been a strong tendency to stress the outstanding importance of discoveries of new properties and relationships, and of new devices for producing motion when they are merely laboratory models or small devices for entertainment or mystification. These attitudes are well grounded, and this distinction between the working model or demonstration and the commercially useful machine is especially important for the history of technology since 1600. New scientific concepts have emerged, new devices have been invented, but practical application has been long deferred.

Transcendentalists and romantic individualists have commonly sought a single inventor in each sequence of achievement. Insofar as emphasis has been placed on sheer priority, the scientist is ranked ahead of the engineer, the achievement of a new principle counted as the true invention, and the practical application of the principle treated as a mere unimpeded act of skill achievable by any competent technician. There has been, thus, a large group who wish to credit Galileo with the 'invention' of the pendulum clock because he perceived the bare principles involved, though no complete clock movement was produced during his lifetime. The work of Huygens has been treated as the explicit achievement of the discoveries and inventions of Galileo. In general, there has been a tendency to emphasize the scientific achievement if there is any single principle definite enough to seem to imply all the subsequent steps in the sequence. In the history of the steam engine the attempts to give primary credit to the early work have not successfully challenged the common appraisal of Watt's work.

This search for the unique inventor is naïve and ill-grounded. All the items specified and many other acts of insight are an integral part of the history of the steam engine. Even if we accept the unduly restricted common meaning of the word invention, there were many inventions and a large number of acts of insight which are entirely ignored by the lay public though fully appreciated by engineers who have any interest in history. Emphasis upon the achievements below the level of practical use must not lead to an underestimate of the high degree of originality involved in practical applications of the general principle.

The scientific and technical achievements below the level of practical commercial use fall into several broad categories that do not lend themselves to comprehensive enumeration in detail. In the field of pure science we can recognize easily the discoveries of new properties of materials, the perception of new relationships expressed as principles or laws, the invention of apparatus for the observation and measurement of natural phenomena. Above this level of generality we find laboratory models and demonstrations which lead directly from primary principles to applications. Otto von Guericke's work on air pressure affords a conspicuous illustration of this phase of scientific work. The primary principles had been worked out by Torricelli and Pascal, but Guericke invented an air pump by which he could produce a significant vacuum. Guericke's work was important both for the development of the atmospheric engine of Newcomen and for the establishment of a technique of experimentation which, with improvements, enabled Boyle to carry the analysis of pressure in gases to the formulation of the famous laws.[1]

If we wish to complete the survey of work below the level of

1. Pascal (translation, 1937), pp. xv–xx, affords a survey of the whole episode. The appendices contain selections from the works of Galileo and Torricelli. See also Schimank (1936), pp. 37–55.

general commercial use, we must include the production of new objects of luxurious consumption. The development of technology has been profoundly influenced by the desire to produce articles of superior quality and special character for ritualistic use and for consumption by dignitaries of church and state outside the limits of explicit ceremonial use. Under such circumstances considerations of cost have been ignored and processes and products developed to gratify the desire to achieve distinction. In the early history of metallurgy, glass-making and textile production, the development of luxury items was an important factor in invention. In the modern period the development of clocks and watches illustrates the importance of ritualistic and luxury demand. The mechanical clock was clearly developed in response to the ritualistic needs of the larger ecclesiastical establishments. Water clocks were not a convenient means of maintaining the schedule of services. Even a crude mechanical clock was superior; it was more accurate and required less attention. With improvements in craftsmanship, clocks and watches became an outstanding item of luxury consumption. In many instances the accuracy of the movement was subordinated to the decoration of the case. But these crafts became the basis of work in light engineering which laid a secure foundation for the heavy-duty power engineering that became important in the eighteenth century.

There is thus a broad field of activity in which costs have been subordinated to the achievement of novelties in science and in the production of luxuries. Activities in this field have fallen somewhat outside any economic calculus. In the early modern period much of the work of scientist-inventors was associated with gainful professional work, especially in the fields of art and engineering. Painting, sculpture, architecture and general engineering were not sharply specialized occupations. Much experimental work was done in the shops or studios, so that science, invention and practice of the arts went hand in hand. The universities also created opportunities for science and invention. In the seventeenth century Galileo and Newton were the most distinguished representatives of the universities, though they were not alone. Boyle, Huygens and Otto von Guericke were the most distinguished men who had personal wealth to use for their work.

Patronage of the wealthy and of chiefs of state was of course a further source of finance for work in primary science and invention.

The sixteenth, seventeenth and early eighteenth centuries were more notable for the accomplishments in science and primary invention than for achievements that added conspicuously to the productivity of industry and agriculture. However, the foundations were laid for the great technical achievements that followed directly upon the development of power engineering that stems from Watt. In the nineteenth century, basic work in the field of electricity was accomplished in a period in which practical achievements lay in the fields of mechanical and civil engineering. It is, therefore, important to recognize the necessity of this underlying work in science and primary invention. In general, the financing of such work came from sources which were essentially the same as in the sixteenth and seventeenth centuries; but the external features were somewhat different.

Since the beginning of the nineteenth century the research programs of the major universities have undergone a remarkable development. The private laboratory has given place to organized laboratory instruction in the universities, and research has become a recognized duty. General university functions have been expanded by the establishment of special research centers in the universities supported by public funds or private endowment, or both. Direct government support of basic research has a longer history in the agricultural field than elsewhere, but the importance of public provision for primary research is increasing in other fields. Even if political pressures require state-supported research to give much time to secondary research of immediate interest, the larger organizations will doubtless contribute much to basic research and primary invention.

The research organizations of many corporations today provide a certain amount of free time for the personal projects and interests of the research worker. In some corporations, too, the general program of the staff includes many basic or primary problems not expected to yield immediate commercial results (Bernal, 1939, pp. 35–70, 126–54, and 261–91; Freedman, 1950; Mees, 1950, pp. 5–16 and 51–149).

Secondary inventions and new investment

In order to clarify distinctions between different types of invention, they may be classified as primary, secondary and tertiary. Underlying inventions not carried to a stage of general commercial use may be classified as primary inventions. Inventions which open up a new practical use may best be considered as secondary inventions, whatever their importance. Any invention which extends a known principle to a new field of use should be so classified. The noncondensing engine and the locomotive should thus be treated as distinct secondary inventions, despite the utilization of some of the principles of the Watt condensing engine. Improvements in a given device which do not clearly extend the field of use can be classed as tertiary inventions. They are not to be ignored, but they stand in a different position in the sequence of technical change, and they have different consequences for the economy. Such inventions may increase the efficiency of the secondary invention or add to its convenience and safe operation; but they remain subordinate in importance if they do not extend the field of use.

In the field of secondary invention the associations and problems of the inventor are profoundly changed. Contacts with science are weakened and contacts with business assume commanding importance. When functions are not fully specialized, the inventor acts as an entrepreneur; this stage is characterized by the inventor-entrepreneur just as the first stage is characterized by the scientist-inventor. These differences in the activities of inventive types were clearly present in Schumpeter's mind, but the problems cannot be adequately analysed in terms of his categories of invention and innovation. All activities at the stage of secondary invention involve close interweaving of acts of skill, acts of insight and inventions. In enterprises which take a lead in the introduction of new inventions and processes, both inventors and administrators are engaged in inventive work of commanding importance. They also achieve great virtuosity in the performance of the associated acts of skill. Schumpeter underestimated the degree of novelty involved in these acts of skill – of both the engineers and technicians and the administrative staff of the enterprise.

The character of the choices to be made and their relation to the financing of the enterprise can be appreciated best if we concentrate attention on particular examples. Three cases are especially significant: the development of the locomotive, the development of interchangeable-part processes of manufacture, and the introduction of the Bessemer process in the iron and steel industry. All three cases exhibit the importance of a period in which technical achievements were imperfect. The early locomotives did not compete decisively with horsepower. Whitney's methods of production were at first limited in scope and used elementary techniques; but though not fully developed for thirty or more years they were commandingly successful from the start. The Bessemer process, in its early form, was restricted to particular ores. The failure to understand these limitations at the outset led to such great disappointment that the whole procedure for the introduction of the process had to be changed.

In histories of these inventions these critical periods of difficulty are frequently ignored or underemphasized. Full analysis is clearly necessary in order to understand the process of investment in new industries. If good judgement is exercised, risks of loss do not exceed the risks in established industries. The beginnings of commercial application precede the full accomplishment of the secondary invention. In many instances even the major invention remains to be achieved; in other cases the process of critical revision is conspicuously incomplete. The implications of Schumpeter's analysis suggest the opposite order of development: the completion of the secondary invention is represented as preceding the entrepreneurial work on application.

If we study the history of the locomotive with a dominant interest in engineering detail, we find three important steps in the achievement: the Pen-y-darran locomotive of Trevithick, 1804; the *Royal George*, built by Timothy Hackworth in 1826; and the *Rocket*, built by Robert Stephenson & Co. in 1829. Trevithick's engine was impractical because its steam capacity was low and it was too heavy for the cast iron rails then in use. Hackworth's *Royal George* was an effective heavy-duty freight locomotive decisively superior to horses, but it was not suitable for passenger service and needed much improvement in details. The *Rocket* was the first locomotive in which all essential features were incorporated in a mature form, and definitely the first locomotive designed to operate at high speeds on rails. It is an oversimplification to stress any single one of these steps as the controlling secondary invention. The vocabulary of common speech does not supply convenient words to express an achievement spread over time in a number of steps. The best we can do is to use the plural form 'secondary inventions'.

The problems of new investment have been dominated by the multiplicity of steps involved at this stage and by the relatively small magnitudes of improvements necessary to justify the expenditures. Trevithick's locomotives were incidental to his work on the high-pressure engine as a stationary source of power. The expenses incurred in making the model and in demonstrations of the road locomotive (1798–1802) were covered by the income from the engineering work that was Trevithick's primary concern. The demonstration at Pen-y-darran was financed by Homfray, the mine-owner for whom Trevithick had built a number of stationary engines. The engine itself was built to work a hammer, so that the special expenditure was restricted to the adaptation of the engine to operation on the tram line at the mine. The test was not intended to open up an application of steam power to the transport work of the mine (Dickinson and Tetley, 1934, pp. 42–65). Although demonstrations of the locomotive on rails were made at London in 1808, Trevithick did not himself, or through associates, develop any project for the systematic operation of a tram line by steam locomotives. His work, however, inspired the projects in the northeastern coal fields which began with Blenkinsop's work at the Middleton Colliery, three miles south of Leeds (1812).

These locomotives were largely the work of Matthew Murray, but royalties were paid to Trevithick for the use of his patents. Blenkinsop added a rack rail, so that this application of steam has been frequently regarded as a diversion of attention from the basic pattern of steam traction. The toothed wheel, however, was not intended to make up for lack of adhesion in a smooth rail; it was a naïve and relatively simple driving mechanism which was good enough to remain in operation for many years around the collieries (Dendy Marshall, 1939, pp. 19–30). These applications, therefore, were in no sense failures.

A locomotive sent to the Royal Iron Foundry in Berlin for the mines at Gleiwitz could not be put into operation because of the opposition of the local engineers and workers. There was similar resistance to the use of a locomotive in the colliery at Saarbrücken (Dendy Marshall, 1939, p. 34; and 1928, pp. 19–21). It may be that these engines failed to achieve all the potentialities that other inventors realized, but they were good enough to compete directly with horses. There are a few statements about the costs and the performances of these engines, but they are not detailed enough to inspire much confidence. Fully loaded, speeds were about 2·5 miles per hour. With light loads, 10 miles per hour was claimed.

The development of the mature type of locomotive and of the civil engineering work on the line was accomplished by Hedley and George Stephenson on the tram lines operated by the Wylam, Killingworth and Hetton Collieries and the famous Stockton & Darlington Railroad, which was a colliery line with supplemental common-carrier functions. All this highly novel work was accomplished in the course of the systematic operation of transport service at these collieries. Use on a restricted scale, but with success, led to a great enlargement of the scale of operation and to a commanding technical superiority over alternative modes of transportation. It is easy to overlook the fact that the early achievements in the restricted field were of material value.

The Stockton & Darlington and the Liverpool & Manchester Railroad represent two successive enlargements of the application of the locomotive. The Stockton & Darlington was an alternative to a canal. It offered common-carrier service, but the company leased the right to operate passenger service to contractors who put horsedrawn coaches on the line. Freight service was operated by locomotives on level stretches and by cable haulage on two inclines. No feature of the project was at that time new, though it did involve a longer line than the colliery tram lines. The Liverpool & Manchester project was an alternative to a line operated by cables and stationary engines. It was planned to offer both freight and passenger service. Passenger service had not previously been attempted on rails, but a number of steam coaches were in operation on highways out of London and the potentialities of the locomotive were understood though still underestimated. Speeds of fifteen to twenty miles per hour were presumed to be achievable. The conditions for the Rainhill competition prescribed a speed of not less than ten miles per hour. The *Rocket* averaged fifteen miles per hour, and ran for a time at a rate of twenty-nine miles per hour. After the accident to Huskisson, the *Northumbrian* of similar design ran fifteen miles at a rate of thirty-six miles per hour. Fifteen or sixteen years later the *Rocket* ran four miles in four and one-half minutes, or at a rate of fifty-three miles per hour (Smiles, 1868, pp. 325 and 327n; Stretton, 1896, p. 36).

The Liverpool & Manchester project marked the culmination of eighteen years of work in the application of the steam locomotive to tram lines. Under progressively exacting conditions of use, both the locomotive and the civil engineering work were greatly improved. Wrought iron rails of improved design replaced the cast iron rails used at Pen-y-darran for Trevithick's demonstration. The locomotive had greater steam capacity, and the differentiation between the freight and the passenger locomotive was understood. The technique had been carried to a point at which it could be used for a generalized system of inland transport. The work of these critical years involved the co-operation of engineers as skilled technicians and as inventors, and of businessmen as colliery managers seeking means of expanding transport services for which the supply of horses was becoming a limiting factor. Novelties were emerging at many levels and in many forms but under conditions which made losses unlikely, though the magnitude of the gains was uncertain. The new technique could be used if it was not less productive than the current alternative. As long as the scale of operations remained small, the dangers of losses were also small. Over-optimism was not likely to develop until the whole group of secondary inventions had been completed, and work began for the generalization of the new technique in the economy as a whole. The railway crisis started in England in 1845.

The general pattern observed in the development of the locomotive and the railroad may be seen also in the history of Eli Whitney's manufacture of muskets on the principle of interchangeable parts. The idea itself was not new. The Swedish engineer Christopher Polhem had perceived the general elements of such a system of manufacture, though he had not attempted to work within the field of precision mechanisms. Work on fire-arms on such a system had been tried in France but had not been pushed to conspicuous and decisive accomplishment.

The significance of Whitney's work lay not only in the bare idea, but also in the progressive mechanization of the process with machine tools capable of great precision. The actual manu-facture of muskets passed through several stages. The first con-tracts were executed by filing the parts of the lock to conform to patterns or jigs. The parts were worked out in soft metal, sub-sequently tempered and hardened. At first, only the locks were made on an interchangeable-part system. Later the stocks were shaped on pattern-turning lathes which reduced the amount of hand labor to a minor fraction of the prior requirements. The lock mechanism, too, was manufactured by methods which re-duced handwork and increased the degree of precision achieved. This was accomplished by developing the technique of die stamping for some parts, and by the development of the milling machine to supplant the laborious processes of filing. These activities covered about twenty years. There was thus a group of interrelated secondary inventions, all of which were essential to the mature accomplishment, though commercial success was assured when the new procedure had not been carried beyond its most elementary form (Mirsky and Nevins, 1952, pp. 128–46 and 177–205).

The array of machine tools ultimately developed were not implicit in or even suggested by the system of jig filing that was first used. The new tools were independent inventions of great merit and importance whose earlier history and background lay in the field of lathes, which worked originally in wood and soft metals below the level of precision required by the system of interchangeable-parts manufacture.

The complexity of the stage of early secondary invention is illustrated also by the history of the refining processes in the iron and steel industries. The transition from an industry dominated by malleable iron to one dominated by steel was brought about by the improvement of refining processes, so that a larger scale of production could be achieved at lower costs. The development was begun by the introduction of the Bessemer converter. By means of intense internal combustion, this device decarburized the cast iron coming from the blast furnace. The treatment of the charge of the converter required about twenty minutes. Puddling, the process then in use, required eight or ten hours to treat a much smaller charge and used highly skilled labor.

When Bessemer's process was announced it was immediately recognized as a potentially controlling factor in the industry. Bessemer proposed to lease the right to use the process on a royalty basis. As soon as attempts were made to apply the process, however, serious difficulties appeared. Much of the iron produced was brittle and poor, its quality far below any standard required in the industry. Bessemer was convinced of the truth of his claims, but careful analysis finally revealed the fact that the process could be used only in the treatment of iron that was free from phosphorus. Many of the ores then in use contained more phosphorus than was allowable if the new process was to be used. Bessemer now found it necessary to join some ironmasters in the organization of a company to apply the process to suitable ores (Bessemer, 1905, pp. 155–77).

Conditions in the iron and steel trades became very complex. The new process was directed to new fields in the market, and the older firms and older processes of production were able to maintain themselves for an interval; but conditions were very unstable, and the value of ores was profoundly affected by the selectivity of the Bessemer process. Balanced use of ores was ultimately made possible by two new secondary inventions: the open hearth process provided a method of refining that was effectively competitive with the Bessemer process commercially and happily suited to an array of ores that could not be treated by the Bessemer process. Neither of these processes was adapted to ores containing much phosphorus, though the open hearth was more tolerant of phosphorus than the Bessemer process. This difficulty was overcome by two chemists, Thomas and Gilchrist. They devised a basic lining for the refining furnaces which removed the phosphorus by a chemical reaction. With the two refining processes and the use of basic linings when required by the ores, it became possible to treat the whole array of commercially important ores. The transformation of the industry thus required all three of these secondary inventions. The full accomplishment took place between 1854, when the Bessemer process was announced, and 1878, when the basic lining of the furnaces was achieved by Thomas and Gilchrist.

The introduction of secondary inventions is profitable from a very early stage once capitalist methods of production have reached a point at which monopoly power can be achieved even for short periods of time. The inventor-entrepreneur cannot secure profit unless the introduction of the device or process can be controlled through patent privileges, secrecy or dependence upon small numbers of specially trained workmen. The textile inventions of the eighteenth century, Whitney's cotton gin, and Oliver Evans's mechanization of flour-milling afford characteristic illustrations of the weakness of the position of the inventor when the machines can be built by any craftsman without drawings or detailed specifications. When machine-building becomes a fully specialized occupation in any given field, new devices can be controlled and protected even without a patent system, though patent privileges are of undoubted social value in creating property rights to invention.

The development of the use of power in transport and manufacture is also an important condition. Increases in the scale of manufacture through the use of power or through the opening of wider markets give new significance to monopoly. The great watchmakers of France and England in the eighteenth century made outstanding contributions to their craft and to the whole field of mechanical engineering. But their most distinguished inventions merely gave them prestige in a craft which at best enabled them to sell their products to distinguished customers willing to ignore the costs of objects of luxury and ostentation. The work of the Dollands in the optical field is perhaps even more striking, because telescopes and microscopes appealed almost exclusively to wealthy amateurs with little serious interest in science. Work on the chronometer was stimulated by the prizes offered in France and England for a means of determining longitude at sea.

The characteristic phenomena of capital formation as we know them since 1800 are clearly attributable to the precision working of iron and steel and the developments in the related field of general power engineering. The textile inventions of the eighteenth century would have had a different and more restricted application if it had not become possible to supplant the machines of wood and soft metals with well-designed machines of iron and steel.

The spatial diffusion of new techniques

When the first cycle of secondary inventions is complete, or even well launched, the diffusion of the new technique throughout the economy presents a new set of problems. At this stage new acts of insight and new inventions are somewhat obscured because such a large part of the total task can be achieved by acts of skill. Furthermore, no particular invention is necessary at any given time to make the process or activity commercially effective. Delays incidental to positive invention cease to be of immediate consequence; improvements that are achievable are desirable, but they merely increase the profit derived from a process that is already profitable. There is thus some justification for the over-emphasis so commonly placed upon particular inventions in the first cycle of secondary invention; but, though understandable, these judgements are misleading.

The usual treatment of Watt's inventions gives too little credit to Newcomen and to the engineers who developed the high-pressure engine. In the textile series the stress on the earliest forms of the inventions is excessive. With the caprice characteristic of incomplete analysis, the importance of the work of the Darbys in the iron industry is ordinarily understated, but the emphasis upon puddling and rolling and upon the steel-making processes in the nineteenth century is sound. The history of the iron industry is perhaps the clearest instance of a succession of secondary inventions whose independence and importance are rarely questioned.

The acts of skill which dominate the spatial diffusion of new techniques appear both in the field of technology and in the field of economic administration. The process of development discloses new characteristics which are most effectively described if we distinguish between meliorative effects and the cumulative accumulation of knowledge, on the one hand, and growth as a quantitative and adaptive phenomenon, on the other. Once the point of secondary invention is reached the scale of the economy and its productivity are affected. These results can be observed in the growth of population and in the changes and net increases in consumption.

Approached from this point of view, the process of growth in the economy is a resultant of the array of accelerating factors represented by technical changes and the decelerating or limiting factors of scarcity of resources. Older concepts of change assumed the intermittent occurrence of particular changes in an essentially stable economy. A mature concept of the process of invention requires us to conceive of changes as essentially continuous in units of small magnitude. Discontinuities are not felt as such because the magnitudes are small. The substantial continuity of the process of growth is clear in the records of the iron and steel industry and in the field of power engineering. When products of specialized use are involved, the case is not as clear. Illustrative material is afforded by the history of the petroleum industry and by the development of heavier-than-air flight. The statistics of the production of illuminating oils are interesting when the whale oils and mineral oils are combined in one series for a substantial period. Adequate analysis of the records, however, would require more space than is here available.

There is enough statistical material to show that growth in the economy proceeds at varying rates: at accelerating rates up to the inflection point; at decelerating rates beyond it. In some fields growth may occur as a continuous process of deceleration, as presumed by Raymond Pearl, but such rigorous conditions are not characteristic of the biological processes in individual organisms or of human economies. It is not possible at this time, with the data available, to identify a specific mathematical formula with the process of social growth. The records we have now, however, are consistent with the suggestions made by R. A. Lehfeldt in 1916. He pointed out that the integral of the normal curve affords a satisfactory formula for much statistical material if we use the logarithms of the items. The formula expresses a cycle of orderly growth between two limiting magnitudes at varying rates of change (Lehfeldt, 1916, pp. 329–32). The framework of the curve does not quite fit the facts of social life, because major technical changes displace the upper asymptote and release new potentialities of growth. The upper asymptote of the preceding cycle is passed and becomes the lower asymptote of the next cycle. Even with good statistical material, the records of actual growth would not conform to the formula in portions of the curve close to the asymptotes implicit in the major portion of the record.

The production of pig iron in Great Britain in the eighteenth and nineteenth centuries exhibits as much conformity to the integral of the normal curve as the quality of our statistical material would warrant us to expect. The magnitude of the growth is most vividly reflected in the figures for *per capita* consumption: 1735, 15 pounds; 1800, 26 pounds; 1830, 77 pounds; 1890, 303 pounds. The record discloses a relatively smooth curve despite the major inventions that were required to make this development possible (Bowden, Karpovich and Usher, 1937, pp. 384–5). It seems strange that there is no evidence of more discontinuity. Some might take the position that the whole development was implicit in the application of coal and coke to iron-making. A better explanation is that the occurrence of inventions and acts of insight does not stand outside the formula of probability. Favorable constellations of events occur as a matter of chance. Over a large field and long periods of time, the emergence of novelty may well conform to the formula of probability. The normal curve may thus be related to all the arrays of phenomena involved in the process of growth.

If these inferences are justifiable, we have means of studying the rate of capital formation in economies undergoing technical change. Even if the statistical record of capital is inadequate, sampling methods will give an indication of the changing relations between product and capital, so that inferences can be drawn from the records of production.

The organic quality of the process of growth has important bearings upon the financial operations underlying new investment. It is not necessary to assume that the funds are deflected from older channels by the higher bids in new industries and undertakings. The credit resources available are sufficient to make the new investments largely dependent upon the expectations of gain in the new activity. New techniques of production affect the value of resources so directly that technological change is reflected in increases in the physical quantity and in the value of basic resources. New techniques of transportation affect site values and the values of agricultural and mineral lands that would otherwise be inaccessible to markets. These facts are obvious, but the timing of these changes in values is sometimes inadequately analysed.

Analysis must be considered from two points of view: that of the statistician and that of the entrepreneur. The statistician may legitimately regard the change in value as implicit in the substantial accomplishment of the secondary inventions. Changes in the coal and iron resources of Great Britain would, therefore, date from the successful work of the Darbys with coke. All these resources would have been revalued at least after 1735. Any survey of physical resources may legitimately distinguish between potential and actual resources. If technical and market conditions do not justify current utilization, particular deposits should be classified as potential. If workable at current costs, resources are actual. In a region whose resources are significantly known, the stocks available from the statistical point of view commonly provide for long periods. Primary minerals are available as 'actual' resources for intervals of 300 to 500 years; and potential resources in mineralized areas extend in many instances to 1000 years.

The entrepreneur must deal with shorter time spans and more immediate market relationships. He is primarily concerned with close analysis of actual resources, though potential resources may be of great moment if the technical or transportation problems admit of some change in technique that would bring the resources into the current cost structure in the industry. The entrepreneur considers three categories of resources: resources that are actual but not proven by detailed surveys, proven resources, and currently developed resources. The time span for entrepreneurial activity would rarely exceed twenty-five years, though acquisitions of title to prospective ores may well be carried further. The essential feature of the situation lies in the emergence of material values well in advance of current exploitation. These values can properly support a structure of long-time loans, and with good judgement should involve no greater risk than in a supposedly established industry. The long-time value structure in the old industry is not insulated against change. The development of petroleum as a primary fuel, for example, has had an adverse effect on the values of coal properties.

The influence of new techniques of transport upon land values is more complex than the effects of technical change on particular fixed resources. Changes in transportation affect site values for all classes of sites, but most particularly for the first-, second- and third-class urban units. These sites increase in size proportionately to the population within a given area and an expansion of their market areas (Zipf, 1949, pp. 374–83). The internal structure of the urban units is affected both by local facilities for transport and by techniques of house construction.

Both mineral lands and agricultural land are sensitive to the market connections that are a function of transport costs. Improvements in transportation, therefore, affect all the values of the land and fixed resources of the economy. The application of power to land transport has completely transformed the world economy. Before the development of the railroad the world economy was in effect the maritime fringe of the great continents. The interiors were open only to the extent that some form of water transport was available, and, as a consequence, contacts with the deeper interiors were very restricted. The railroad opened all these interiors to world markets, and today the emergence of a truly global economy is painfully evident. Its substantial emergence can perhaps be dated as early as 1878, when wheat from the prairies of the United States entered the European markets with such devastating effect.

No problems of technical diffusion and expansion are as complex as the agrarian dislocations created by the introduction of new transport facilities extending the areas devoted to particular crops. Such areas of cultivation bear no significant relation to the development of demand for the product. The extension of the sugar culture to the New World presents a complex episode of this type that cannot be accurately analysed because of the inadequacy of our statistical material. On the other hand, the expansion of wheat culture in the prairie regions of the United States, Russia and Australia can be studied in detail. Large portions of these areas were suitable only to wheat culture, as the rainfall pattern was unfavorable to general farming. Beyond the margins of wheat culture, specialized grazing became important.

Competition with these frontier producing areas was disastrous to the established agricultural areas in Europe dependent upon general farming; both mixed and cereal farming areas suffered, though the latter were more severely affected than the former. Wheat prices fell, even in relation to other commodities. Since non-human uses could be brought into the market only at prices that were unremunerative to growers, primary reliance lay in the expansion of human consumption. Over the period of fifty-five years studied by Malenbaum, acreage expanded in excess of the requirements for human consumption. In the decade 1929–39, this excess acreage averaged thirty-five million acres per year (Malenbaum, 1941).

The problem of maintaining equilibrium in the spatial economy appears in a special form in the construction of the railroad network. In Great Britain, public policy became committed at an early date to a principle of requiring presumptive evidence of cost recovery as a condition for the grant of a charter. In France and in the United States it was recognized that some railroad construction could advisedly be authorized even if there were no prospect of recovering all the costs immediately – or perhaps even ultimately. The grounds for this position are different in different circumstances. Some areas of France were provided with more service than they could themselves support in order to avoid losses from major displacement of population in such areas. In the United States it was presumed that losses would be temporary, but this optimism has not always been justified, though the operation of the network in the United States has been affected by a pattern of ownership that results in the segregation of strong and weak lines.

The diffusion of new techniques throughout the world economy is certainly not a smooth process that admits of the continuous maintenance of equilibrium conditions. The difficulties come from many sources: the magnitude of the units of investment, the length of the time intervals needed to produce effective responses in the economy, the number of different regions affected by a single change in technology, and the divergent influences upon particular kinds of resources or activities. Coal deposits and coalmining have been profoundly affected by the development of petroleum as a major fuel and by large-scale hydroelectric installations. Furthermore, the demand for coal in transportation has been influenced by the development of pipeline transport of oil and gas, and by long-distance transmission of electricity. Economy in the use of coal was recognized as an undoubted necessity as late as 1920; but when the changes came, the magnitudes of the reductions in demand could not be absorbed without drastic reorganization of the industry.

For all these reasons it is desirable to think of these spatial problems as growth problems. All such changes are not necessarily positive and advantageous. Even the most important technical advances affect adversely many activities and even entire regions.

The time intervals involved in the general process of technical advance are certainly much longer than current economic literature is ready to recognize. Some broad entrepreneurial policy decisions most certainly recognize intervals of twenty-five to thirty years, but the technical changes themselves do not disclose all their influence within such intervals. The development of the mineral economy reaches maturity with the development of petroleum and a generalized system for the distribution and use of energy in the form of electricity. The beginning of the change can be dated from the Newcomen engine and Darby's use of coal and coke, in the years 1708–12. A period of more than 200 years must, therefore, be considered. With long time intervals the achievement of equilibrium becomes more difficult.

Technology and centralized administration

It is tempting to presume that centralized administration might achieve a more adequate equilibrium in the economy than has been achieved in the past. Analysis of technical change certainly creates the presumption that important increases in the size of administrative units are likely to emerge as a consequence of changes in the techniques of communication. Increased speed of communication, new techniques of collecting and organizing information, and larger-scale units of production all lead us to presume that administrative units will become larger both in private and in public administration. Comprehensive control of the economy from some highly centralized policy committee raises a different issue. The precise form of the central authority is perhaps less important than the concept of the direction of investment in terms of statistical aggregates. If we assume that these aggregates can become the sole basis of primary economic policy, the economy must be a closed organic system.

The organic concept of the state and of the economy has exerted a powerful influence upon the social sciences since the early nineteenth century. No problem presents more sharply the issues between idealistic and empirical interpretations of history and of social structure. Idealists assert that the social structure is a closely knit and comprehensive whole. Empiricists hold that society consists of a loosely related array of structures, none of which is comprehensive. Individual elements may belong to more than one structure, and the parts may be combined in different ways in different periods of history. To the empiricist the wholes possess less vitality than their parts – though the parts cannot function except in some broader structure.

We therefore face a significant issue of analysis when we consider the aggregative records of an economy: do these aggregates indicate the presence of an organic whole that can be directly administered as a comprehensive entity? The idealist assumes that society is such a monolithic whole. However, a concept of emergent novelty and a theory of invention presented as cumulative synthesis are inconsistent with such a position. We should think of the events in our social life as a relatively large number of systems of events, some of which have contacts with each other; some of them, however, are positively opposed to each other. It is possible to show readily that many of these systems of events have no genetic relationship; though, over long periods of time, events which have not been related in the past converge toward a new synthesis. This pluralistic concept does not exclude the concept of organic structure, but it attributes organic relationships to smaller units and presumes that these relationships are created by social process. The organic wholes recognized by the empiricist are never comprehensive.

The statistical universes which are found in social life disclose systems of order that can be analysed in terms of probability. We must think of these aggregates merely as representations of arrays of events disclosing a multiplicity of causes which afford no clues to the degrees or degree of interdependence or independence. The empiricist thinks of them as statistical aggregates which reflect large numbers of individual items. In the field of investment, multiplicity is manifested as a large number of separate judgements – judgements about the values of resources and new materials, about costs of production and of marketing, and about consumer responses to different price policies. Even in fields dominated by well-organized markets, values are not independent of individual judgements; they do not come to the entrepreneur from the outside as a final fact. He is concerned with a future that is not fully calculable, so that even in the best market economy individual judgement must be exercised. The market registers an array of judgements; it does not make judgements.

Centralization of fiscal and monetary authority can serve usefully in guiding many of the value judgements to be made, but fiscal and monetary policies do not make independent judgements unnecessary. Wise policies can direct the process of investment, but they cannot change the primary character of the process as a summation of individual items proceeding upward from a large number of separate decisions. It is not necessary to assume that all decisions are made by entrepreneurs charged with the administration of independent corporations operated on a profit basis. Government corporations or departments may also be making judgements about investment in which cost factors may be given primary importance or subordinated to other social interests. The form of the process would not be altered by the introduction of new criteria for decisions as long as it is recognized that many independent decisions are being made.

The fully administered economy may seem to achieve unity, but this is not necessarily the case. Are the totals for the comprehensive plan aggregates of items that have been added together or are they allocations of funds that are divided after the total has been determined by a judgement of a central executive officer? No final answer can safely be given to such a question. In the past, concentration of authority has frequently been carried beyond the limits of the actual and effective making of executive decisions. Orders that emanate from the highest levels of authority are actually based on information and texts that are collected and prepared by lower levels of the bureaucratic hierarchy. Since the operation of bureaucracies is capricious, it is dangerous to generalize about the effective limits of the authority of central executive officers. Practical limits of centralization are determined by techniques of communication. We may now recognize the desirability of a much greater concentration of administrative power than was possible before the introduction of the telegraph and telephone, but even after all these developments, and with ancillary inventions in electrical tabulation, it is probable that important limitations to centralization should be recognized. If there is to be any innovation, there must be opportunity for independent judgements about the use of time and of uncommitted funds. Novel achievements emerge most freely beyond the limits of the pressures of convention and authority. The existence of many different authorities is the ultimate safeguard of the individual and the basis for the continuity of the process of invention in all its forms and fields. Major innovations in technology would probably encounter more resistance in an authoritarian society than in a society without full concentration of control over property and choice of occupation.

The adaptation of social and economic administration to changing circumstances will probably be quicker and more successful in a society without full centralization. But social and economic structures will undoubtedly function even if centralization is carried beyond the point of full efficiency.

References

BERNAL, J. D. (1939), *The Social Function of Science*, Routledge & Kegan Paul.

BESSEMER, SIR H. (1905), *Autobiography*, Offices of Engineering.

BOWDEN, W., KARPOVICH, M., and USHER, A. P. (1937), *An Economic History of Europe Since 1750*, American Book.

DAVIS, M. M. (1906), 'Gabriel Tarde: an essay in sociological theory', thesis, Columbia University

DENDY MARSHALL, C. F. (1928), *Two Essays*, Locomotive Publishing, London.

DENDY MARSHALL, C. F. (1939), *Early British Locomotives*, Locomotive Publishing, London.

DICKINSON, H. W., and TETLEY, A. (1934), *Richard Trevithick: The Engineer and the Man*, Cambridge University Press.

FOX, H. G. (1947), *Monopolies and Patents*, University of Toronto Press.

FREEDMAN, P. (1950), *The Principles of Scientific Research*, Public Affairs Press.

HIRSCH, N. D. M. (1931), *Genius and Creative Intelligence*, Sci-Art Publishers.

KOFFKA, K. (1935), *Principles of Gestalt Psychology*, Routledge & Kegan Paul.

KÖHLER, W. (1929), *Gestalt Psychology*, Liveright.

LANGE-EICHBAUM, W. (1931), *The Problem of Genius*, Routledge & Kegan Paul.

LEHFELDT, R. A. (1916), 'The normal law of progress', *J. Royal Stat. Soc.*, vol. 79.

MALENBAUM, W. (1941), 'Equilibrating tendencies in the world wheat market', thesis, Harvard University; substantially the same analysis was published by Malenbaum in 1953 as *The World Wheat Economy, 1885–1939*, Harvard University Press.

MEES, C. E. K. (1950), *The Organization of Industrial Scientific Research*, McGraw-Hill.

MIRSKY, J., and NEVINS, A. (1952), *The World of Eli Whitney*, Macmillan.

MORGAN, C. L. (1930), *The Criminal Mind*, Longman.

PASCAL, B. (1937), *The Physical Treatises of Pascal: The Equilibrium of Liquids and the Weight of the Mass of the Air*, Columbia University Press, trans. by I. H. B. and A. G. II. Spiers.

PASSER, H. C. (1953), *The Electrical Manufacturers, 1875–1900*, Harvard University Press.

SCHIMANK, H. (1936), *Otto von Guericke, Burgermeister von Magdeburg*, Magdeburg.

SMILES, S. (1868), *The Life of George Stephenson and of His Son Robert Stephenson*, Harper & Row.

STONE, C. P. (ed.) (1951), *Comparative Psychology*, Prentice-Hall, 3rd edn.

STRETTON, C. E. (1896), *The Locomotive Engine and Its Development*, Lockwood.

TINBERGEN, N. (1951), *The Study of Instinct*, Clarendon Press.

USHER, A. P. (1929), *A History of Mechanical Inventions*, McGraw-Hill; revised edn, 1954, ch. 4.

WELFORD, A. T. (1951), *Skill and Age: An Experimental Approach*, Oxford.

WHEELER, W. M. (1928), *The Social Insects*, Routledge & Kegan Paul.

ZILSEL, E. (1926), *Die Entstehung des Geniebegriffes*, J. C. B. Mohr, Tübingen.

ZIPF, G. K. (1949), *Human Behaviour and the Principle of Least Effort*, Addison-Wesley.

3 V. Ruttan

Usher and Schumpeter on Invention, Innovation and Technological Change[1]

V. Ruttan, 'Usher and Schumpeter on invention, innovation and technological change', *Quarterly Journal of Economics*, November 1959, pp. 596–606.

Most social scientists would probably accept the sequence in which the three terms – invention, innovation and technological change – are ordered in the title of this paper as representing a logical sequence; that is, invention in some manner is antecedent to innovation, and innovation is in turn antecedent to technological change. The distinction between exactly what is meant by invention in contrast with innovation, and innovation in contrast with technological change, is usually less clear.[2] This absence of any clear-cut analytical distinction among concepts which have been assigned such important places in current economic discussion is particularly disturbing.

In this paper, I attempt to clarify the meaning of the three concepts. A comparison of Schumpeter's treatment of the role of innovation and the innovator in the process of economic development with Usher's discussion of the emergence of strategic inventions provides a useful focus around which to carry out this attempt. This is particularly true in view of the fact that neither the long association of Usher and Schumpeter as colleagues at Harvard nor their common interest in economic development has resulted in any published attempt to explore the relationship between their work.[3]

1. Research for this paper was conducted under the auspices of the Purdue Agricultural Experiment Station and was financed by grants from the National Science Foundation and Resources for the Future. The author has benefited from critical comments on an earlier draft of this paper by Rondo Cameron of the University of Wisconsin and Robert Wolfson of Michigan State University.

2. For illustrations of this point, see the discussion of 'Innovation' by H. M. Kallen and of 'Invention' by Carl Brinkman in the *Encyclopedia of the Social Sciences*, vol. 8, pp. 58–60 and 247–51.

3. References by either Schumpeter or Usher to the works of the other are

The discussion of Schumpeter's (1939) treatment of the concept of innovation will center mainly on chs. 3 and 4 of *Business Cycles*, where his discussion of the role of innovation in economic growth is stated in its most developed form (see also Schumpeter, 1934, esp. ch. 2, for an earlier and less well-developed discussion). Usher's discussion of the emergence of strategic inventions is presented most fully in chapter 4 of the revised edition of his classic *A History of Mechanical Inventions* (1954).

Innovation and technological change

The concept of innovation has played a more important role in economics than the concept of invention. It was not, however, until Schumpeter identified innovation as the essential function of the entrepreneur and then went on to make the innovator and the process of innovation one of the three elements, along with credit and profit maximization, out of which he constructed a theory of economic development[4] that the concept achieved its greatest vogue.

Innovation and the innovator were quite distinct from invention and the inventor to Schumpeter. He repeatedly emphasized the distinction.

Innovation is possible without anything we should identify as invention, and invention does not necessarily induce innovation, but produces of itself . . . no economically relevant effect at all (Schumpeter, 1939, vol. 1, p. 84).

Schumpeter not only rejected the idea that innovation is

exceedingly rare. Schumpeter's limited references to Usher's work are no doubt due in part to the fact that Usher's theory of the process of invention was less well developed in the first edition (1929) of *A History of Mechanical Inventions* than in the second edition (1954), which appeared after Schumpeter's death. We have, of course, Usher's (1951) article on the 'Historical implications of the theory of economic development', which was dedicated to Schumpeter. Usher's only critical comment on Schumpeter's work is, so far as I know, contained in two brief paragraphs in an article on 'Technical change and capital formation' (1955; reprinted as Reading 2 of the present selection).

4. 'Schumpeterian development rests on (1) the innovation process and on the innovator, (2) on the credit mechanism, and (3) on the drive for profit maximization' (Wolfson, 1958, p. 45).

dependent on invention in any direct manner; he also asserted that the social process which produces innovations is distinctly different 'economically and sociologically' from the social process which produces inventions.

Refusal to identify innovation with invention does not in itself sufficiently identify the concept of innovation. Schumpeter's definition is in terms of a change in the form of the production function.

We will now define innovation more rigorously by means of the production function. . . . This function describes the way in which quantity of products varies if quantity of factors vary. If, instead of quantities of factors we vary the form of the function, we have an innovation (pp. 87–8).

Again:

. . . we will simply define innovation as the setting up of a new production function. This covers the case of a new commodity as well as those of a new form of organization or a merger, or the opening up of new markets . . . (pp. 87–8).

Schumpeter's formulation of the production function differed from the neo-classical formulation in that capital was excluded and only labor and land were included as inputs. Furthermore, Schumpeter explicitly rejected the possibility of measuring the effect of innovations through changes in the production function. He argued that price changes and non-neutrality of innovation would effectively limit the possibilities of measurement.

In spite of the difference in the production function concept, Schumpeter's definition of innovation bears a remarkably close resemblance to the definition of technological change currently used by students of productivity and technological change. Compare, for example, a recent definition by Solow with the above quotation from Schumpeter.

If Q represents output and K and L represent capital and labor in 'physical' units, then the aggregate production function can be written as:

$$Q = F(K, L; t);$$

the variable t . . . appears in F to allow for technical change. I am using the phrase 'technical change' as a shorthand expression for *any kind of a shift* in the production function (Solow, 1957, p. 312).[5]

5. For a similar definition, see Ruttan (1956, p. 61).

Fellner (1956a, pp. 35–41; and 1956b, pp. 195–6) discusses the same concept under the heading of technological-organizational change.

It seems fairly clear that current interest in technological change and growth of total productivity is focused on the same problem which Schumpeter treated under the heading of innovation. That problem is the effect of technological and organizational change, operating through the production function, on economic growth. Schumpeter was primarily interested in changes in the production functions of the technological leaders – the innovating firms – because of the growth forces which adoption of new methods of production set in motion. Recent students of growth and productivity have, on the other hand, given major attention to the production function which describes the average performance of the economy or industry.[6]

Neither Schumpeter nor the growth economists have given explicit attention to the *process* by which innovation – technological and organizational change – is generated. Schumpeter *used* his definition of innovation referred to above, three simplifying assumptions,[7] and a set of historical observations regarding the time sequence of cycles of innovation and business activity as one cornerstone of his theory of economic development.

Neither in *Business Cycles* nor in Schumpeter's other works is there anything that can be identified as a theory of innovation. The business cycle, in Schumpeter's system, is a direct consequence of the appearance of clusters of innovations. But no real explanation is provided as to why innovations appear in clusters or why the clusters possess the particular type of periodicity which Schumpeter identified in terms of Kitchen (forty months), Juglar (ten years), and Kondratieff (sixty years) cycles. Indeed, in discussing the cyclical pattern of innovation that emerges from his review of economic history, Schumpeter specifically eschewed any theoretical basis for the observed pattern of behavior (1939, p. 173).

6. This distinction is based on Brozen (1951).
7. These assumptions are that innovations: (1) entail construction of new plants and equipment, (2) are introduced by new firms, and (3) are always associated with the rise to leadership of new men. (Schumpeter, 1939, pp. 93–6.)

If Schumpeter did not present a theory of innovation, where do we go for such a theory? I will argue, in the next section, that the basis for such a theory is presented by Usher in *A History of Mechanical Inventions*.

Invention and innovation

The concept of invention has played a rather important role in the literature of applied technology, of sociology and of history. It has occupied only a peripheral, though expanding role, in the literature of economics.[8]

One of the problems economists have had to face in using the concept of invention has been the difficulty of providing a generally acceptable analytical definition of the term in contrast with the legal-institutional definitions emphasized by national patent offices.

Usher's solution to this problem is to define invention in terms of the emergence of 'new things' which require an 'act of insight' going beyond the normal exercise of technical or professional skill.

Acts of skill include all learned activities whether the process of learning is an achievement of an isolated adult individual or a response to instructions by other individuals. *Inventive acts of insight* are unlearned activities that result in new organizations of prior knowledge and experience . . . (Usher, 1955, p. 526).

Such acts of insight frequently emerge in the course of performing acts of skill, though characteristically the act of insight is induced by the conscious perception of an unsatisfactory gap in knowledge or mode of action (Usher, 1955, p. 523).[9]

Usher's interest in the study of inventions goes beyond definition and description. In chapter 4 of the revised edition of *A History of Mechanical Inventions*, he faces up to the question: how does one explain the emergence of inventions in contrast with the performance of acts of skill? He identifies three general approaches to this problem: the transcendentalist, the mechanistic process, and the cumulative synthesis.

Usher rejects both the transcendentalist and the mechanistic

8. For an excellent survey of the literature on the economics of invention, see Nelson (1959, pp. 101–27).

9. See Reading 2, above.

theories. *The transcendentalist approach*, which attributes the emergence of invention to the inspiration of the occasional genius who from time to time achieves a direct knowledge of essential truth through the exercise of intuition, is rejected as unhistorical. He argues that, in fact, the act of insight has not been the rare, unusual phenomenon assumed by the transcendentalists. Furthermore, the act of insight which results in the perception of a new relationship requires a highly specific conditioning of the mind within the framework of the problem that is to be solved. It is not an accident, in other words, that a bicycle mechanic contributed to the development of the automobile.

Usher also rejects *the mechanistic process theory* of the Chicago sociologists, although he stresses the importance of the empirical results achieved by Ogburn and Gilfillan (Gilfillan, 1935; 1952; Ogburn, 1922, pp. 80–102; 1938; Ogburn and Gilfillan, 1937; Ogburn and Nimkoff, 1940, pp. 775–810). By demonstrating that the process of invention typically represents a new combination of a relatively large number of individual elements accumulated over long periods of time, the sociologists have effectively refuted the claims of the transcendentalists. Usher rejects, however, the mechanistic hypothesis that this process proceeds under the stress of necessity and that the individual inventor is merely an instrument of historical processes.[10] He argues that such an approach overlooks the significance of discontinuities inherent in the process of invention and that the 'acts of insight' required to overcome a particular discontinuity or resistance are possible for only a limited number of individuals operating under conditions which bring both an awareness of the problem and the elements of a solution within their frame of reference. Even under these conditions, it is not certain that the specific act of insight required for a solution to the problem will occur.

In formulating *the cumulative synthesis approach*, which he offers as an alternative to the transcendentalist and mechanistic

10. According to Gilfillan (1935, p. 10), 'With the progress of . . . invention, apparently a device can no longer remain unfound when the time for it is ripe. . . '. And 'There is no indication that any individual's genius has been necessary to any invention that has any importance. To the historian and social scientist, the progress of invention appears impersonal.'

process theories of invention, Usher draws on the insights into mental and social processes provided by Gestalt psychology.[11] Within this framework, major inventions are visualized as emerging from the cumulative synthesis of relatively simple inventions, each of which requires an individual 'act of insight'.

In the case of the individual invention, four steps are outlined:

1. *Perception of the problem*, in which an incomplete or unsatisfactory pattern or method of satisfying a want is perceived.

2. *Setting the stage*, in which the elements or data necessary for a solution are brought together through some particular configuration of events or thought. Among the elements of the solution is an individual who possesses sufficient *skill* in manipulating the other elements.

3. *The act of insight*, in which the essential solution of the problem is found. Usher stresses the fact that large elements of uncertainty surround the act of insight. It is this uncertainty which makes it impossible to predict the timing or the precise configuration of a solution in advance.

4. *Critical revision*, in which the newly perceived relations become fully understood and effectively worked into the entire context to which they belong. This may again call for new acts of insight.

A major or strategic invention represents the cumulative synthesis of many individual inventions. It will usually involve all the separate steps that may be found in the case of the individual invention. Many of the individual inventions do no more than set the stage for the major invention, and new acts of insight are again essential when the major invention requires substantial critical revision to adapt it to particular uses. A schematic presentation of the elements of the individual act of insight and

11. 'The Gestalt analysis presents the achievements of great men as a special class of acts of insight, which involves synthesis of many items derived from other acts of insight. In its entirety, the social process of innovation thus consists of acts of insight of different degrees of importance and at many levels of perception and thought. These acts converge in the course of time toward massive synthesis. Insight is not a rare, unusual phenomenon as presumed by the transcendentalist; nor is it a relatively simple response to need that can be assumed to occur without resistance and delay' (Usher, 1954, p. 61).

the cumulative synthesis as visualized by Usher are presented in Figures 1 and 2.

Usher's cumulative synthesis theory is appealing on other grounds in addition to its basis in Gestalt psychology. It provides a unified theory of the social processes by which 'new things' come into existence that is broad enough to encompass the whole range of activities characterized by the terms science, invention and

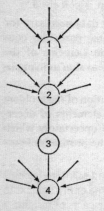

Figure 1 The emergence of novelty in the act of insight. Synthesis of familiar items: (1) perception of an incomplete pattern; (2) the setting of the stage; (3) the act of insight; and (4) critical revision and full mastery of the new pattern

innovation. One is no longer forced to maintain, as Schumpeter did, the increasingly artificial distinction between the processes of invention and innovation nor to explain away the association between scientists, inventors and entrepreneurs as 'merely a chance coincidence'.[12]

One may wish to reserve the term innovation to cover the emergence of elements of novelty in the area of economic organization and invention to cover the emergence of elements of

12. For a discussion of the difficulty of maintaining this distinction, see C. S. Solo (1951). For an excellent case study illustrating the close association of scientists, inventors and entrepreneurs in a modern scientific industry, see Maclaurin (1950); also, Nelson (1959).

novelty in the area of technology. I see no particular point in such a distinction, and I can see many advantages in eliminating a distinction for which there is no real conceptual basis. Indeed, it would be more in line with both popular usage and the termin-

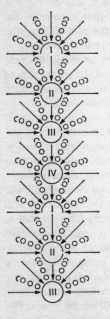

Figure 2 The process of cumulative synthesis. A full cycle of strategic invention and part of a second cycle. Large figures I–IV represent steps in the development of a strategic invention. Small figures represent individual elements of novelty. Arrows represent familiar elements included in the new synthesis

ology of other disciplines to use the term innovation to designate any 'new thing' in the area of science, technology or art.[13] When

13. See, for example, the use of the term innovation in the series of articles in the *Scientific American*, vol. 199, September 1958, devoted to 'Innovation in science'. The individual articles include: J. Bronowski, 'The creative process', pp. 58–65; Paul R. Halmos, 'Innovation in mathematics', pp. 66–73; Freeman J. Dyson, 'Innovation in physics', pp. 74–82; George Walk, 'Innovation in biology', pp. 107–15; John R. Pierce, 'Innovation in

greater precision is required, innovation can be preceded by an appropriate adjective – 'scientific innovation', 'technical innovation', 'organizational innovation', or something even more precise. Invention then becomes that special sub-set of technical innovation on which patents can be obtained.[14]

Another advantage of Usher's theory is that it clarifies the points at which conscious efforts to speed the rate or alter the direction of innovation can be effective. The possibility of affecting the rate or direction of innovation is obscured by the transcendentalist approach with its dependence on the emergence of the 'great man' and is denied by the mechanistic process approach with its dependence on broad historical trends or forces.

The focus of conscious effort to affect the speed or direction of innovations centers around the second and fourth steps in the process as outlined by Usher – in *setting the stage* and in *critical revision*. By consciously bringing together the elements of a solution – by creating the appropriate research environment – the stage can be set in such a manner that fewer elements are left to chance. It would be inaccurate to suppose that we yet, or perhaps ever, will find it possible to set the stage in such a manner as to guarantee a breakthrough in any particular area. As we learn more about the effectiveness of various research environments, the probability that breakthroughs will be achieved should be increased. For the present, we can only emphasize how little we know about the administration of basic research (see Villard, 1958; US National Science Foundation, 1958; Nelson, 1959, pp. 119–29).

At the level of *critical revision*, we have already made considerable progress in bringing economic and administrative resources to bear. Many of the elements of critical revision require 'acts of skill' in contrast with 'acts of insight'. The effectiveness of modern research procedures in shortening the time span from the test

---

technology', pp. 116–34; Frank Barron, 'The psychology of imagination', pp. 150–69; Warren Weaver, 'The encouragement of science', pp. 170–78.

14. This same distinction is also emphasized by Nelson (1959), p. 103: 'Granted that the term [invention] can be used to include all human creative activity, from composing a poem to developing a chemical process, for our purpose it is well to define the term narrowly so as to include only patentable inventions.'

tube to the production line testifies to our ability to exert conscious direction at the applied research level.

One limitation of Usher's theory should also be noted. If the operational criteria against which one tests the theory is that of predicting the course of technical (and other) innovations in the sense that the Chicago sociologists have attempted, the approach is obviously inadequate. Indeed, Usher (1954, p. 66) asserts that prediction in this sense is impossible.

The theory is operational in another sense, however. It does permit one to predict those points in the innovation process where economic resources can be brought to bear most effectively – in 'setting the stage' and at the level of 'critical revision'. The effective institutionalization of applied research at the 'critical revision' level during the last two hundred years, and the last fifty in particular, together with the growing interest in the problem of creating an institutional environment favorable to 'basic research', does provide an operational test that is consistent with Usher's theory.

Summary and conclusions

The implications of the preceding discussion can be summarized in terms of their implications for the problems of terminology which were raised in the introduction.

Firstly, I suggest that we abandon the attempt to provide an analytically meaningful definition of the term invention. The term is most useful in a descriptive sense when confined to its institutional content and is used to refer only to that sub-set of technical innovations which are patentable. As such, the term has relatively little historical or spatial continuity.

Secondly, I suggest that we extend the concept innovation to cover the entire range of processes to which Usher's theory applies, that is, to the process by which 'new things' emerge in science, technology and art. In this context, inventions become an institutionally defined sub-set of technical innovations.

Thirdly, I suggest that we employ the term technological change in a functional sense – to designate changes in the coefficients of a function relating inputs to outputs resulting from the practical application of innovations in technology and in economic organization.

It should be emphasized that, while acceptance of Usher's theory to cover the process of innovation, as defined above, eliminates the need for Schumpeter's distinction between the processes of invention and innovation, no violence is done to Schumpeter's theory of economic growth. Schumpeter's major interest was not in explaining the process of innovation but in discovering the effect of variations in the rate of both technological and organizational changes on economic growth and development.

And Usher's theory does suggest how two defects in Schumpeter's system can be overcome. It eliminates the need for an artificial distinction between the process of invention and the process of innovation. Furthermore, it directs attention to those stages in the process of innovation where conscious effort can be brought to bear to alter or change the direction of innovation.

References

BROZEN, Y. (1951), 'Invention, innovation and imitation', *Amer. econ. Rev.*, vol. 41, pp. 239–57.

FELLNER, W. J. (1956a), 'Discussion' [of Abramovitz's paper on 'Resource and output trends in the United States since 1870'], *Amer. econ. Rev.*, vol. 46, pp. 35–7.

FELLNER, W. J. (1956b), *Trends and Cycles in Economic Activity*, Holt, Rinehart & Winston.

GILFILLAN, S. C. (1935), *The Sociology of Invention*, Follett Publishing.

GILFILLAN, S. C. (1952), 'Prediction of technical change', *Rev. Econ. Stat.*, vol. 34, pp. 368–85.

MACLAURIN, W. R. (1950), 'The process of technological innovation: the launching of a new scientific industry', *Amer. econ. Rev.*, vol. 60, pp. 90–112.

NELSON, R. R. (1959), 'The economics of invention: a survey of the literature', *J. Bus.*, vol 32, pp. 101–27.

OGBURN, W. F. (1922), *Social Change*, Viking Press.

OGBURN, W. F. (1938), *Machines and Tomorrow's World*, Public Affairs Committee, pamphlet no. 25.

OGBURN, W. F. and GILFILLAN, S. C. (1937), several articles in W. F. Ogburn (ed.), *Technological Trends and National Policy*, US National Resources Committee.

OGBURN, W. F., and NIMKOFF, D. F. (1940), *Sociology*, Houghton Mifflin.

RUTTAN, V. W. (1956), 'The contribution of technological progress to farm output, 1950–75', *Rev. Econ. Stat.*, vol. 38, pp. 61–9.

Schumpeter, J. (1934), *The Theory of Economic Development*, Harvard University Press.

Schumpeter, J. (1939), *Business Cycles*, McGraw-Hill.

Solo, C. S. (1951), 'Innovation in the capitalist process: a critique of the Schumpeterian theory', *Quart. J. Econ.*, vol. 65, pp. 417–28.

Solow, R. M. (1957), 'Technical change and the aggregate production function', *Rev. Econ. Stat.*, vol. 39, pp. 312–20. [Reprinted as Reading 16 of the present selection.]

Usher, A. P. (1951), 'Historical implications of the theory of economic development', *Rev. Econ. Stat.*, vol. 33, pp. 158–162.

Usher, A. P. (1954), *A History of Mechanical Inventions*, Harvard University Press.

Usher, A. P. (1955), 'Technical change and capital formation', in *Capital Formation and Economic Growth*, Universities – National Bureau Committee for Economic Research; [Reprinted as Reading 2 of the present selection.]

US National Science Foundation (1958), *Research and Development and Its Impact on the Economy*, US Government Printing Office.

Villard, H. H. (1958), 'Competition, oligopoly and research', *J. polit. Econ.*, vol. 66, pp. 483–97.

Wolfson, R. J. (1958), 'The economic dynamics of Joseph Schumpeter', *Econ. Devel. cult. Change*, vol. 7, pp. 31-54.

4 M. Blaug

A Survey of the Theory of Process-Innovations

M. Blaug, 'A survey of the theory of process-innovations', *Economica*, February 1963, pp. 13–32.

The topic considered here is innovations, not inventions: the entrepreneur is viewed as facing a list of known but as yet unexploited inventions from which he may select. How this list is itself drawn up and continuously augmented is an issue that deserves separate treatment.[1] Any analysis of the *rate* at which techniques improve in an economy cannot ignore the pace and scope of inventive activity. But my subject is not what Kaldor has aptly called 'the degree of technical dynamism' in an economy – the flow of new ideas and readiness of the system to absorb them – but rather the *pattern* of technological progress when the economy is in fact technically dynamic.

Innovations fall into two classes: process-innovations and product-innovations. The terms are self-explanatory. The distinction is to some extent an artificial one: the introduction of a cost-reducing process is sometimes accompanied by a change in the product mix, while new products frequently require the development of new equipment. In practice the two are usually so interwoven that any distinction between them is arbitrary. Nevertheless, in principle, novel ways of making old goods can be distinguished from old ways of making novelties. Since the index-number problem has so far doomed all theoretical analysis of innovations which alter the quality of final output, the refusal to discriminate between product- and process-innovations would close the subject of technical progress to further analysis.

A process-innovation is defined as any adopted improvement in technique which reduces average costs per unit of output despite the fact that input prices remain unchanged. The new

1. See Jewkes *et al.* (1958). For a valuable review of the entire field, see Nelson (1959).

technique may involve drastic alterations in equipment, but this is not a necessary feature of the definition: the mere reorganization of a plant may be as factor-saving as the introduction of new machines. We have to guard ourselves against a widespread misunderstanding at this point. An innovation represents an addition to existing technical knowledge. Since the production function already takes account of the entire spectrum of known technical possibilities – known, in the sense of being practised somewhere in the system – innovating activity ought to denote the adoption of hitherto untried methods. But few indeed are the successfully adopted innovations which do not have a long history of unsuccessful trials, and even the imitation of previously tried techniques almost always involves a 'creative response'. This difficulty has caused some authors to define technical progress as any change in the production methods of an enterprise, irrespective of whether the new method has been tried before. But this compromise blurs the distinction between a movement along, and a shift of, the production function. In the interest of theoretical clarity, we will adhere for the time being to the traditional definition.

Process-innovations as defined above have been under serious discussion for a generation or more, although older contributions go back to Pigou (1920, pt 4, ch. 4), Schumpeter (1911, trans. 1934, ch. 2), Wicksell (1901, trans. 1934, vol. 1, pp. 163–4), Marx (1894, vol. 3, chs. 4 and 5), Mill (1848, bk 1: ch. 6, sects. 2 and 3; ch. 7, sect. 4; bk 4: ch. 4, sect. 6) and even Ricardo (1821, chs. 2 and 31). In the 1930s attention was chiefly focused on the problem of classifying innovations into mutually exclusive boxes such as labour-saving, capital-saving and neutral. Some attempt was made to interpret the history of technical change in terms of these classifications. This aspect of the discussion has come to dominate the stage in the post-war period. Recently, interest has centered on the mechanism which accounts for systematic bias or lack of it in the factor-saving slant of technical change. Nothing like a consensus has yet developed with respect to any of these questions. There is in fact much confusion about the analytical status of the theory. Indeed, some readers may object to the very title of this article as an attempt to confer scientific dignity on what is nothing more than a hodge-podge of suggestions. In the words of a student of growth theory: 'contemporary economics does not offer a "theory of innovations" in the sense of a systematic theory accounting for the rate and "slant" of innovations through time', and 'the lamentable state of our understanding of the origin and process of technical change' constitutes 'the most important deficiency' in contemporary theorizing on economic growth (Bruton, 1956b, pp. 287, 297). Nevertheless, there is a vast literature on technical change which reveals certain lines of agreement. It is hoped that this survey may indicate where we stand and how much more needs to be done.

Taxonomy

Assuming there are only two factors of production, it seems natural to classify an innovation as labour-saving when it raises, and capital-saving when it lowers, the capital–labour ratio. The factor-proportions criterion, however, remains ambiguous unless a level of output or a time-period is specified. The introduction of an innovation cannot fail to affect a firm's output-decision and, unless the firm operates at constant returns to scale, a larger output will in the long run imply a different capital–labour ratio even at constant factor prices. Furthermore, innovations take time to install, and the results of newly installed equipment do not accrue immediately: this is the basis of Lange's (1945, p. 74) distinction between the gestation-period and the operation-period of an innovation. An innovation which is capital-saving once in place may nevertheless absorb capital relative to labour during the period of its construction. Conversely, an innovation which reduces initial capital costs may ultimately prove to be capital-using if it accelerates the rate of replacement through obsolescence.

It would seem that the way out of these difficulties is to classify innovations with reference to their effect on the capital–labour ratio utilized in the production of a *given* volume of output. Technical change is represented graphically by a movement towards the origin of each and every isoquant. Since factor prices are given to the individual firm, this leads to a straightforward interpretation of different innovations (see Figure 1).

The classificatory scheme which has gained the widest adherence is that associated with the works of Hicks and Robinson: it will be recognized immediately that their definitions differ from those just laid down. They define a labour-saving innovation as one which raises the marginal product of capital relative to that of labour *at a given capital–labour ratio* employed in producing a given output, and conversely for a capital-saving innovation (see Figure 2).

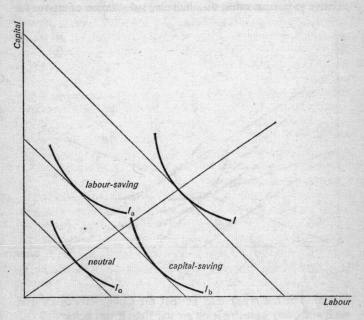

Figure 1

On the face of it, the Hicksian definition does not conflict with our earlier one: an innovation which is labour-saving at constant factor prices in terms of the capital–labour ratio will certainly be labour-saving at constant factor-proportions in terms of the relative marginal products. The earlier definition, however, is applicable to individual firms or industries facing given factor prices, while the Hicksian definition is geared to the economy as a whole. In aggregative analysis we are interested in the effect of innovations on relative factor prices. Changing factor-prices induce factor-substitution, and the Hicksian definition is designed to distinguish the latter from technical change proper. For the economy as a whole, the capital–labour ratio by itself is obviously an inadequate criterion for classifying innovations. An innovation which is labour-saving on the industry level may, if it is widely adopted in other industries, lower wage rates relative to interest rates, thus inducing substitution of labour for

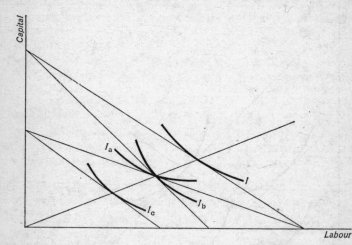

Figure 2

capital. The net result may be that the capital–labour ratio in each and every industry is no greater than it was before.

For a given capital–labour ratio, a labour-saving innovation reduces labour's share of total income while a capital-saving innovation increases it. The ultimate effect on relative shares depends of course on the ease of substitution between capital and labour in response to changing factor prices. With the aid of Hicks's concept of the elasticity of substitution (σ), we can add that, say, a labour-saving innovation will tend to raise labour's relative share if $\sigma > 1$.[2]

2. The elasticity of substitution is defined for a given output as the percentage change in the relative amount of the factors employed divided by the proportionate change in their relative marginal products or relative

Mrs Robinson both extended and simplified Hicks's argument by showing[3] that σ varies with the elasticity of the average productivity of a factor (η). For any given amount of the factors, an innovation necessarily raises the average productivity of each factor taken separately in the same proportion as output. It also raises the marginal productivity of each factor, but not necessarily in equal proportions. If a labour-saving innovation raises the marginal product of labour (M_l) less than the marginal product of capital (M_k), M_l must have risen less than A_l. Graphically expressed, a labour-saving innovation increases the percentage gap between A_l and M_l; it reduces η_l. Similarly, a capital-saving innovation is 'gap-increasing' with respect to the productivity functions of capital: it reduces η_k. Neutral innovations, on the other hand, raise the average and marginal productivity curves of each factor iso-elastically. Since the elasticity of the average productivity curve of a factor determines its relative share,[4] Mrs Robinson's classification agrees with that given by Hicks.

But whereas Hicks adhered to the premise of given factor

---

prices; to put it graphically, the relative change in the slope of a vector drawn through the origin up to the isoquant divided by the tangent gradient of the isoquant. It follows that σ is inversely proportional to the degree of the curvature of the isoquant. Thus a labour-saving innovation which lowers the wage rate relatively to the interest rate at a given capital–labour ratio by 1 per cent may ultimately lower the capital–labour ratio by more than 1 per cent if it increases the scope of factor-substitution. See Hicks (1932, pp. 117–30); Lerner (1933; 1947, ch. 13); and Allen (1938, pp. 340–43).

3. J. Robinson (1938); also in W. Fellner and B. F. Haley (eds.) (1946). Robinson (1934, p. 330n) offered a definition of σ identical to Hicks but applied to single industries.

4. With $Q = $ output, the elasticity of A_l can be expressed as

$$\eta_l = \frac{L}{Q/L} \frac{d(Q/L)}{dL} = \frac{L^2}{Q} \frac{d}{dL} \frac{(Q)}{(L)} = \left(L \frac{dQ}{dL} - Q\right) = \frac{L}{Q} \frac{dQ}{dL} - 1 = a - 1.$$

But $a = \dfrac{L}{Q} \dfrac{dQ}{dL} = \dfrac{M_l}{A_l}$,

whereas a is the elasticity of the production function with respect to labour. Under perfect competition, $dQ/dL = w/p$, where w/p is the real wage rate. Hence $a = M_l/A_l = Lw/Qp = $ labour's relative share. Likewise, capital's share is equal to β, the ratio M_k/A_k. If, following the usual formula for the elasticity of demand, we define η not as $M/A - 1$, but as $1 - A/M$, then $a(\beta) = 1 - 1/\eta$. Hence, a 'gap-increasing' innovation, implying a reduction in η, reduces the share of output imputed to the factor in question.

inputs, supplemented by some loose remarks on the possible magnitude of σ, Mrs Robinson's argument ran in terms of comparative statics: she assumed full employment of all factors before and after the introduction of the innovation at a constant rate of interest. Once an innovation has raised the yield of capital, investment is adjusted to the new technique so that equilibrium is once again attained, the rate of interest having remained unchanged throughout; that is, M_k is the same under the new technique as under the old.[5] If the rate of profit is determined

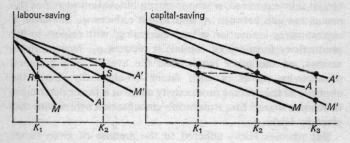

Figure 3

by M_k, this in turn implies that labour-saving innovations raise, and capital-saving innovations lower, the average capital–output ratio – the reciprocal of A_k. This follows simply from the algebraic fact that the rate of profit is a quotient of the profit share and the capital–output ratio.

We are thus provided with two alternative formulations, both of which look at the effect on relative shares. Drawing linear capital-productivity functions, Hicks's version of the left-hand diagram (in Figure 3) is that the innovation is labour-saving because capital's relative share for K_1 amounts of capital has increased in consequence of the innovation; only $\sigma < 1$ can account for the fact that factor-substitution, involved in the move from K_1 to K_2, has not reversed this result. Mrs Robinson's version is that the innovation is labour-saving because the

5. At a constant market rate of interest, the notional rate entering into the cost of capital goods is equal to the ruling rate of profit. The assumption that the rate of interest remains constant therefore avoids most of the difficulties of measuring capital, and seems to have been adopted for that reason. See Robinson (1956, pp. 118–23, 418–20).

elasticity of A' at K_2 (determined by the condition that $K_1 R = K_2 S$) is greater than the elasticity of A at K_1; since the percentage gap between the two curves has decreased, capital's relative share as well as the capital–output ratio have risen. The argument for capital-saving innovations is, of course, perfectly symmetrical.[6]

The Hicks–Robinson classification came in for considerable criticism.[7] Firstly, it was argued that the scheme breaks down

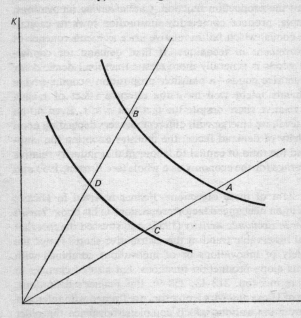

Figure 4

when the underlying production functions do not obey constant returns to scale: it is only for linear homogeneous production functions that the marginal productivity of a factor is determined solely by the ratios of the factors employed and not by their absolute amounts. We can always apply Hicks's definition to a given amount of the factors, but nothing much can be said about

6. The same reasoning is illustrated graphically in terms of total product functions by Meade (1961, pp. 25–6, 40–41).

7. Hicks's (1936) reply to his critics cites all the relevant contributions.

the new equilibrium level of inputs without knowledge of the shape of the production function.[8]

A backward-falling supply curve for labour and imperfect competition further complicate matters. But the fatal objection to Hicks's criterion, and Mrs Robinson's extension of it, is that it takes no account of commodity-substitution. The effect of an innovation cannot be inferred solely from the physical characteristics of the production function. Capital-saving innovations, for example, produce commodity-substitution towards capital-intensive goods, which fall in relative price as a consequence of the improvement in technique. If final demand for capital-intensive goods is generally more elastic than final demand for labour-intensive goods – a plausible proposition – capital-saving improvements might well have the ultimate effect of raising capital's relative share despite the fact that $\sigma < 1$. Even on an industry level, we emerge with different answers depending upon the elasticity of demand facing the industry adopting the innovation and the ratio of capital to labour in that industry relative to the average for the economy as a whole (see Brozen, 1953 and 1957).

In the light of these objections, Fellner reverted to Hicks's emphasis upon unchanged factor-proportions in his book, *Trends and Cycles in Economic Activity* (1956), and stressed the fact that the actual observable trend in the distributive shares is not the result solely of innovations or of innovations combined with movements along production functions, but also of changes in the product mix (pp. 212–13, 258–9). But Fellner's definitions, like those of Hicks and Mrs Robinson, are framed with reference to once-over innovations which completely displace the older techniques. For a classification explicitly designed to apply to the steady indivisible stream of improvements we must turn to Harrod's schema which runs in terms of the capital coefficient: labour-saving technical change raises and capital-saving technical change lowers the capital coefficient, while neutral progress

8. σ refers only to the same isoquant; but so long as the production function is linear and homogeneous, a move from A to D gives the same value for σ as a move from A to B since $\sigma = 0$ for the change from B to D. But the moment the production function departs from linearity, σ cannot be employed to measure the effects of changes both in ratio and in the scale of factors employed. Even for the same isoquant, σ is no longer determinate because σ_{lk} is no longer equal to σ_{kl}.

leaves the coefficient undisturbed, 'at a constant rate of interest' (Harrod, 1948, pp. 22–7). Harrod's criterion has met with the objection that all technical change, insofar as it prevents M_k from falling as capital accumulates through time, tends to reduce the average capital–output ratio. The capital coefficient appears to be an index of the rate of innovating activity, rather than of its factor-saving bias; a non-increasing capital–output ratio suggests that diminishing returns to capital along given production functions are being adequately offset by innovations (see Fellner, 1961a, pp. 59–60; and Kaldor, 1961, pp. 324, 340). But, Harrod might reply, a labour-saving innovation, because it is 'gap-reducing', is capable of raising M_k without raising A_k, or lowering the capital–output ratio. A steady stream of labour-saving innovations, therefore, is not likely to produce a perversely falling capital–output ratio. Thus the trend in the capital coefficient can be made to yield rough insights about the direction of technical change.

It remains true, however, that Harrod's definition is rigorous only if the rate of interest ($= M_k$) stays constant, in which case it is Mrs Robinson's definition travelling in disguise.[9] Harrod accepts the test of relative shares but rejects Mrs Robinson's comparative static analysis as 'inappropriate'. But if it is indeed inappropriate, keeping the rate of interest constant – a purist's argument at best – does not tell us what will happen to M_k. Hence, the capital coefficient criterion cannot be precise: an innovation which is capital-saving in terms of relative shares may be labour-saving in the technical coefficient sense if it fails to prevent a fall in M_k (see, e.g., K_3 in Figure 3).

For purposes of explaining the history of technical change, the capital coefficient suffers from the same shortcoming as the capital–labour ratio. The capital stock has been growing faster than the labour supply in all developed economies for a century or more. How much of this change in relative factor supplies has been induced by the rate and direction of technical change,

9. In *Essays in the Theory of Employment* (1937, pp. 96–8), Mrs Robinson in fact defined innovations in terms of the capital coefficient, given equilibrium adjustments to a constant rate of interest. For a neat demonstration that Harrod's neutral technical change comes to the same thing as Hicks's definition of neutrality, see Kennedy (1962).

and how much has been due to autonomous influences upon saving-propensities and upon birth and death rates? If we knew that the impulse to change came from the side of factor supplies, that autonomous forces were raising the ratio of capital to labour, we could infer from the observed stability of the capital–output ratio and from the failure of returns to capital to fall significantly through time, that innovating activity had been plentiful and not seriously biased in a labour-saving direction. For if the aggregate production function obeys non-increasing returns to scale, an increase of capital per man must in the absence of innovations raise capital requirements per unit of output and lower interest rates relative to wage rates.[10] But we have no way of separating autonomous from induced capital formation, and no direct knowledge of the shape of the aggregate production function. Not knowing what is the cause and which is effect, we cannot discriminate between arguments that account for the changes in the capital–labour and capital–output ratios by the character of technical change and those that explain the nature of technical change by the fact that capital for whatever reason grew faster than labour (see Solow, 1959, pp. 272–3, 283). As soon as we leave the abstract world of given factor-proportions, classifications of innovations in terms of capital–labour ratios, capital–output ratios or relative shares are all equally arbitrary.

The shortcomings of available classificatory schemes are not inherent in their logic. The crux of the difficulty lies in the inability of static-equilibrium theory to analyse continuous processes through time. Still, we must interpret real-world technology. Loose definitions are better than none. The test of relative shares, combined with the capital coefficient, does provide a rough apparatus for thinking consistently about the character of technical change, keeping in mind Mrs Robinson's dictum that 'a craven scruple of thinking too precisely on definition [of innovations] must not prevent us from trying to analyse them'.

The myth of a labour-saving bias

No one has ever been hard pressed to find examples of labour-saving or land-saving innovations, or to describe in general what

10. For a review of neo-classical theory on this point, see Ara (1958).

we have in mind when we talk about them. Capital-saving improvements, however, are often said to be difficult to exemplify because they represent exceptions to the general trend of technical change. And, indeed, if we confine ourselves to spectacular inventions it is difficult to find examples other than those usually mentioned, such as radio, telegraphy, airplanes and explosives for mining. But *any* change in the production function involves technical change. In this sense, examples of capital-saving innovations are perfectly easy to come by. Improvements extending the scope of auxiliary instruments, reducing the floor space needed to accommodate a given set of machines, or lengthening the physical life of a plant, fall into the class of innovations which economize fixed capital. Economies of working capital, on the other hand, always take the form of reducing the stock of goods-in-process which must be carried for a given output; typically, operating funds are released by the speeding-up of machinery, by faster handling of materials, by reductions in delivery time, and by fuel savings through recovery and use of waste products. On the face of it, there is no reason to believe that such improvements are any less frequent or pervasive than labour-saving improvements, or that they become important only when an economy is already richly endowed with capital. Some of the crucial innovations of the Industrial Revolution released rather than absorbed capital, and improvements in production-layout and in machine-design were as important then as now.[11]

Until quite recently, however, the prevailing view among economists was that technological progress has always been dominated by labour-displacing inventions. The usual rationalization for this viewpoint is that given by Hicks in *The Theory of Wages*. Drawing a distinction between innovations induced by changes in relative factor prices and innovations dependent upon autonomous progress in scientific and technical knowledge, he argued that the more rapid increase of capital relative to labour over the last century has caused induced innovations to be labour-saving. Autonomous innovations, however, might be

11. See Ashton (1948, pp. 91–2; 1953, pp. 90, 100, 108–13); Blaug (1961). (Habakkuk, 1962, pp. 162–9, advances some tentative arguments for a labour-saving bias under certain circumstances at early stages of development. The book appeared too late to take into account in this survey.)

assumed to be randomly dispersed without bias in either direction. Hence, the two types taken together yielded a preponderance of labour-saving improvements; this, he noted, seemed to be in conformity with observed facts. Apart from the troublesome notion of innovations induced by changes in factor prices – this would seem to involve factor-substitution, not technical change – it is clear from the reference to random dispersion that Hicks treats each innovation as having equal weight. But particular innovations differ radically in economic significance. What does the frequency distribution of the *number* of innovations of one kind or another have to do with the factor-saving slant of a continuous stream of improvements? To argue that, in the absence of specific knowledge, 'neutrality' is the best assumption is to commit the 'fallacy of equi-probabilities'. Hicks's entire discussion begs the question whether technical progress has in fact been biased in any economically meaningful sense.[12]

It is difficult nowadays to appreciate how quickly and how recently economists have changed their minds on this question. In 1937 Mrs Robinson declared: 'It appears obvious that the development of human methods of production, from the purely hand-to-mouth technique of the ape, has been mainly in the direction of increasing "round-aboutness", and that the discovery of short cuts, such as wireless, are exceptions to the general line of advance'. But in 1956 she concluded: 'There is no reason to expect technical progress to be exactly neutral in any one economy, but equally there is no reason to expect a systematic bias one way or the other. Capital-using innovations raise the cost of machines in terms of commodities and give entrepreneurs an extra motive to find ways to cheapen them. Capital-saving innovations tend to produce scarcity of labour in the consumption sector and give entrepreneurs an extra motive to increase productivity. Each type of bias tends to get itself compensated by the other' (Robinson, 1937, p. 135; 1956, p. 170; see also 1952, p. 43).

The myth of an inherent labour-saving bias in technical change derived its appeal from knowledge of the familiar facts of industrial history, strengthened by the tendency to discuss technical

12. See Bloom (1946). For a recent endorsement of Hicks's argument, see K. W. Rothschild (1954, pp. 117–19).

progress as if it consisted solely of inventions in the narrow sense of the word, rather than of any change whatever in the technical horizon of producers.[13] Since most capital-saving innovations consist of relatively minor improvements in the utilization of machinery, they tend to escape recognition. Indeed, awareness of the very existence of capital-saving improvements came late in the history of economic thought. The classical economists realized that time-saving improvements raise the rate of profit by increasing the turnover of money capital, but such ideas were not systematized, and economies of fixed capital were never seriously contemplated (see Blaug, 1960).

Sidgwick in *The Principles of Political Economy* of 1883 seems to have been the first to question the traditional idea that technical change is necessarily capital-using. Taussig in *Wages and Capital* (1896) suggested that the inventions of the future might save capital by 'shortening the period of production', and J. B. Clark, a decade later, made the point that many capital-using innovations do ultimately release capital after their gestation-period is over. As Jevons put it: 'The first cost of a steam collier is greater than for sailing colliers of equal tonnage. But then capital invested in the steam vessel is many times as efficient as the sailing vessel.' But none of them doubted that technical change had been overwhelmingly labour-saving in the past. The growing influence of the Austrian theory of capital around the turn of the century, emphasizing as it did capital formation which increases the durability of plant and equipment, further encouraged the belief that economic development, even when technical change is allowed for, typically entails not only an increase in capital per man but also a steady rise in capital requirements per unit of output.[14]

13. For example, when Douglas sought to justify neutral technical change as implied by a Cobb–Douglas production function fitted to aggregate data, he cited a study of the factor-saving distribution of '120 inventions which are commonly agreed upon as having been the most important of the last generation', implicitly assuming that technical change consists of nothing more than mechanical inventions (Rothschild, 1934, pp. 214–15).

14. Böhm-Bawerk maintained that while some inventions do reduce roundaboutness, the capital so released tends to be applied to lengthening the period of production elsewhere. Only if the innovation is both capital-saving and product-replacing will the average period of production be

The break with the labour-saving myth can be traced to the stagnation thesis of the 1930s: the increasing importance of capital-saving innovations was said to be one of the factors that contributed to the existence of a chronic deflationary gap. In the 1940s, empirical findings, demonstrating the long-run stability of the aggregate capital–output ratio in advanced economies, made economists more receptive to the idea of neutral technical change.[15] Neutrality came to be assumed in model-building, the earliest and most notable example being Harrod's growth model. As might be expected from a neo-Keynesian growth model, Harrod employs a two-sector breakdown of the economy, and it is a peculiarity of two-sector models that overall neutral technical progress need not imply neutral innovations in each sector. All technical advance, whatever its factor-saving bias, in the sector making investment goods is capital-saving inasmuch as it reduces the capital costs of real investment in every sector even at a constant rate of interest. But cheaper machines in turn induce the substitution of capital for labour. In the same way, all technical advance in the consumption-goods sector cheapens labour and encourages the substitution of labour for capital. Hence, overall neutrality for Harrod means either unbiased technical change proceeding at a uniform rate at all 'stages of production', or else a situation in which technical change in the investment-goods sector is just offset by a labour-saving bias in the consumption-goods sector, leaving the average capital–output ratio constant at an unchanged rate of interest.[16]

---

shortened. This he dismissed as exceptional, citing the secular increase in physical capital per head as presumptive evidence of the greater frequency of time-increasing inventions: 'Industrial experience will verify two propositions . . . first that with the larger capitalistic equipment, the product per unit of labour increases; and second, that this increase in product does not go on *pari passu* with the capitalistic equipment' (Böhm-Bawerk, 1896, p. 150). It is worth noting that Wicksell's famous discussion of the effect of inventions upon wages in the *Lectures on Political Economy* does not consider the possibility of capital-saving improvements.

15. The first writer to draw attention to the constancy of the capital–output ratio in a variety of industries over long periods of time was Carl Snyder, in an article published in 1936. See Stern (1945).

16. The former interpretation seems to be the one favoured by Harrod himself (1948, p. 23), and by Mrs Robinson (1956, p. 87).

Whatever interpretation we accept, it remains true that the trend in the aggregate capital–output ratio, even when calculated at a constant rate of interest, furnishes only presumptive evidence of neutrality. Leaving aside the consideration that only strongly biased technical change will show up clearly, the aggregate capital–output ratio is influenced not only by the direction of technical change but also by saving propensities, inter-industry shifts in investment, expectations about the future rate of technical advance, and the cumulative influence of the rate of growth of output. Moreover, the denominator of the ratio includes the end-product of spending on education, health and training, while the numerator refers solely to non-human capital. It is always possible to argue that technical change has shown no tendency towards 'capital deepening' because of the rise in the ratio of human to non-human capital over the last century. The fact that capital has grown faster than labour may itself be a statistical illusion, the result of measuring labour in man-hours instead of efficiency-units (see Solow, 1958, p. 630). Furthermore, it is not at all clear how capital is to be measured for purposes of verifying the neutrality hypothesis (see Green, 1960, pp. 59–62; Harrod, 1961; Kennedy, 1961, pp. 292–9). And what is still more serious is that there appears to be no agreement as to whether the average capital–output ratio has in fact been relatively stable: at a recent conference on capital theory, Barna, Fellner and Kaldor found that the ratio exhibited surprising stability for long periods of time while Domar, Lutz and Solow thought that the ratio had shown considerable variability (Lutz and Hague, 1961, pp. x–xi, 342–4, 369–70).

The argument for neutrality gains something when facts about the capital coefficient are combined with other empirical findings. Neglecting the possibility of increasing returns to scale, the sharp rise that has been witnessed in the capital–labour ratio should have led, in the absence of innovations, to a falling yield on capital as well as a falling share of profits in total income. In the last 100 years the rate of technical change has in fact been very high, judged by the upward trend in the average productivity of labour. Yet the rate of return on capital in the UK and the USA has shown a mild downward trend. This suggests that innovations, however plentiful, were not sufficiently labour-saving. The shift in the distribution of income from property to labour that took place over the same period likewise suggests that technical progress was not strongly biased in a labour-saving direction.[17] A somewhat stronger version of this impressionistic argument relies upon the Hicksian proposition that the elasticity of substitution between machinery and labour must eventually fall below unity in an economy where the capital stock is always growing faster than the labour supply; without innovations, the fact that $\sigma < 1$ implies that the profit rate and the profit share will fall as the capital–labour ratio increases. Economic growth in the advanced economies has, therefore, depended upon sufficient labour-saving technical change to prevent the chronic excess of saving over investment that would be produced by a falling yield on capital. In the light of 'the great depression' of the 1930s, it would seem that technology has not been adequately labour-saving: neutral technical change or even a mild capital-saving bias in the last three or four decades best account for the course of events.[18]

This Hicksian line of reasoning, valid as it may be, is not central to the argument for neutrality;[19] but the static micro-economic notion of diminishing returns to capital is. The latter rests on the treacherous assumption that it is meaningful to contemplate a temporally homogeneous structure of capital when the capital–labour ratio is rising. Inputs do not in reality remain homogeneous over time when factors are substituted for each other even with given technical knowledge; the mere process of capital

17. For a discussion of the evidence see Fellner (1956, pp. 246–57); Kravis (1959).
18. If $\sigma < 1$, innovations have to be labour-saving to produce constancy in relative shares. Since labour's relative share has in fact risen, neutral innovations with $\sigma < 1$ do account for the observed facts. Similarly, if $\sigma = 1$, as implied by a constant-elasticity Cobb–Douglas production function, neutrality is consistent with the historical record. But if $\sigma > 1$ only a capital-saving bias could explain how a rising capital–labour ratio produces an increase in the ratio of wages to profits. Unfortunately, it is not possible to infer $\sigma \leqslant 1$ or even $d\sigma/dt < 0$ from diminishing returns to the faster-growing factor without knowledge of the shape of the aggregate production function. See Hamberg (1959).
19. Hamberg asserts the contrary, citing Bruton (1956a) and several works by Fellner. But the essential part of Fellner's *Trends and Cycles* does not rest upon the assumption that $\sigma < 1$.

formation gives rise to changes in the composition of the capital stock and in the quality of the labour force.[20] It is misleading to base any 'strong' conclusions on the historical trend in the ratio of non-human capital to physical man-hours employed. At best we can say that the fact that reproducible capital has nearly tripled in quantity since 1900 with only a mild secular decline in the yield of capital, while man-hours rose by less than 50 per cent despite a threefold rise in hourly wage rates, is compatible with neutral technical change provided the elasticity of supply of capital was greater than that of labour (see Kravis, 1959, pp. 940–45). The assumption of neutrality does not violate the evidence. It has proved possible to fit a first degree homogeneous production function to American data, containing a multiplicative factor expressing neutral shifting in the function (Hogan, 1958; Solow, 1957).[21] By itself this proves nothing, but it does succeed in placing the burden of proof on those who argue that technical change is necessarily labour-saving.

If the data are compatible with the assumption of neutral but also with that of somewhat biased technical change, why the eagerness to uphold neutrality? In good part, no doubt, for model-building conveniences, but also because the idea of a persistent bias in the stream of innovations is itself difficult to swallow. It supposes the absence of any market mechanism that would counteract the bias. The standard view, however, best represented by Schumpeter's *Theory of Economic Development*, is that there is indeed no reason to believe that technical change is directly responsive to market pressures. The proof of this assertion is tautological: any market mechanism involves reactions to changing prices, and technical change is defined as consisting of cost-reducing improvements at constant factor prices; by definition, a market response is excluded. The traditional attitude, however, has been challenged in recent years. Perhaps it is possible to

20. This point recalls the age-old debate about the validity of *historically* diminishing returns. See, e.g., Knight (1944); Wright (1944). For a recent book on capital theory, emphasizing the heterogeneity of capital, see Lachman (1956).

21. For Solow's method applied to a broader set of data, see Massel (1960); also Reder (1959). In these studies, 'technical change' includes not only improvement of techniques within industries, but also improvements in labour skills and returns to investment in education and research.

account for the direction of technical change in terms of rational optimizing behaviour.

The inducement mechanism

An individual firm under perfect competition, facing given wage and interest rates over which it has no control, is not concerned with the factor-saving character of improvements. Its aim is to reduce total costs irrespective of whether the saving is made in the operating costs of labour or in the total cost of capital, provided they are of equal magnitude. The activities of all producers taken together may not be consonant with relative factor scarcities in the economy, with the result that future factor prices diverge from current ones. But no individual firm can take account of these macro-economic repercussions. Hence, the competitive market seems to provide no signals to induce 'appropriate' factor-saving innovations.

Given the orthodox view of the competitive firm, something like the relative constancy of the aggregate capital coefficient can only be explained on strictly technological grounds. Major changes in productive technique are labour-saving and capital-using, but these occur only sporadically in most countries. Once the new methods are in use, routine day-to-day modifications raise the capacity of equipment without additional expenditures. As a consequence, the capital–output ratio tends after a time to return to its technically-determined normal level. Since spurts of investment do not occur simultaneously in all industries, the net effect is to maintain a roughly constant trend in the aggregate capital coefficient.[22]

This sequence, it has been suggested, is best observed in 'social overhead' industries where indivisibility of capital makes for a phase of inaugural build-up ahead of the market. The more complex and expensive the capital installation, the more diverse the opportunities for improving its efficiency; hence, the strong correlation between the initial height of the capital–output ratio and the rate of its subsequent fall which has been observed in all

22. This argument is found, to give only a single reference, in Bruton (1955, pp. 326–7). It has been generalized by Kuznets (1960) who suggests that the history of all industries shows a definite 'life-cycle of capital–output ratios'.

regulated industries in the US (see Ulmer, 1960, pp. 65–71, 93–8, 101–10). This argument may be relaxed to take account of the influence of relative factor prices upon standards of obsolescence and so upon decisions to scrap and replace with new equipment. But even here proponents of the view that the capital coefficient is technologically determined would stress the fact that a high capital–output ratio need not imply proportionately higher fixed capital charges because it may be associated with long-lived equipment, and in general they emphasize the accidental appearance of new inventions promoting unforeseen type of investment. In Harrodian language, the thesis is that there is nothing in innovations that will automatically produce the warranted rate of growth (see Bruton, 1956a; and Power, 1958).

In opposition to this viewpoint, Fellner has reasserted Hicks's concept of induced innovations, notwithstanding the textbook theory of the competitive firm. The argument runs as follows (Fellner, 1956, pp. 220–22). Firstly, strongly biased technical change even under perfect competition creates conditions akin to monopsony in factor markets, and a monopsonist is made directly aware of relative factor scarcities in the economy by the gap between his average and his marginal factor costs. For example, persistent 'overshooting' in the labour-saving direction is bound in the short run to raise the price of capital[23] and in the long run to reduce the elasticity of the supply curve of capital to a firm below infinity; some form of capital rationing is instituted and firms are induced to seek out capital-saving improvements. On the other hand, 'overshooting' in the capital-saving direction leads to tightness in the labour market reflected in a 'wage drift'; once again, a situation of quasi-monopsony provides incentives to introduce labour-saving innovations. Secondly, although producers may face perfectly elastic supply curves in all factor markets, they become conditioned by experience to avoid disappointment by choosing improvements which save the relatively scarcer factor. Even a perfectly competitive firm 'learns' to adapt itself to a persistent and hence discernible trend in the shifting of factor-

23. The concept of the price of capital which corresponds to the wage rate as the cost of employing labour is that of annual capital costs per unit of real investment, taking into account both depreciation and interest charges and allowing (somehow) for expectations of obsolescence.

supply curves. In this way sharp cumulative changes in factor rewards, relative shares, and the capital coefficient are offset by appropriately slanted technical change. In other words, the warranted rate of growth does tend to be automatically maintained.

In effect, Fellner's two arguments come to the same thing: because of a 'learning process', firms behave *as if* they were monopsonists (Fellner, 1961b). The idea that firms learn from experience may strike some readers as alien to the static theory of competitive price. But, in fact, the competitive firm cannot rationally decide upon a particular level of output without some estimate of future product and factor prices. It is not sufficient to have 'primary information' about technical production possibilities, consumers' preferences and current input prices. To determine the profitability of investment, firms must also be acquainted with 'secondary information', which permits them to form expectations of future sales and prices. In the real world this type of 'secondary information' becomes available in the form of known restraints on rival behaviour which makes it possible to estimate the minimum or maximum competing supply in some future period. Thus, the competitive firm is by necessity 'forward looking' and driven to adjust behaviour in the light of expected events.[24] But even if individual firms lack foresight and act simply on the expectation that prices will remain unchanged, a Darwinian selection-process will produce an automatic adjustment mechanism (see Alchian, 1950; Heflebower and Stocking, 1958). Firms which persist in adopting capital-saving devices when wage rates are rising and interest rates falling will not prove viable. The successful innovator, alert to the signals transmitted by the price system, will be saving labour and absorbing capital, and the economist looking on will find the system as a whole adapting technical change to relative factor scarcities.

The theory of market-induced innovations is not required to explain why every individual firm adopts certain inventions rather than others. There is almost always a large gap between the average and the best-practice technique in an industry because machines are not scrapped until their operating costs equal the total costs (including capital charges) of a new machine. The

24. This argument is borrowed from Richardson (1959).

delay in the adoption of the best-practice technique is itself conditioned by relative factor prices. As Salter has shown, when real investment is cheap relative to labour, standards of obsolescence are stringent and the spread between the best and the average-practice technique is narrow. A fall in the interest rate or in equipment prices lowers the capital costs of adopting the best-practice technique, while a rise in wage rates increases the operating costs of average-practice techniques; both tend to induce scrapping of old machines. On the other hand, when real investment is dear relative to labour, the capital structure of an industry consists largely of outmoded equipment (Salter, 1960, pp. 66–70).

Thus, as soon as one firm in an industry has established the best-practice technique by introducing an innovation, the problem of the speed with which the new technique is diffused throughout the industry falls outside the domain of the theory of technical change. It is the adoption of the best-practice technique itself which alone constitutes genuine innovating activity. At least, so much is implied by the traditional definition of technical change as involving a shift in the spectrum of known techniques. This is not to deny that much of the literature on technical change is concerned with the problem of inter-firm diffusion of innovations (see, e.g., Mansfield, 1961, and the references cited there), but simply that diffusion raises no theoretical problems outside the corpus of received doctrine.

Whatever the validity of the theory of market-induced innovations, it is subject to important qualifications. The problem of indivisibilities of capital in certain industries restricting the scope of adjustments to relative factor prices has already been mentioned. Then there is the fact that improvements which save working capital are frequently the result of external economies generated by the growth of social overhead facilities which accrue to firms independently of their own actions and in no direct relationship to relative factor scarcities in the economy. Furthermore, while there is considerable evidence that inventive activity is itself responsive to perceived profit opportunities, it is far from true that inventions nowadays can be simply 'manufactured' in research laboratories to suit economic needs. Although we have no reason to believe that the available pool of inventions is itself systematically biased – labour-saving technical knowledge is not easier to come by than capital-saving technical knowledge – this is a possibility which should not be overlooked. On the other hand, the presence of monopoly in the real world may not prove to be as serious an objection to the thesis as appears at first glance. No clear picture has emerged on the relationship between market structure and innovations; the rate of growth of firms, as related to their age, seems to be a more important influence than size or power to set the price (see Fellner, 1951; for a general view of the literature, see Hennipman, 1954). Lastly, there is the possibility that wage–price rigidities will permit 'overshooting' in the short run. And if overshooting of a capital-saving nature occurs in an economy in which capital is already the relatively abundant factor, the resulting fall in the yield of capital may lead to Keynesian under-employment equilibrium. Instead of increasing labour scarcity inducing corrective labour-saving technical change, there is an excess supply of labour encouraging further 'overshooting'. Thus, the presence of a deflationary gap can put the adjustment-mechanism out of commission (Fellner, 1962). On the other hand, market-induced innovation is one of the forces preventing Keynesian unemployment. Without flexible wages and prices, however, it is powerless to act.

It is evident from casual impression that industrialized countries have not suffered over the last 100 years from excessively biased technical change. If technical progress is not something that happens wholly by chance, this suggests that some kind of adjustment-mechanism has been at work. But proof by way of *post hoc, ergo propter hoc* is inherently unsatisfactory. The theory of market-induced innovations, however, could be verified directly. Unfortunately, most of the available empirical material on technical change is useless for this purpose: it has not been gathered systematically to test any hypothesis and is rarely available in a suitably disaggregated form. What is needed to test the notion of the 'learning firm' are detailed case studies of innovating activity in particular industries.[25] Until then we

25. Very little has yet been done in this area, but see Bright (1947), Maclaurin (1949) and Scoville (1948). On the industry level, see the suggestive study by Melman (1958) and Salter (1960). Even the potentialities of the questionnaire-approach have not been fully explored. See Bloom (1951), and Carter and Williams (1958).

shall not be able to choose decisively between the concept of a technically determined 'life-cycle of capital–output ratios' and the theory of market-induced innovations.

Conclusion

The *rate* of technical advance influences its *direction* via changing standards of obsolescence. In this way, the flow of *inventions* plays a role in determining the pattern of *innovations*. New machines represent *product-innovations* for the machine-goods industries but *process-innovations* for the consumer-goods industries. Technical change, by producing cheaper machines or cheaper workers, generates its own pressures to bring about *factor-substitution*. None of these water-tight distinctions is realistic. Nevertheless, all fruitful theoretical analysis of changing technology has to this date proceeded upon the basis of these distinctions.

Recently we have been told to abandon all this and in particular to dispense with the fundamental neo-classical concept of the production function. In Kaldor's growth model, we are presented instead with a 'technical progress function', linking the growth of capital per man to the growth of output per man, which is defined for a given *rate* of change, rather than a given *level* of technical knowledge (Kaldor, 1957, p. 596). Since this function increases at a decreasing rate, it is possible to show that a high rate of capital accumulation 'appears' to produce labour-saving innovations in the capital coefficient sense, not because of the technical characteristics of innovations but simply because a larger volume of investment is competing for a constant flow of innovations. No one has yet managed to measure the state of technical knowledge, much less the rate of change of technological knowledge.[26] But the value of Kaldor's approach lies in the assumption that technical knowledge increases at a reasonably stable rate so that the technical progress function does not shift about. Since spurts of investment are often the occasion for innovative activity, it may well be that the technical progress function shifts systematically with the capital–labour ratio. This

26. The neo-classical idea of a given state of knowledge is admittedly an abstract one. But the concept of a given rate of change of knowledge is almost metaphysical.

deprives Kaldor's function of any meaningful interpretation. Whatever the other virtues of Kaldor's model, it throws no new light on the dynamics of technical progress and does not even provide meaningful categories for its analysis. Having postulated a particular investment function, he eliminates, by assumption, the problem of bias in technical change (Kaldor, 1957, p. 604).[27]

The neo-classical conception of technical change as involving shifts in the production function is full of difficulties, and the very notion of a production function as something that is purely technically determined, showing no traces of the influence of factor prices, tends to break down once we accept the idea that current output decisions are influenced by expectations of the future. The case for the neo-classical approach is that it provides a meaningful framework for organizing our knowledge of technical progress, and, to provide a more decisive consideration, that no satisfactory alternative approach is in view.

Judging from the findings of Solow, Abramovitz and Kendrick (see the references to Solow, 1960), the real-world importance of technical change in contrast to factor-substitution is the inverse of the attention their respective analysis now receives in economic textbooks. The Ricardo Effect – will a rise in wages induce substitution of capital for labour? – is still debated at the expense of discussing the much more significant question whether a tight labour-market will influence the selection of inventions and innovations. In the past, neglect of technical change has been justified by its intractability to the traditional tools of economics. So much the worse for the traditional tools, we might reply. But a survey of the literature on technical change suggests that the potentialities of orthodox theory in this connection have not yet been exhausted.

References

ALCHIAN, A. A. (1950), 'Uncertainty, evolution and economic theory', *J. polit. Econ.*, June.
ALLEN, R. G. D. (1938), *Mathematical Analysis for Economists*, Macmillan.

27. In a new paper, 'Capital accumulation and economic growth', *The Theory of Capital*, Kaldor meets some of the objections that were raised against his investment function, but adds nothing to the original statement of the technical progress function.

ARA, K. (1958), 'Capital theory and economic growth', *Econ. J.*, September.

ASHTON, T. S. (1948), *The Industrial Revolution, 1760–1830*, Home University Library.

ASHTON, T. S. (1953), *An Economic History of England: The Eighteenth Century*, Methuen.

BLAUG, M. (1960), 'Technical change and Marxian economics', *Kyklos*, vol. 13, no. 4.

BLAUG, M. (1961), 'The productivity of capital in the Lancashire cotton industry during the nineteenth century', *Econ. Hist. Rev.*, April.

BLOOM, G. F. (1946), 'A note on Hicks's theory of inventions', *Amer. econ. Rev.*, March.

BLOOM, G. F. (1951), 'Wage pressure and technological discovery', *Amer. econ. Rev.*, September.

BÖHM-BAWERK, E. (1896), 'The positive theory of capital and its critics', *Quart. J. Econ.* January.

BRIGHT, A. A., Jr (1947), *The Electric Lamp Industry: Technological Change and Development from 1800 to 1947*, Macmillan Co.

BROZEN, Y. (1953), 'Determinants of the direction of technical change', *Amer. econ. Rev.*, May.

BROZEN, Y. (1957), 'The economics of automation', *Amer. econ. Rev.*, May.

BRUTON, H. J. (1955), 'Growth models and underdeveloped countries', *J. polit. Econ.*, August.

BRUTON, H. J. (1956a), 'Innovations and equilibrium growth', *Econ. J.*, September.

BRUTON, H. J. (1956b), 'Contemporary theorizing on economic growth', in B. F. Hoselitz *et al.* (eds.), *Theories of Economic Growth*, Free Press.

CARTER, C. F., and WILLIAMS, R. B. (1958), *Investment in Innovation*, Oxford University Press.

FELLNER, W. (1951), 'The influence of market structure on technological progress', *Quart. J. Econ.*, November; also in Heflebower and Stocking (1958).

FELLNER, W. (1956), *Trends and Cycles in Economic Activity*, Holt, Rinehart & Winston.

FELLNER, W. (1961a), 'Appraisal of the labour-saving and capital-saving character of innovations', in F. Lutz and D. C. Hague (eds.), *The Theory of Capital*, International Economic Association.

FELLNER, W. (1961b), 'Two propositions in the theory of induced innovations', *Econ. J.*, June. [Reprinted as Reading 10 of the present selection.]

FELLNER, W. (1962), 'Does the market direct the relative factor-saving effects of technological progress?', in *The Rate and Direction of Inventive Activity: Economic and Social Factors*, Princeton University Press.

FELLNER, W., and HALEY, B. F. (eds.) (1946), *Readings in the Theory of Income Distribution*, Irwin.

GREEN, H. A. J. (1960), 'Growth models, capital and stability', *Econ. J.*, March.

HABAKKUK, H. J. (1962), *American and British Technology in the Nineteenth Century*, Cambridge University Press.

HAMBERG, D. (1959), 'Production functions, innovations and economic growth', *J. polit. Econ.*, June.

HARROD, R. F. (1948), *Towards a Dynamic Economics*, Macmillan.

HARROD, R. F. (1961), 'The "neutrality" of improvements', *Econ. J.*, June.

HEFLEBOWER, R. B., and STOCKING, G. W. (eds.) (1958), *Readings in Industrial Organization and Public Policy*, Irwin.

HENNIPMAN, P. (1954), 'Monopoly: impediment or stimulus to economic progress?', in E. H. Chamberlin (ed.), *Monopoly and Competition and their Regulation*, St Martin's Press.

HICKS, J. R. (1932), *The Theory of Wages*, Macmillan.

HICKS, J. R. (1936) 'Distribution and economic progress: a revised version', *Rev. econ. Stud.*, October.

HOGAN, W. P. (1958), 'Comments', *Rev. Econ. Stat.*, November.

JEWKES, J., *et al.* (1958), *The Sources of Invention*, St Martin's Press.

KALDOR, N. (1957), 'A model of economic growth', *Econ. J.*, December.

KALDOR, N. (1961), 'Capital accumulation and economic growth', in F. Lutz and D. C. Hague (eds.), *The Theory of Capital*, International Economic Association.

KENNEDY, C. (1961), 'Technical progress and investment', *Econ. J.*, June.

KENNEDY, C. (1962), 'Harrod on "neutrality"', *Econ. J.*, March.

KNIGHT, F. H. (1944), 'Diminishing returns from investment', *J. Polit. Econ.*, March.

KRAVIS, I. B. (1959), 'Relative income shares in fact and theory', *Amer. econ. Rev.*, December.

KUZNETS, S. (1960), Preface to D. Creamer *et al.* (eds.), *Capital in Manufacturing and Mining*, Princeton University Press.

LACHMAN, L. M. (1956), *Capital and Its Structure*, Bell, for the LSE.

LANGE, O. (1945), *Price Flexibility and Employment*, Principia Press.

LERNER, A. P. (1933), 'The diagrammatic representation of elasticity of substitution', *Rev. econ. Stud.*, October.

LERNER, A. P. (1947), *The Economics of Control*, Macmillan Co.

LUTZ, F., and HAGUE, D. C. (eds.) (1961), *The Theory of Capital*, International Economic Association.

MACLAURIN, W. R. (1949), *Invention and Innovation in the Radio Industry*, Macmillan Co.

MANSFIELD, E. (1961), 'Technical change and the rate of imitation', *Econometrica*, October.

MARX, K. (1894), *Capital*, vol. 3, Meissner, Hamburg.

MASSEL, B. F. (1960), 'Capital formation and technological change in US manufacturing', *Rev. Econ. Stat.*, May.

MEADE, J. E. (1961), *A Neo-Classical Theory of Economic Growth*, Allen & Unwin.

MELMAN, S. (1958), *Dynamic Factors in Industrial Productivity*, Blackwell.

MILL, J. S. (1848), *Principles of Political Economy*, Parker & Co.

NELSON, R. R. (1959), 'The economics of invention: a survey of the literature', *J. Bus.*, April.

PIGOU, A. C. (1920), *The Economics of Welfare*, Macmillan.

POWER, J. H. (1958), 'The framework of a theory of economic growth', *Econ. J.*, March.

REDER, M. W. (1959), 'Alternative theories of labour's share', in M. Abramovitz *et al.* (eds.), *Allocation of Economic Resources*, Stanford University Press.

RICARDO, D. (1821), *Principles of Political Economy*, Dent.

RICHARDSON, G. B. (1959), 'Equilibrium, expectations and information', *Econ. J.*, June.

ROBINSON, J. (1934), *Economics of Imperfect Competition*, Macmillan.

ROBINSON, J. (1937), *Essays in the Theory of Employment*, Macmillan.

ROBINSON, J. (1938), 'The classification of inventions', *Rev. econ. Stud.*, February.

ROBINSON, J. (1952), *The Rate of Interest and Other Essays*, Macmillan.

ROBINSON, J. (1956), *The Accumulation of Capital*, Macmillan.

ROTHSCHILD, K. W. (1954), *The Theory of Wages*, Blackwell.

SALTER, W. E. G. (1960), *Productivity and Technical Change*, Cambridge University Press.

SCHUMPETER, J. A. (1911), *The Theory of Economic Development*, trans. 1934, Harvard University Press.

SCOVILLE, W. C. (1948), *Revolution in Glassmaking: Entrepreneurship and Technological Change in the American Industry, 1880–1920*, Harvard University Press.

SIDGWICK, H. (1883), *The Principles of Political Economy*, Macmillan.

SOLOW, R. M. (1957), 'Technical change and the aggregate production function', *Rev. Econ. Stat.*, August.

SOLOW, R. M. (1958), 'A sceptical note on the constancy of relative shares', *Amer. econ. Rev.*, September.

SOLOW, R. M. (1959), 'Comment' [on L. L. Pasinetti's 'On concepts and measures of changes in productivity'], *Rev. Econ. Stat.*, August.

SOLOW, R. M. (1960), 'Investment and technical progress', in K. J. Arrow *et al.* (eds.), *Mathematical Methods in the Social Sciences*, Stanford University Press.

STERN, E. H. (1945), 'Capital requirements in a progressive economy', *Economica*, August.

ULMER, M. J. (1960), *Capital in Transportation, Communications and Public Utilities*, Princeton University Press.

WICKSELL, K. (1901), *Lectures on Political Economy*, vol. 1, trans. 1934, Kelley.

WRIGHT, D. McC. (1944), 'Professor Knight on limits to the use of capital', *J. polit. Econ.*, May.

Part Two
The Determinants of Technological Change

What is the role of market forces in generating technological change? How exactly does technology respond to the pressures of the market place? What peculiar aspects of the distribution of its costs and benefits are essential to an understanding of the process of technological change? The Readings in this section explore such questions from a variety of perspectives. The late Jacob Schmookler's seminal paper forcefully argues the case that the allocation of resources to inventive activities can be substantially explained in terms of individuals acting in response to expectations of future profit, as these expectations are shaped and influenced by the complex of socio-economic changes. Far from being the product of autonomous forces, therefore, technological progress can be studied and explained as the outcome of society's demand, as expressed in the market place, for specific categories of invention.

Parker, in his short but wide-ranging paper, examines the timing and the sequence of technological changes over the past two centuries. In so doing he offers some provocative suggestions concerning the manner in which the underlying state of fundamental science and differences in the inherent complexity of classes of natural phenomena may have shaped the historical pattern of industrial change.

Since, in a growing number of sectors of the economy, technological advances are dependent upon the expansion of scientific knowledge, the mechanisms by which society allocates resources to basic research are a matter of increasing importance and concern. Nelson examines the conditions of knowledge production and the external economies generated by it which cause business firms, responding to ordinary financial incentives of private profit maximization, to allocate a quantity of resources to such activities which is clearly suboptimal from society's point of view.

Arrow, in a closely related paper, considers why, even under perfect competition, society will fail to allocate an optimal amount of resources to inventive activity. In so doing he interprets invention as the production of information and stresses the peculiar characteristics of information when treated as a commodity. Since it is never possible to predict the output to be produced by a given combination of inputs, inventive activity is inherently a risky process and devices which will permit the shifting of risks have been only imperfectly developed. As a consequence, society under-invests in such risky enterprises. Furthermore, in creating property rights in information – e.g., through the patent system – there is an unavoidable conflict between strengthening the incentive to engage in information production, on the one hand, and assuring a full utilization of information which has already been produced, on the other.

The paper by Griliches on hybrid corn attempts to quantify what is regarded as one of the necessary conditions justifying public investment in support of research: a sizeable divergence between the social and the private rates of return. Griliches in fact estimates that the social rate of return over the period 1910–55 to the public and the private resources committed to agricultural research which resulted in the development of hybrid corn was of the order of at least 700 per cent.

Finally, Fellner's note addresses itself to the question of whether, and under what circumstances, market forces prompt individual firms to seek out inventions which are not merely cost-reducing but have specific factor-saving biases. Is there a mechanism which induces firms, in response to changing relative scarcities at the macro-economic level, to direct inventive resources in a specifically labour-saving or capital-saving direction?

5 J. Schmookler

Economic Sources of Inventive Activity[1]

J. Schmookler, 'Economic sources of inventive activity', *Journal of Economic History*, March, 1962, pp. 1–20.

The fundamental conclusion of this paper is that technological progress is intimately dependent on economic phenomena. The evidence suggests that society may indeed affect the allocation of inventive resources through the market mechanism somewhat as it affects the allocation of economic resources generally. If this is true, then technological progress is not an independent cause of socio-economic change, and an interpretation of history as largely the attempt of mankind to catch up to new technology is a distorted one. Cultural lags undoubtedly exist in social history. The automobile – to use an obvious example – rendered obsolete many pre-existing social arrangements and behavior patterns. But the reverse is also true. New goods and new techniques are unlikely to appear, and to enter the life of society without a pre-existing – albeit possibly only latent – demand. Even a long-standing demand may have been intensified shortly before a technique to satisfy it is invented. In addition to cultural lag, there exists technological lag – a chronic tendency of technology to lag behind demand (Gilfillan, 1935, esp. ch. 1).

The problem considered in this paper may be viewed in still another way. If one charts over time a firm's research and

1. An earlier version of this paper entitled 'Growth, cycles and invention' was read at the meetings of the American Association for the Advancement of Science, Section K, New York City, 26 December 1960.

I wish to acknowledge the helpful criticisms of Simon Kuznets and of my colleagues at the University of Minnesota; the financial assistance of the National Science Foundation, the Ford Foundation, the John Simon Guggenheim Foundation, and the University of Minnesota; the co-operation of the United States Patent Office in the derivation of the patent statistics used; and the research assistance of Manmohan S. Arora, Heinrich H. Bruschke, A. Luis Darzins, Irwin Feller, Sushila Gidwani, Donald W. Katzner and Allan L. Olson. The opinions expressed are mine.

development expenditures leading to a given product (for example, a given model of railway car) and the volume of sales of the resulting product, one would expect in the nature of the case to find – if the product were marketed at all – that the research expenditures were made largely before the product was marketed (see Johnson and Turner, 1956, pp. 705–10). What would one find if total research and development expenditures of all firms on a given *class* of product, such as railroad passenger cars, were compared over a period of time with sales of all products in the class? On this question, the data examined in this paper throw some light. Inventive effort, it would appear, usually varies directly with the output of the class of goods the inventive effort is intended to improve, with invention tending to lag slightly behind output. The explanation of this pattern, in my judgement, is that variations in invention are a consequence of economic conditions with which output is also positively correlated. While the discussion and evidence relate to classes of products already in existence, the conclusion may well apply with almost equal force to basic, industry-establishing inventions.

A few remarks about the basic data used – statistics of patents granted, classified by industry – are in order at the outset. Beginning with 1874, the data represent patents counted as of the time the application for the patent was filed. This means, according to members of the patent bar, that generally the inventions represented were reduced to patentable, although not necessarily commercially useful, form on the average about six months beforehand. For the period prior to 1874 the patents are counted as of the time of granting. During this earlier period the interval between application and grant averaged only a few months.

Such data are subject to criticism because their use entails three unproved assumptions: (a) that the ratio of patents to inventions within the field is constant over time, (b) that the ratio of the number of patented inventions in the field to the number of patents covered by the particular series is constant over time,[2] and (c) that the average invention in one period in the

2. The statistics used in this paper were compiled by taking all the patents in those Patent Office subclasses in which two-thirds or mo e of the patents included in the subclass pertained to the industry. For the fields discussed

field is of the same importance, in some relevant sense, as the average invention in any other period. The validity of these assumptions cannot be established. Moreover, since 1940, there may have been a marked decline in the proportion of inventions patented. But these data provide the most effective means for a comprehensive historical investigation of the problems involved. The fact that the data seem to behave reasonably, albeit remarkably, suggests that these assumptions are tolerable, at least for the series discussed here.

Two further preliminary points need to be made about the data:

1. The proportion of patented inventions used commercially is apparently large. Studies by the Patent Foundation in Washington, which are consistent with considerable independent evidence, indicate that between 40 and 50 per cent of inventions patented by independent inventors and between 55 and 65 per cent of those patented by corporations are used commercially. In short, the data do not reflect the vaporings of disordered minds.

2. In the two industries for which we have been able to compile the most reliable chronologies of major inventions – railroading and petroleum refining – the number of major inventions tends to vary directly with the number of patents (see Schmookler, 1960). This suggests that the inferences drawn from the behavior of the minor inventions, which unquestionably dominate the patent statistics, probably apply in considerable measure to major inventions too.

Consider first the hypothesis that variations in the output of a class of products are associated with corresponding variations in invention. Figure 1 shows the annual number of patents in railroading from 1846 to 1950, with the patents subdivided into two groups: those pertaining to railroad track, and all other railroad patents. These two comprehensive components of railroad invention show remarkably similar behavior. They reach their

in this paper, at least 95 per cent of the patents pertain to the industries to which they have been assigned. However, patents pertaining to the industry assigned to other subclasses are not included. The data will be published and discussed more fully in a forthcoming book.

all-time peaks within three years of each other, their major swings are similarly timed, and even minor fluctuations in each tend to be duplicated in the other. This pervasive similarity suggests that common external forces shaped the course of invention in both railroad fields.

We get an inkling of what those common forces may be from

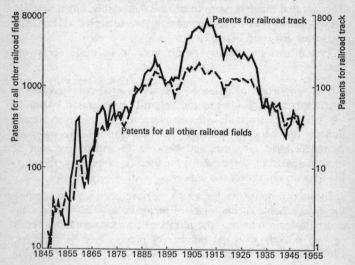

Figure 1 Railroad patents, track and non-track, United States, 1846–1950, annual data*

*Beginning with 1874 patents are counted as of the year of application. For earlier years they are counted as of the year of granting.

Figure 2, which shows seventeen-year moving averages of four variables – total railroad patents, net changes in miles of road, gross capital formation in 1929 prices, and the real price of railroad shares. The variables displayed in the graph suggest at first glance that the railroad industry became increasingly profitable until 1908 when, with the end of railroad expansion and the coming of highway transportation, its profitability began to decline. As profits rose and fell, so did investment in the industry – as indicated by changes in miles of road and in the industry's gross capital formation. Invention in the industry rose and fell, too, along with profits and the output of railroad equipment, which the capital formation series indicates.

The next question is whether this apparent long-run relationship holds also within shorter periods. Figure 3 reveals that it does, and with considerable fidelity. This graph shows deviations of seven-year moving averages from the seventeen-year moving

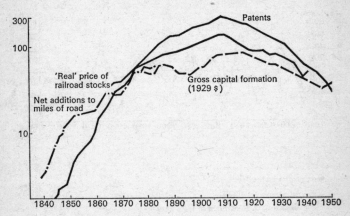

Figure 2[3] Railroad patents and railroad investment, United States, seventeen-year moving averages*

* Beginning with 1874 patents are counted as of the year of application. For earlier years they are counted as of the year of granting.

averages of the preceding graph. The major turning points in the various series, summarized in Table 1, usually come close together, with patents usually lagging behind the economic indicators. Even when the major turning points are spread out,

3. Sources, Figures 2 and 3. Railroad stock prices: Macaulay (1938), Appendix Table 10, and *Moody's Transportation Manual* (1959), p. 30. These were adjusted for changes in the BLS wholesale price index of all commodities, Department of Labor (1941), p. 715 and (1947), p. 126, and subsequent supplements thereto.

Net additions to miles of road: calculated by successive subtraction from statistics on miles of road operated given in *Poor's Manual of Railroads*.

Gross capital formation in 1929 prices: annual data used for calculating the seventeen-year moving average from Ulmer (1960), Appendix Table C-1; nine-year moving averages from Appendix Table K-2.

minor irregularities in the individual series reveal the persistence of considerable conformity between the series.

The apparent correspondence between invention and investment indicators for the railroad industry as a whole is also found between invention and output of specific kinds of railroad equipment. Each of the next three graphs shows the annual output of

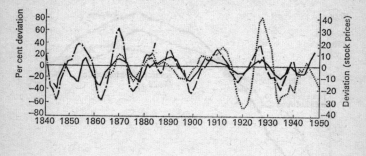

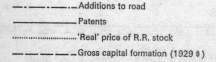

—·—·—·—· Additions to road

————————— Patents

·····················'Real' price of R.R. stock

— —— — —— Gross capital formation (1929 $)

Figure 3 Railroad patents and railroad investment, United States, deviations of seven- or nine-year moving averages from seventeen-year moving averages*

* Beginning with 1874 patents are counted as of the year of application. For earlier years they are counted as of the year of granting.

a major variety of railroad equipment coupled with the annual number of patents pertaining to that equipment.

Figure 4 shows railroad rail output and patents annually from 1860 to 1950. Inspection of the graph reveals the existence of substantially parallel major and minor fluctuations in the two variables. Output reaches its all-time high in 1906, and patents in 1911. A tentative though somewhat arbitrary list of timing of major troughs in each series is provided in Table 2, columns (1) and (2). The corresponding columns of Table 3 offer similar estimates of major peaks. Rail output generally leads patents at the troughs, but not necessarily at the peaks.

Figure 5 compares railroad passenger car output and patents annually from 1871 to 1949. Again both variables move in marked harmony over both the long and the short run. Railroad passenger car output reaches its all-time peak in 1907, railroad passenger car patents in 1908. At the major troughs tentatively

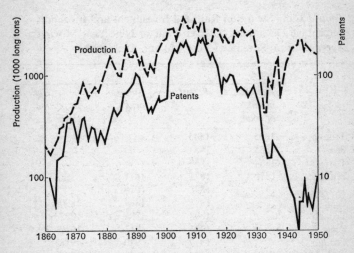

Figure 4⁴ Railroad rails: output and patents, United States, 1860–1950, annual data*

*Beginning with 1874 patents are counted as of the year of application. For earlier years they are counted as of the year of granting.

identified in Table 2, output generally leads patents, but again, leads and lags are roughly even at the peaks shown in Table 3. Figure 6 offers a similar comparison for freight car output and freight car patents, with much the same results. In this case, the all-time peak in patents occurs in 1907, the same year as that in output. At the major troughs patents again usually lag, but again honors are even at the major peaks.

The examples cited thus far all relate to the railroad industry; does the same pattern exist in other industries? Figure 7 shows

4. Sources: Railroad rail output: 1860–1914: Frickey (1947), pp. 10–11; 1915–29: Burns (1934), pp. 294–5; for subsequent years, *Annual Statistical Reports* of the American Iron and Steel Institute.

that the all-time peak of American invention in another part of
the transport industry – horseshoes – was reached around 1900,
at the time that the steam traction engine on the farm, and the
auto, truck and bus began to displace the horse and just before
the output of horse-drawn vehicles reached its peak around 1904.

Table 1 Long Swings in Railroad Investment and Invention –
Percentage Deviations of Seven-Year or Nine-Year Moving
Averages from Seventeen-Year Moving Averages*

Cycle phase	Series			
	Net change in miles of road	Patents	Stock prices	Gross capital formation (1929 $)
Trough	1845	1845		
Peak	1855	1857		
Trough	1864	1863		
Peak	1870	1871	1871	
Trough	1876	1878	1876	
Peak		1891	1884	1890
Trough		1898	1897	1899
Peak		1905	1912	1910
Trough		1919	1920	1918
Peak		1928	1928	1927
Trough		1935	1935	1935

* The deviations for gross capital formation are those of a nine-year
moving average from a seventeen-year moving average. (The original
source provided a nine-year average.) The rest are based on seven-year
moving averages.

The output of the product and inventions pertaining to it apparently moved together, just as in railroads. The graph also shows
that inventions pertaining to the horseshoe calk, a device
attached to the shoe of workhorses to prevent slipping, continued rising until the First World War, perhaps because of the
increased paving of roads. This pattern of invention casts doubt
on the notion that invention declines in an industry because the
possibilities for further improvement are exhausted, and that an
industry's rate of growth declines because technical progress
slackens (Burns, 1934, ch. 4, esp. pp. 141–5; Kuznets, 1930, ch. 1,

esp. pp. 30–2; Merton, 1935, pp. 454–74; Salter, 1960, ch. 10, esp. pp. 133–4). Quite the reverse – the decline in the industry's rate of growth apparently induced the decline in invention and technical progress.

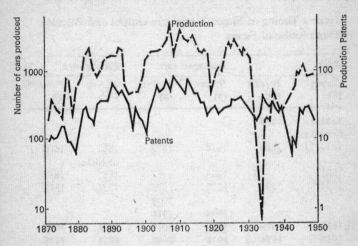

Figure 5[5] Railroad passenger cars: output and patents, United States, 1871–1950, annual data*

* Beginning with 1874 patents are counted as of the year of application. For earlier years they are counted as of the year of granting.

Figure 8 presents a more important series on petroleum refining investment and invention. These investment data are less complete than those for railroads. Beginning with Drake's well in 1859, investment in the industry probably underwent two major growth cycles, one associated with the kerosene phase of the industry and the other with the gasoline phase. Petroleum refining patents traversed a pair of similar growth cycles about the same time. Thus the same long-run relation between investment, or investment goods output, and invention exists in petroleum as in railroads.

5. Sources, Figures 5 and 6. Railroad passenger and freight car output: 1871–1914: Frickey (1947), pp. 14–15; 1915: American Railway Car Institute (1939), p. 83, and subsequent issues. Beginning in 1920 the data include production in railroad repair shops; before 1920 they do not.

In the building industry, too, I have been able to compile a rough but usable long-term record of annual economic data. Here the indicator of building activity has been provided by splicing series which are not wholly comparable; trend com-

Table 2 Timing of Major Troughs in Output and Patents – Three Railroad Fields

Rails		Passenger cars		Freight cars	
Output (1)	Patents (2)	Output (3)	Patents (4)	Output (5)	Patents (6)
1861 or earlier	1863				
1874–77	1880	1878	1879	1887 or earlier	1878
1885		1885	1885	1885	1886
1894	1894	1895	1896–1900	1894	1898
		1908	1911		
1914	1918	1919	1918	1921	1918
1932	1934	1933	1932	1933	1935
1938	1943	1940	1942	1938	1943

Total leads, each series over companion (output *v.* patents):
Rails: 5 to 0   Passenger cars: 4 to 2   Freight cars: 5 to 1
Total ties: Rails: 1   Passenger cars: 1   Freight cars: 0

Summary for all three fields combined, major troughs:
Patents lead: 3 times
Output leads: 14 times
The two tie: 2 times
Total   19

parisons are therefore not warranted, and Figure 9 depicts only the behavior of deviations of seven- from seventeen-year moving averages for building patents and building activity. The picture here is much the same as in railroading. Patents on building components – doors, walls, roofs, chimneys, etc. – also exhibit long swings similar to those observable in building activity itself.

The turning-point dates, listed in Table 4, come within two years of each other nine times out of twelve, with patents lagging slightly more often than they lead.[6] Just as in railroads, the patenting patterns in individual subfields within the building

Table 3 Timing of Major Peaks in Output and Patents –
Three Railroad Fields

Rails		Passenger cars		Freight cars	
Output (1)	Patents (2)	Output (3)	Patents (4)	Output (5)	Patents (6)
1872	1869–73	1876	1876		
1881		1883	1882	1881	1880
1887	1890	1892	1890	1890	1889
1906	1911	1907	1908	1907	1907
		1910	1912		
1926	1921	1926	1928	1923	1929
1937	1935	1937	1934	1937	1937
1944	1946	1946	1948	1948	1949

Total leads, each series over companion (output *v.* patents):
Rails: 3 to 2 Passenger cars: 4 to 3 Freight cars: 2 to 2
Total ties: Rails: 1 Passenger cars: 1 Freight cars: 2

Summary for all three fields combined, major peaks:
Patents lead: 7 times
Output leads: 9 times
The two tie: 4 times
Total   20

6. The substantial lead of building activity over patents at the start of the series may reflect deficiencies in the economic data. This portion of the Riggleman series is regarded as particularly unrealistic. See US Census Bureau (1960), p. 376, discussion of series N64. The frequency with which patents lead building activity (four times out of twelve) may be an illusion induced by aggregation. The long swings in building activity have somewhat different timing in different sections of the country. Since inventive activity may be more heavily concentrated in the leading sections, if the inventions from each section were compared with the building activity in that section, patents might be found to lead even less often than appears to be the case.

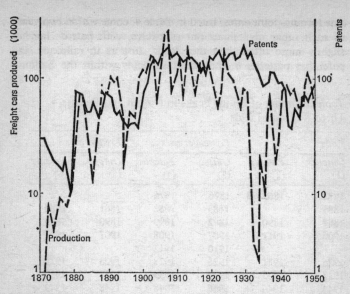

Figure 6 Railroad freight cars: output and patents, United States 1871–1950, annual data*

*Beginning with 1874 patents are counted as of the year of application. For earlier years they are counted as of the year of granting.

industry, not shown here, bear a strong family resemblance, suggesting again that common external forces determined their shape.

In the industries examined, there is a marked tendency for output and invention to move together in the long run, in the

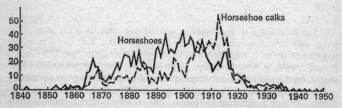

Figure 7 Horseshoe patents, United States, 1840–1950*
*Beginning with 1874 patents are counted as of the year of application. For earlier years they are counted as of the year of granting.

short run, and in intermediate periods. Since it seems implausible, despite the lead–lag relationship, that output variations directly produce variations in invention, two possible explanations for this phenomenon exist: (a) the variations in invention cause those in output; or (b) both variables fluctuate in response to a third set of forces as yet unspecified.

The first explanation conforms to popular belief and to the

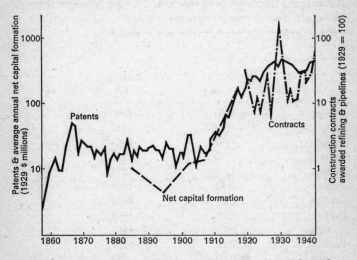

Figure 8[7] Petroleum refining: patents and investments, United States*
*Beginning with 1874 patents are counted as of the year of application. For earlier years they are counted as of the year of granting.

indisputable fact that research on any given product must precede its production and sale. None the less, three reasons count heavily against this as a general explanation. Firstly, it offers no explanation for the parallel behavior of invention in the different subfields within an industry. Further, the lead–lag relations between output and invention are generally in the wrong order.

7. Source: Annual rate of net capital formation: estimated from Creamer, Dobrovolsky and Borenstein (1961), Table A-8; index of contracts awarded in petroleum refining and pipelines: data supplied by F. W. Dodge Corporation.

If the fluctuations in invention cause those in output, the all-time peaks in invention should come before those in output. Instead, the all-time peaks in railroad equipment invention usually follow, or at least coincide with, those in equipment output. Secondly, if the fluctuations in invention cause those in output, then the

Table 4 Long Swings in Building Activity and Invention – Percentage Deviations of Seven-Year Moving Averages from Seventeen-Year Moving Averages

Cycle phase	Series	
	Patents	Building activity
Trough	1850	1842
Peak	1857	1854
Trough	1862	1863
Peak	1870	1870
Trough	1879	1878
Peak	1887	1889
Trough	1897	1899
Peak	1906	1913
Trough	1920	1918
Peak	1928	1926
Trough	1933	1933
Peak	1938	1938

Summary:	Patents lead	4 times
	Building activity leads	5 times
	The two tie	3 times

major troughs in invention should precede those in output by a few years – to allow time for inventions to reach a commercial stage, for plants to get tooled up, and for the necessary marketing effort to be made. Instead, inventions usually increase after rises in output, or occasionally tie or lead by only a very short period. Finally, the correspondence between invention and output is so close that if we are to argue that the fluctuations in invention cause those in output, we are in danger of explaining too much. We in effect assign only a minor role to the state of the economy, the rate of population growth, and other such factors as determinants of the output of an industry. Schumpeter (1939), who made so much of the role of innovation as a determinant of economic development, thought investment in most industries was ordinarily governed by system-wide determinants. He argued only that the state of the whole system was primarily determined by innovations in a few industries – industries which differed from one period to another.

If the line of causality runs neither from production of the good in question to invention nor vice versa, then the parallel movement of both variables must mean that both are effects either of the same or of correlated causes. The causes of fluctuations in the output of commodities have long been studied by economists and economic historians, and we cannot hope to shed further light on the matter here. It is sufficient for our purposes to state that in an economic system, the whole tends to dominate in its parts, and that more often than not variations in the output of a given industry are traceable to developments elsewhere in the system.[8] On this basis, our problem is not to account for the fluctuations in output, but to understand the inventive process and its socio-economic connections well enough to explain why inventive and economic activity are as closely correlated as they appear to be. While this problem cannot yet be definitely solved, it does seem possible at this stage, on the basis of private discussions with and published accounts of inventors, applied

8. This general theoretical premise finds support, not surprisingly, in a comparison of the long swing turning points in railroad investment and building activity. The interested reader who compares the turning points given in Table 1 for railroad investment with those in Table 4 for building activity will find that out of eleven occasions, the two variables turn in the same year three times, within a year of each other four times, within two years of each other twice, and within three years twice. Since the trough-to-trough length of a long swing in each industry is about nineteen years, the presence of common external influences on both industries is strongly indicated. (For a fascinating suggestion as to the possible nature of the causal mechanism at work, see Kuznets, 1958.) It is perhaps also pertinent to direct attention to the substantial, but again not surprising, similarity in the timing of major peaks and troughs in the output of major varieties of railroad equipment (Tables 2 and 3).

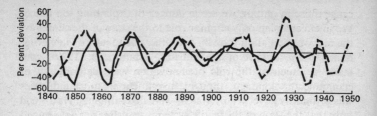

Figure 9[9] Building activity and building patents, United States, deviations of seven-year from seventeen-year moving average*

*Beginning with 1874 patents are counted as of the year of application. For earlier years they are counted as of the year of granting.

research scientists, and research directors, to suggest what may prove to be some of the key factors at work.[10]

Even if the germinal ideas for inventions occur costlessly and at random, which is perhaps seldom the case, it often takes talent, hard work and money to carry them out. These are not likely to be forthcoming in volume without commensurate prospective

9. Sources: Riggleman's index of *per capita* value of building permits given in Colean and Newcomb (1952), Appendix N, Table 2, was multiplied by the US population (US Census Bureau, 1960), Series A2, to provide an estimate of total building in current prices for 1830–1920. By splicing the 1913-based building cost index given in Colean and Newcomb, Appendix N, Table 4, to the American Appraisal Company index of construction costs (US Departments of Labor and Commerce), a continuous index of building costs (1947–9 = 100) from 1830 to 1950 was secured. The latter was then used to convert the index of total building from 1830 to 1920 in current prices into an index of building in 1947–9 prices. The post-1920 index of building activity was derived by subtracting public utility construction in 1947–49 prices from total construction in 1947–9 prices. Public utility construction was eliminated to make the series more comparable with that derived from the Riggleman data and with the patent statistics.

10. The obvious possibility that variations in the proportion of inventions patented occur in response to variations in economic prospects will be ignored in what follows. This is justified by the fact that while such variations may explain minor year-to-year variations in patents, the economic incentives to invent are the same as the incentives to patent. Hence, the major movements in patenting, which have been the focus of our discussion, can be reasonably construed as reflecting movements of invention in the same direction, though not necessarily of the same magnitude.

rewards in income or prestige. Clearly, prospective rewards will be favorably affected by high volume of sales of the commodity to be improved. According to Harrel (1948, p. 116), 'The choosing of projects must be as precisely timed as a well-planned military maneuver. Not only must they be precisely timed – they must also be the correct project properly developed at a utilizable place,' and quoting R. P. Soule (p. 117): 'The history of research indicates that the odds in favor of the project are most favorable when, first, the product to be developed is riding a rising rather than a declining industrial trend, and, second when it is started in the earlier rather than the later phases of that trend.'

However, the relation of invention to sales may be non-linear in the short run; if sales get too high relative to plant and personnel, potential inventors may be kept too busy in current production, and potential innovators may lack both the required slack resources and the pressure to innovate.[11]

To the degree that inventions are made either by producers or consumers of a commodity, more money will be available for invention when the industry's sales are high than when they are low. Increased sales imply that both the producing firms and their employees will be in a better position than before to bear the expenses of invention. Large purchases of a product also suggest that its buyers are better able to finance invention. Finally, the current business practice of setting research budgets at a fixed percentage of sales tends to assure, in recent years, the relation we have observed.

Certain aspects of the psychology of invention are noteworthy in this context. Effective inventive activity apparently requires an intense, almost obsessive preoccupation with its object. Buyers, for example, are more likely to show such a preoccupation when they are shopping for a product, with standards and expectations about performance and cost beginning to form in their minds. The number of potential inventors among the buyers or producers of a product, then, might vary with sales and employment.[12]

11. This consideration may explain the frequency with which railroad patents declined cyclically before output.

12. This may explain why invention in some branches of agriculture has declined with the number of farmers, rather than with the sales of farm implements.

Another psychological requirement for invention, which is also involved in new purchases, is a certain dissatisfaction with existing possessions. Every new purchase indicates such dissatisfaction. New buildings are erected because old ones are crowded or inadequate. New machines are purchased because old ones are breaking down or cannot handle an increased volume or produce unsatisfactory products. The same conditions which lead some men to buy existing goods may lead others to try to improve them. Such dissatisfaction with existing products may stem from several sources whose influence usually varies directly with sales. Firstly, in periods of general economic expansion and rising sales, when proportions of many kinds are changing, products which were satisfactory under earlier conditions may be unsatisfactory under the new. To meet the changed requirements of customers, the producers or the customers themselves may invent new or modified products. Secondly, increased sales often mean that some people have begun to use a product for the first time. Since the circumstances and preferences of new customers may differ from those of old ones, either the producers or the new customers themselves may invent modifications in the product to suit the altered requirements.

Finally, newly purchased goods embodying recent improvements are also likely to have the 'bugs' which seem almost the inevitable accompaniment of progress. These emerge after use of a product has begun, and provide the occasion for corrective invention by the user or the producer. Such invention will tend to vary directly with, but lag behind, sales of the product involved.

To summarize, the evidence strongly suggests that the output of a commodity and invention relating to it vary together, with invention tending to lag. The relation cannot be explained by the hypothesis that the variations in invention cause those in output.[13] Rather, it seems probable that expected profits from invention, the ability to finance it, the number of potential inventors, and the dissatisfaction which invariably motivates it – are all

13. It is essential that the meaning of this statement should be clear. Of course, as noted earlier, inventions affect production. It is only asserted here that their effect does not take the form of inducing fluctuations in production which are synchronized with fluctuations in invention.

likely to be positively associated with sales. That major and minor inventions in an established industry tend to go together is a further indication that inventive effort is responsive to economic pressures and opportunities.

One may suggest further that even basic inventions which establish new industries are often, and perhaps usually, induced by economic forces like those which appear to operate in established industries. Even when a scientific discovery underlies an invention, the discovery may contain the seeds of many potential inventions, and economic factors may then determine which potential applications are selected for exploitation. These economic factors may take the form either of new, or newly intensified, latent demands, or of a greater intensity of the latent demands satisfied by the inventions actually made, as compared to that of the demands which remain unsatisfied. For example, while the invention of wireless telegraphy cannot be understood without reference to prior scientific discoveries in electromagnetism, it cannot be understood with reference to those discoveries alone. Other inventions based on the same discoveries could presumably have been made, and a substantial and probably increased demand for rapid and low-cost communication techniques probably influenced Marconi's efforts.

The essential point is that the incentive to make an invention, like the incentive to produce any other good, is affected by the excess of expected returns over expected costs. Scientific progress may reduce expected costs and so increase the probability that a given invention will be sought and made. However, every invention represents a fixed cost, and the expected benefits from it vary with circumstances. Those circumstances, arising from changes in the prospective market for a commodity or a process, depend not on scientific discovery, but on socio-economic change – urbanization, declining family size, changing status of women, changes in relative factor costs, increases in population and *per capita* income, etc. Antecedent scientific discoveries are sometimes necessary, but seldom sufficient, conditions for invention. The historical shifts in inventive attention appear to reflect the interplay of advancing knowledge, which opens up new inventive opportunities for exploitation, and the unfolding economic needs and opportunities arising out of a changing social order. The evidence adduced in this article suggests that the influence of the latter has been substantial, at least in established industries.

References

BURNS, A. F. (1934), *Production Trends in the United States since 1870*, National Bureau of Economic Research.

COLEAN, M. L., and NEWCOMB, R. (1952), *Stabilizing Construction*, McGraw-Hill.

CREAMER, DOBROVOLSKY, and BORENSTEIN (1961), *Capital Formation in Manufacturing and Mining*, Princeton University Press.

DEPARTMENT OF LABOR (1941), *Handbook of Labor Statistics*, Bureau of Labor Statistics.

DEPARTMENT OF LABOR (1947), *Handbook of Labor Statistics*, Bureau of Labor Statistics.

GILFILLAN, S. C. (1935), *The Sociology of Invention*, Follet Publishing Co.

HARREL, C. G. (1948), 'Selecting projects for research', in C. C. Furnas (ed.), *Research in Industry: Its Organization and Management*, Van Nostrand.

JOHNSON, G. K., and TURNER, I. M. (1956), 'Use of transfer functions for company planning', *Op. Res.*, vol. 5, December.

KUZNETS, S. (1930), *Secular Movements in Production and Prices*, Houghton Mifflin.

KUZNETS, S. (1958), 'Long swings in the growth of population and related economic variables', *Proceedings of the American Philosophical Society*, February.

MACAULAY, F. R. (1938), *Some Theoretical Problems Suggested by Interest Rates, Bond Yields and Stock Prices in the United States since 1856*, National Bureau of Economic Research.

MERTON, R. K. (1935), 'Fluctuations in the rate of industrial invention', *Quart. J. Econ.*, vol. 39.

SALTER, W. E. G. (1960), *Productivity and Technical Change*, University Press, Cambridge.

SCHMOOKLER, J. (1960), 'Changes in industry and in the state of knowledge as determinants of industrial invention', *Proceedings of the Conference on the Economic and Social Factors Determining the Rate and Direction of Inventive Activity*, Universities–National Bureau Committee of Economic Research and Committee on Economic Growth of the Social Science Research Council.

SCHUMPETER, J. (1939), *Business Cycles*, McGraw-Hill, vols. 1 and 2.

ULMER, M. J. (1960), *Capital in Transportation, Communications and Public Utilities*, Princeton University Press.

US CENSUS BUREAU (1960), *Historical Statistics of the United States: Colonial Times to 1957*, US Census Bureau.

6 W. Parker

Economic Development in Historical Perspective

W. Parker, 'Economic development in historical perspective',
Economic Development and Cultural Change, October 1961, pp. 1–7.

This article is divided into three parts. In part 1, changes in techniques of communication are examined for their effects on the rate of diffusion of modern industry from Europe over the past two centuries. In part 2, changes in production and transport techniques are examined for their effects on the shifting geographical distribution of opportunities for economic development within the zone of diffusion of those techniques. In part 3, some of the reasons for the historical sequence of techniques are considered. The core of the argument is that (a) the sequence of technological change over the past two centuries has controlled both the rate of diffusion of modern industrial culture and the shifting production opportunities at specific locations within that zone of diffusion; and (b) that sequence has been governed by human capabilities in examining external nature. Together, these propositions pose a technological determinism, operating at a level of human nature beneath and beyond the influences of pre-existing social environment. To the extent that this thesis is valid, the national society and the national economy lose a portion of the importance with which they are invested by much current work of historians and economists in the study of economic development.

Trade in commodities transmits knowledge as well as goods, and techniques may be reproduced by direct copying of the objects incorporating them. But apart from some striking cases of imitation, the diffusion of technology in the modern world has been largely limited by techniques not unfamiliar to St Paul or Mohammed: the movement of persons and the transmittal of written documents. The latter immensely proliferated after the invention in the fifteenth century of the printing press, and the flow has been maintained by mechanical and chemical improvements in the supply of paper and by the successive applications of steam power and electricity to the mechanism (Coleman, 1957, chs. 2 and 7). Yet the spread of technological knowledge, narrowly considered, is not a matter of mass education, but of the training of a small elite. The Greek and Arabic techniques were transmitted to the medieval world by means of scribes and manuscripts, after all, and no enormous quantity of printed paper is required to transmit modern techniques. Instructions must usually be accompanied by teachers and engineers in person and by models or examples of the equipment. Technological change has been of relatively minor importance in the movement of such persons, carrying printed materials with them and knowledge in their heads. Unlike politicians, technicians need not shuttle rapidly back and forth. The drastic reduction of travel time through the steamship and airplane has permitted the emergency movement of technicians to trouble spots; it has also permitted a wider use of the corps of personnel available for foreign assignment. But communication is not simply, or even mainly, a matter of rapid movement of persons from place to place. Formidable barriers of language and environment have persisted, and it is the slow erosion of these, rather than the more rapid movement of technicians, that has permitted the diffusion of modern technology.

The erosion of the barriers to utilization of new technology is itself partly the result of technological change. One must speak here of the transmission of social traits, or of certain underlying modern attitudes toward nature and human values, rather than of specific bits of knowledge about production. In the regions of European settlement, these barriers have been at a minimum. The movement of migrants on a scale large enough to re-establish fundamental ingredients of European society, in some one of its national variants, was sufficient in the eighteenth and nineteenth centuries to insure a level of receptivity to new techniques that was at least as high as that of the mother country.[1] The movement of a set of tastes and traits suitable to modern industrial

1. No higher, perhaps, in Latin America, but higher in the United States, relative to the regions in Europe from which immigrants came.

and commercial activity into established tribal and peasant societies was very slow. In the nineteenth century, it depended almost entirely upon the work of Christian missionaries and a few merchants and colonial administrators. The techniques involved distribution of the written word – which presented formidable tasks of translation and assumed high levels of literacy – or direct personal contact between persons who had overcome, at least in speech, the barrier of language. It is not surprising that, even between the industrial and rural areas of a single nation-state of Europe or North America, separate subcultures with strongly differing characteristics persisted. The railroad, telegraph and telephone were essentially techniques of commercial communication over long distances; except where the railroad encouraged migration, these techniques could not drastically affect popular tastes or thinking. Beyond the 'Atlantic community', and to a degree even within it, substantial penetration of local cultures had to await the development of still and moving photography, radio and television. For movies, in particular, the production of the image required a much higher level of technology than the showing of it, and is still largely the monopoly of a few Western countries. With these techniques, and also with the closer intermingling of peoples during the Second World War, the spread of Western tastes and attitudes has been greatly accelerated. With the spread of these tastes and attitudes at the popular level, the level of receptivity to new production technology has been raised. The hearts of the heathen have been softened; in the language of the Christian missionaries, the seed is now sown on fertile rather than stony ground.

The rate of flow of Western thought, popular taste and technology has been strongly affected by the development of these techniques of social communication. But the direction in which culture traits and knowledge have flowed over this communication net is a more puzzling and profound question. What does knowledge consist of, and why has it moved out from Europe, rather than in from Asia and Africa? Why is it that the things the Europeans have brought have been the good, the penetrating things, and why have local cultures proved so brittle before their advance? These questions cannot be answered by reference to the culturally conditioned elements in human character. Western ideas have on occasion been spread by force, just as Islam spread in the seventh and eighth centuries. Before the advent of mass communication techniques, the most common way was through the conversion of a small group at the top of the existing social order. Given this sort of revolutionary overthrow of the old society, the education of a new generation may be sufficient to alter radically its character. But the spread of Western ideas is not due wholly to a combination of power and brainwashing; one must postulate instead some universal characteristics of human nature to which scientific thought and its accompanying technology, and even the curious combination of values placed on goods, enterprise and the saving of human labor, make an irresistible appeal.

Under nineteenth-century techniques of communication, European society was transported to the Western hemisphere and Australia; elsewhere, the familiar phenomenon of export enclaves was thrust upon peasant and tribal societies. These regions were all bound by a network of commercial communication, by factor movements and by flows of commodity trade. At the centers of invention in Europe, a transport and production technology was evolving, and knowledge of these techniques was – with some imperfections – diffused over the 'world economy'. At any point in space, the employment of these techniques depended upon factor and materials prices relative to demand – demand being affected by supplies at competing locations, and materials prices by transport costs from their points of origin. Factors moved with considerable freedom, and as between the nation-states of Western Europe, factor immobility was compensated for by the existence of alternative native sets of behavior pattern that insured adequate response to economic opportunity. This world economy, it is well known, arranged itself with heavy concentrations of industry in Northwestern Europe and the Northeastern United States. One of the most important questions for modern economic history is whether this arrangement was rooted in nature and technology, or whether it was culturally and historically determined by a relative lack of mobility of ideas, capital and enterprise from their points of origin. It is contended here that this arrangement was essentially a natural and rational one, given the limits of diffusion of Western society as set by the feeble means of social communication.

The analysis of the shift, in Europe and the United States, from local industry to the concentrated pattern of the nineteenth century, on the basis of changing techniques, is familiar to most economic historians. The reasoning runs as follows:

1. In agriculture, techniques remained land-using; rises in land yields were not very great, and the main drift of technological changes was labor-saving.

2. In transport, the railroad cheapened costs in specific directions; it produced concentrations of activity at ports and rail junctions, and it was used most cheaply where long-distance hauls of bulk commodities were possible with considerable backhaul.

3. In manufacture, economies of scale were present to permit considerable concentration; apart from the scale of plant, imposed by the use of power-driven equipment, there were external economies arising from the scale of an industrial complex – economies in communication, and in the use of social overhead capital and the overhead of a relatively immobile labor force.

4. In the location of manufactures and their agglomerations, the substitution of coal for wood and the heavy fuel requirements of the steam engine and smelting technology made the opportunity for industry widest at coal fields.

Growth opportunities based upon the European price structure, and the sets of demand patterns, techniques and conditions of factor supply that underlay it, were spread to specific locations in the 'world economy' by movements of goods, gold, credit, men and ideas. Beyond the regions of massive European settlement, this spread was imperfect for the reason already advanced. Some regions technically suited to industry remained undeveloped, simply because of the absence of cultural prerequisites – mass consumption patterns, disciplined labor force, strong native enterprise, a native engineering tradition, and continuing sources of capital. In these locations, European society had not been reproduced on a large enough scale to generate European production conditions. Specialization in export crops and reductions in international transport costs smothered the opportunities for industrial growth.

Improvement in communications techniques after 1920 may have brought industry within sight of a few such naturally well-endowed regions, even using nineteenth-century techniques of production and transport. But as modern industrial culture has spread, its techniques have altered. In the late eighteenth and early and mid-nineteenth-centuries, technical change moved in the direction of geographical concentration, and rural industry crumbled before the factory system. Since that time, the drift of technological change has been almost wholly in the other direction. The modern developments that have produced a diffusion of industrial opportunity in formerly agricultural areas are by now fairly evident. Briefly, they include the following:

1. Improvements in land yields, which have undermined agricultural areas' comparative advantages based on natural conditions of soil and climate.

2. Reduction in transport costs at minor industrial locations, through the development of more flexible modes of transport, e.g. the truck and airplane; reduction in materials transport costs by improved economy of materials and fuels utilization.

3. Maintenance of economies of scale in single plants, but the break-up of economies of scale in industrial complexes, through the improvements in communications, transport and power transmission.

4. Proliferation of usable fuels and raw materials, through utilization of oil, gas, electricity and atomic power, and through the development of synthetic materials.

Without these developments, improvements in social communication between industrial and agricultural areas might still ultimately have equalized factor returns; but the opportunities for industrial growth at a variety of locations around the world would be narrow indeed. Since technological change does not appear likely to reverse itself in these respects in the near future, these developments offer the strongest hope that – despite the persistence of pre-industrial behavior patterns and their unpredictable responses to new knowledge – the underdeveloped regions will ultimately be able to industrialize.

Nearly all the production and transport techniques developed between 1770 and 1870 promoted the geographical concentration of industry. Those developed since that time have, in contrast, favored deconcentration. At first sight, it appears odd that the drift of technological change should have been so strongly in one direction in one period and in the reverse in another. It suggests that some relatively simple principle may underlie the variety of changes in each period, and that a fundamental shift occurred between, say, 1870 and 1920, in the principle on which technology advanced. Our ignorance of the history of modern technology is so profound that it is not possible to work out the nature of this shift in detail from readily available materials. The most appealing hypothesis is that it involved the relative decline in purely mechanical inventions and a redirection of invention into chemistry and sub-atomic physics.

That such a shift would have produced the locational results observed seems likely. Mechanical invention involves the transmission of power by rigid bodies; it immensely multiplies the advantage of plants located at fairly ample power sources. The steam engine added coal-beds to waterfalls as sources of power, and a few simple but drastic changes in the metallurgical industries – coke smelting and the new steel processes – were sufficient to confirm the dominance of coal sites. In contrast, three principal bases of modern technology stand out: the internal combustion engine, the chemical breakdown and transformation of substances, and the increasing control over electrical phenomena. The internal combustion engine itself involves a chemical principle – that of an explosion; but the principle was known in the late eighteenth century, and its substitution for the steam engine as a power source involved considerable achievements in mechanical engineering. But chemical and electrical engineering involve knowledge of quite a different kind. The developments in the generation of electricity and in electronics have produced changes in communications techniques and offered alternative locations to coal-beds as sources of large-scale power. Chemical technology has, by its nature, been concerned with the materials basis of economy, and its goal is unchanged from the days of the alchemists: to transform common substances into 'gold'.

We must now inquire why the history of technology should have proceeded from the mechanical to the chemical and the electronic. This is a problem well beyond the limits of economics or of economic history. Economists' recent suggestions of an economic direction to the course of technological change are relevant in a narrower context (Fellner, 1960; Schmookler, 1960). Within specific established industries, need – as shaped by markets and relative factor prices – may have produced inventions with certain economic characteristics. Even as between industries, the course of invention – whether mechanical or chemical – may have been guided by prospective returns. But in the broad terms considered here, the drift of invention from mechanical into other lines must surely be controlled by the state of fundamental science. The processes by which scientific knowledge and interests transmit themselves to relatively unsophisticated inventors and engineers are not apparent. There are inventions in basic science, for example, in experimental apparatus, that help to shape its direction. But one must observe that the proliferation of mechanical inventions followed hard upon the formulation of Boyle's law of gases and Newtonian physics. And Edison, his most recent popular biography shows (Josephson, 1959, pp. 25, 42, 52, 53, 61, 93 and 94), was not unfamiliar with the rapid advances in the scientific understanding of electricity that were occurring in the mid-nineteenth century. The development of chemical engineering would have been impossible without a science of chemistry; and the relationship to fundamental science of the newly hatched field of nuclear engineering is so close at hand that it is perfectly obvious.

Possibly the century-long lag between the steam engine and the internal combustion engine was not due to inherent technical or scientific causes. Had petroleum been as plentiful in England as coal, or had the Arabic or the Apache culture developed a strong interest in technology, an oil-using industrial structure might have antedated a coal-using one.[2] But the whole sequence of technological development, from the mechanical to the other

2. Forbes (1958) has re-examined the Arabic knowledge and use of petroleum; see also Forbes (1936), esp. pp. viii and 28 *et seq.*

forms, can hardly be explained in such terms. Explanations based on local economic or cultural characteristics appear pathetically inadequate. So long as historians of science pursue Galileo and his predecessors, with an occasional excursion in the direction of Newton, the modern development remains shrouded in mystery. One suggestion may be advanced from a fragment of C. S. Pierce's, written in 1891:

The first step taken by modern scientific thought – and a great stride it was – was the inauguration of dynamics by Galileo. A modern physicist on examining Galileo's works is surprised to find how little experiment had to do with the establishment of the foundations of mechanics. His principal appeal is to common sense and *il lume naturale*. He always assumes that the true theory will be found to be a simple and natural one . . . the straight line appears to us simple, because, as Euclid says, it lies evenly between its extremities; that is, because viewed endwise it appears as a point. That is, again, because light moves in straight lines. Now, light moves in straight lines because of the part which the straight line plays in the laws of dynamics. Thus it is that our minds having been formed under the influence of phenomena governed by the laws of mechanics, certain conceptions entering into those laws become implanted in our minds, so that we readily guess at what the laws are. Without such a natural prompting, having to search blindfold for a law which would suit the phenomena, our chance of finding it would be as one to infinity. The further physical studies depart from phenomena which have directly influenced the growth of the mind, the less we can expect to find the laws which govern them 'simple', that is, composed of a few conceptions natural to our minds (Pierce, 1891, in Weiner, 1958, pp. 145–6).

If this tantalizing hint could be pursued, it might lead to the unfolding of a history of modern science and technology based on the characteristics of the human mind and senses when facing external nature. The sequences of this history – determined simply by the inherent intellectual difficulties in unravelling the secrets of nature – have governed both the rate of diffusion of Western culture patterns and techniques and the shifting patterns of industrial opportunity within the zone in which that culture has spread. Technological determinism of this sort need employ pre-existing cultural factors only to explain impediments to the spread of Western techniques, and how those techniques and behavior patterns originated from the pre-industrial culture of the West. Once arrived in the world, they appear to have power exceeding that of a new religion, to transcend the culture of any place and time. As Western science finds universal technical applicability, so Western materialism – in its Communist or capitalist versions – appears to have a universal human appeal. Perhaps some fundamental, cross-cultural uniformities in human psychology account for this. One who shares fully in Western culture may continue to hope that these uniformities include some portions of the Western heritage beyond curiosity and greed. But a non-Westerner may possibly have valid reason to hope that cultural change will not extend too far into his existing system of values.

Appendix

Locational Characteristics of Advanced Techniques

	Period	Distant communication	Agriculture	Transport		Manufacture	
				Water	Land	Scale	Materials localization
1	Pre-1750	nil	land-using	cheap	costly	small-scale	little
2	1750–1900	limited	land-using	cheap	cheap-rigid	large-scale	great
3	1900–1950	extensive	land-using	cheap	cheap-flexible	large-scale	less
4	1960–	extensive	less land-using	cheap	cheap-flexible	large-scale	little

Note: In each case, the comparison is made as between periods, except that water transport is called 'cheap' in comparison with the land transport of period 1.

References

COLEMAN, D. (1957), *The British Paper Industry, 1495–1860*, Oxford University Press.

FELLNER, W. J. (1960), 'Does the market direct the relative factor-saving effects of technological progress?', National Bureau of Economic Research Conference on Inventive Activity, Social Science Research Council.

FORBES, R. J. (1936), *Bitumen and Petroleum in Antiquity*, Leiden.

FORBES, R. J. (1958), *Studies in Early Petroleum History*, Leiden.

JOSEPHSON, M. (1959), *Edison: A Biography*, McGraw-Hill.

PIERCE, C. S. (1891), 'The architecture of theories', in P. Weiner (ed.) (1958), *Values in a Universe of Chance*, Stanford University Press.

SCHMOOKLER, J. (1960), 'Changes in industry and in the state of knowledge as determinants of industrial invention', National Bureau of Economic Research Conference on Inventive Activity, Social Science Research Council.

7 R. Nelson

The Simple Economics of Basic Scientific Research

R. Nelson, 'The simple economics of basic scientific research', *Journal of Political Economy*, June 1959, pp. 297–306.

Basic economic framework

Recently, orbiting evidence of un-American technological competition has focused attention on the role played by scientific research in our political economy. Since Sputnik it has become almost trite to argue that we are not spending as much on basic scientific research as we should. But, though dollar figures have been suggested, they have not been based on economic analysis of what is meant by 'as much as we should'. And, once that question is raised, another immediately comes to mind. Economists often argue that opportunities for private profit draw resources where society most desires them. Why, therefore, does not basic research draw more resources through private-profit opportunity, if, in fact, we are not spending as much on basic scientific research as is 'socially desirable'? In order to answer some of these questions, it seems useful to examine the simple economics of basic research. How much are we spending on basic research? How much should we be spending? Under what conditions will these figures tend to be different? Is basic research marked by these conditions? If so, what can we do to eliminate or reduce the discrepancy?

How much are we spending on basic research? In 1953, the latest date for which relatively sophisticated estimates are available, total expenditure on research and development was about $5·4 billion. Of that total, much more than half was for engineering development, much less than half for scientific research. Even less of the total, about $435 million in 1953, was spent on 'basic research'. All evidence indicates that since 1953 expenditure on research and development has increased markedly; $10 billion seems a reasonable estimate for 1957. Expenditure on basic research has also increased at a rapid rate, perhaps at a faster rate than total research and development expenditure. But basic-research expenditure today is probably under $1 billion, less than one-quarter of 1 per cent of gross national product (National Science Foundation, 1956, 1957a, 1957b, 1957c).

How much should we spend on basic research? Replacing the X_i of the familiar literature on welfare economics with 'basic research' provides the theoretical answer. From a given expenditure on science we may expect a given flow, over time, of benefits that would not have been created had none of our resources been directed to basic research. This flow of benefits (properly discounted) may be defined as the social value of a given expenditure on basic research. However, if we allocate a given quantity of resources to science, this implies that we are not allocating these resources to other activities and, hence, that we are depriving ourselves of a flow of future benefits that we could have obtained had we directed these resources elsewhere. The discounted flow of benefits of which we deprive ourselves by allocating resources to basic research and not to other activities may be defined as the social cost of a given expenditure on basic research. The difference between social value and social cost is net social value, or social profit. The quantity of resources that a society should allocate to basic research is that quantity which maximizes social profit.

Under what conditions will private-profit opportunities draw into basic research as great a quantity of resources as is socially desirable? Under what conditions will it not? If all sectors of the economy are perfectly competitive, if every business firm can collect from society through the market mechanism the full value of benefits it produces, and if social costs of each business are exclusively attached to the inputs which it purchases, then the allocation of resources among alternative uses generated by private-profit maximizing will be a socially optimal allocation of resources. But when the marginal value of a 'good' to society exceeds the marginal value of the good to the individual who pays for it, the allocation of resources that maximizes private profits will not be optimal. For in these cases private-profit opportunities do not adequately reflect social benefit, and, in the absence of positive public policy, the competitive economy will tend to spend less on that good 'than it should'. Therefore, it is in the interests of society collectively to support production of that good.[1]

Society does, in fact, collectively support a large share of the economy's basic research. About 60 per cent of our basic-research work is performed by non-profit institutions, predominantly government and university laboratories. And a portion of the basic research performed in industrial laboratories is government financed. (This flow of funds is about equal to the flow of funds from industry to non-profit laboratories in the form of grants and contracts for basic research.) (National Science Foundation, 1957a.) Much, though not all, of the government contribution to basic research is national defense-oriented. But defense-oriented expenditure aside, the American political economy certainly treats basic research as an activity that creates marginal social value in excess of that collectable on the free market.[2] Is this treatment justified? If so, since, in fact, society is collectively supporting much basic research and hence resources directed to basic research do exceed the quantity drawn by private-profit opportunity, is present social policy adequate?

Scientific research and economic value

What are the social benefits derived from the activity of science? It is sometimes argued that most of our great social and political problems would simply evaporate if all citizens had a scientific point of view and, hence, that the benefits derived from scientific research are only in small part reflected in the useful inventions generated by science, for science helps to make better citizens.

1. Of course, the resources supplied to the industry must be withdrawn from other industries which generate no external economies or less external economies. Although significant external economies are probably rare, they almost certainly exist in education and preventive medicine as well as in basic research. Though the burden of this paper is that more resources should be allocated to basic research, the argument is probably invalid if these resources are taken, for example, from education or preventive medicine.

2. This paper will not consider the vital question of whether the Department of Defense is spending enough on defense-oriented basic research. It probably is not, but the analysis of the paper is independent of this.

And many scientists and philosophers take the point of view that the very activity of science – considered as the search for knowledge – is itself the highest social good and that any other benefits society might obtain are just by-products of the activity of science – social gravy. Dissents on both of these points are often sharp. The economist, after the usual perfunctory statement that he is fully aware that his definition does not capture everything, might define the benefits derived from the activity of science as the increase resulting from scientific research in the value of the output flow that the resources of a society can produce. In order to examine the extent to which a private firm can capture through the market the increased value of output resulting from the scientific research, in particular the basic scientific research, that it sponsors, it is necessary to examine the link between scientific research and the creation of something of economic value.

Scientific research may be defined as the human activity directed toward the advancement of knowledge, where knowledge is of two roughly separable sorts: facts or data observed in reproducible experiments (usually, but not always, quantitative data) and theories or relationships between facts (usually, but not always, equations). Of course, no strict line can be drawn between scientific research and all other human activities. Men have always experimented and observed, have always generalized and theorized; thus all men have always been, at least in a limited way, scientists. And knowledge has often (usually?) been acquired in activities in which pursuit of knowledge was of no, or negligible importance. But even fuzzy definitions often have value. Scientific knowledge rests on reproducible experiments, but science is more than experimentation leading to new observations of facts which are believed observable by any other scientist undertaking the identical experiment. Science is most fruitful when it leads to ability to predict facts about phenomena without, or prior to, experimentation and observation. Scientific knowledge has economic value when the results of research can be used to predict the results of trying one or another alternative solution to a practical problem.

Scientific research has increasingly been coupled to invention, where invention is defined as the human activity directed toward the creation of new and improved practical products and processes. But though many inventions occur as a result of a reasonably systematic effort to achieve a particular goal, many other inventions do not. They are a by-product of activity directed in a quite different direction, often a scientific research project directed toward solving an unrelated problem. Mauve, the first analine dye, was discovered by W. H. Perkin while he was attempting to synthesize quinine, and calcium carbide and the acetylene gas that it produces were invented by a group attempting to develop a better way to extract aluminum from clay. And though many inventions are made possible by closely preceding advances in scientific knowledge, many others require little knowledge of science or occur long after the relevant scientific knowledge is available: scientific knowledge certainly had little to do with the development of such useful inventions as the safety razor and the zipper; scientists have long known that expanding gases absorb heat, thus cool whatever they contact, but the gas refrigerator is an invention of the twentieth century. But particularly in the institution of the industrial research laboratory, applied science and invention are closely linked, and inventions usually result from a systematic attack on a problem.

In the activity of invention, as in most goal-directed activities, the actor has a number of alternative paths among which he must choose. The greater his knowledge of the relevant fields, the more likely he will be eventually to find a satisfactory path, and the fewer the expected number of tried alternatives before a satisfactory one is found. Thus, the greater the underlying knowledge, the lower the expected cost of making any particular invention.

A rationally planned inventive effort will be undertaken only if the expected revenue of the invention exceeds the expected cost. In many instances the economic utility of a particular invention is so great that an inventive effort is economically rational, even though the underlying scientific knowledge is scanty and hence the expected cost of making the invention is great. Edison's attempt to develop an incandescent lamp and Goodyear's attempt to improve the characteristics of rubber are cases in point. In these cases, since there was little useful underlying scientific knowledge, the invention procedure was trial and error, the next trial being roughly – but only roughly – indicated by a very loose theory formulated as the research proceeded. But though the inventors knew that it would probably prove costly to achieve their objective, they believed that the gains, if they were successful, were sufficiently great to make the effort profitable.

But often, though the inventor believes that there is great demand for a particular invention, it is not rational for him to attempt the invention, given the state of scientific knowledge. Expected cost will exceed expected revenue unless additional scientific knowledge can be obtained. If the expected cost of acquiring the relevant scientific knowledge is low, an organization interested in making a particular invention may undertake an applied scientific research project. A profit-maximizing firm will undertake a research project to solve problems related to a development effort if the expected gains – for example, reduction in development costs, or improvement in the final developed product – exceed expected research costs and if total research and development cost is exceeded by the expected net value of the invention. To the extent that the results of applied research are predictable and relate only to a specific invention desired by a firm, and to the extent that the firm can collect through the market the full value of the invention to society, opportunities for private profit through applied research will just match social benefits of applied research, and the optimum quantity of a society's resources will tend to be thus directed.

However, by no means all scientific research is directed toward practical problem-solving, though the line between basic scientific research and applied scientific research is hard to draw. There is a continuous spectrum of scientific activity. Moving from the applied-science end of the spectrum to the basic-science end, the degree of uncertainty about the results of specific research projects increases, and the goals become less clearly defined and less closely tied to the solution of a specific practical problem or the creation of a practical object. The loose defining of goals at the basic research end of the spectrum is a very rational adaptation to the great uncertainties involved and permits a greater expected payoff from the research dollar than would be possible if goals were more closely defined. For commonly, not just sometimes, in the course of a research project unexpected possibilities not closely related to the original objectives appear, and concurrently it may become clear that the original objectives are unobtainable or will be far more difficult to achieve than originally expected. While the direction of an applied research project must be closely constrained by the practical problem which must be solved, the direction of a basic research project may change markedly, opportunistically, as research proceeds and new possibilities appear. Some of the most striking scientific breakthroughs have resulted from research projects started with quite different ends in mind.

Pasteur's discovery of the value of inoculation with weakened disease strains is one of the more famous cases in point, but what is important is that the case is so similar to many others. While studying chicken cholera, Pasteur accidentally inoculated a group of chickens with a weak culture. The chickens became ill but, instead of dying, recovered. Since Pasteur did not want to waste chickens, he later reinoculated these chickens with fresh culture – one that was strong enough to kill an ordinary chicken – but these chickens remained healthy. At this point Pasteur's attention shifted to this interesting and potentially very (socially) significant phenomenon, and his resulting work, of course, brought about a major medical advance.

Applied research is relatively unlikely to result in significant breakthroughs in scientific knowledge save by accident, for, if significant breakthroughs are needed before a particular practical problem can be solved, the expected costs of achieving this breakthrough by a direct research effort are likely to be extremely high; hence applied research on the problem will not be undertaken, and invention will not be attempted. It is basic research, not applied research, from which significant advances have usually resulted. It is seriously to be doubted whether X-ray analysis would ever have been discovered by any group of scientists who, at the turn of the century, decided to find a means for examining the inner organs of the body or the inner structure of metal castings. Radio communication was impossible prior to the work of Maxwell and Hertz. Maxwell's work was directed toward explaining and elaborating the work of Faraday. Hertz built his equipment to test empirically some implications of Maxwell's equations. Marconi's practical invention was a simple adaptation of the Hertzian equipment. It seems most unlikely that a group of scientists in the mid-nineteenth century, attempting to develop a better method of long-range communication, would have developed Maxwell's equations and radio or anything nearly so good.

The limitations of an applied-research project constrained to the solution of a specific practical problem, and the practical value of many research projects where the goal is simply knowledge, not the solution of a practical problem, is well illustrated by the development of hybrid corn. During the latter half of the nineteenth century several attempts were made to improve corn yields. Many of the researchers directed their attention, at one time or another, to the inbreeding of corn to obtain a predictable and profitable strain. But as corn plants were inbred, though they tended to breed true, they also tended to deteriorate in yield and in quality. For this reason, applied researchers attempting to improve corn dropped this seemingly unpromising approach. But George Harrison Shull, a geneticist working with corn plants and interested in pure breeds not for their economic value but for experiments in genetics, produced several corn strains that bred true and then crossed these strains. His project was motivated by a desire to further the science of genetics, but a result was high-yield, predictable hybrid corn.

Basic research and private profit

It is clear that for significant advances in knowledge we must look primarily to basic research; the social gains we may expect from basic research are obvious. But basic research efforts are likely to generate substantial external economies. Private-profit opportunities alone are not likely to draw as large a quantity of resources into basic research as is socially desirable.

Significant advances in scientific knowledge, the types of advances that are likely to result from successful basic-research projects, very often have practical value in many fields. Consider the range of advances resulting from Boyle's gas law or Maxwell's equations. On Gibb's law of phases rests the design of equipment in fields as diverse as petroleum refining, rubber vulcanization, nitrogen fixation and metal-ore separation. Few firms operate in so wide a field of economic activity that they are able themselves to benefit directly from all the new technological possibilities opened by the results of a successful basic research effort. In order to capture the value of the new knowledge in fields which the firm is unwilling to enter, the firm must patent the practical applications and sell or lease the patents to firms in the industries affected.

But significant advances in scientific knowledge are often not directly and immediately applicable to the solutions of practical problems and hence do not quickly result in patents. Often the new knowledge is of greatest value as a key input of other research projects which, in turn, may yield results of practical and patentable value. For this reason scientists have long argued for free and wide communication of research results, and for this reason natural 'laws' and facts are not patentable. Thus it is quite likely that a firm will be unable to capture through patent rights the full economic value created in a basic-research project that it sponsors.

A firm with a narrow technological base is likely to find research profitable only at the applied end of the spectrum, where research can be directed toward solution of problems facing the firm, and where the research results can be quickly and easily translated into patentable products and processes. Such a firm is likely to be able to capture only a small share of the social benefits created by a basic-research program it sponsors. On the other hand, a firm producing a wide range of products resting on a broad technological base may well find it profitable to support research toward the basic-science end of the spectrum.

A broad technological base insures that, whatever direction the path of research may take, the results are likely to be of value to the sponsoring firm. It is for this reason that firms which support research toward the basic-science end of the spectrum are firms that have their fingers in many pies. The big chemical companies producing a range of products as wide as the field of chemistry itself, the Bell Telephone Company, General Electric, and Eastman Kodak immediately come to mind. It is not just the size of the companies that makes it worthwhile for them to engage in basic research. Rather it is their broad underlying technological base, the wide range of products they produce or will be willing to produce if their research efforts open possibilities. (Eastman Kodak entered the vitamin business when a research project resulted in a new way to synthesize vitamin B.) Strangely enough, economists have tended to see little economic justification for giant firms not built on economies of scale. Yet it is the many-product giants, not the single-product giants, which have been most technologically dynamic, and, to the extent that we wish the private sector of the economy to support basic research, we must look to these firms.

The importance of a broad technological base as a factor permitting a company to engage profitably in basic research is clearly illustrated by Carothers's famous research project for Du Pont. Carothers's work in linear superpolymers began as an unrestricted foray into the unknown with no particular practical objective in view. But the research was in a new field of chemistry, and Du Pont believed that any new chemical breakthrough would probably be of value to the company. The very lack of a specific objective, the flexibility of the research project, was an important factor behind its success. In the course of research Carothers obtained some superpolymers which at high temperatures became viscous fluids and observed that filaments could be obtained from these materials if a rod were dipped in the molten polymer and then withdrawn. At this discovery the focus of the project shifted to these filaments. Nylon was the result, but at the start of the project Carothers could not possibly have known that his research would lead him to the development of a new fiber.

A wide technological base (usually involving a diversified set of products) does not imply a position of monopoly power in any or all of the product markets, nor does a monopoly position in a market imply a wide technological base. Focusing attention on market position, a business firm operating in a competitive environment will seldom find it profitable to engage in a research project which is not likely to result quickly in something patentable, even if the firm can predict the nature of the research results, unless the firm keeps tight secrecy. For if the results of research cannot be quickly patented and are not kept secret, other firms producing similar products using similar processes will be free to use the results as an input of a development program of their own, designed to achieve a similar patentable objective. If competing firms develop a patentable product first, or develop a competing product, these firms will in effect steal from the research-sponsoring firm, through price and product competition, a large share of the social utility created by research. In fact, many companies engaging in research keep their research findings secret until the new knowledge is put to practical use and the results are patented.

Many industries have attempted to reconcile their need for new knowledge with the lack of incentives to individual private firms to produce that new knowledge by establishing cooperative industry research organizations. To the extent that an industry rests on a field of science that is likely to get little attention in the absence of sponsorship by the firms in the industry, it may be in the interests of all the firms that research in this field be pushed, though each firm would prefer the others to do the financial pushing. An industrial cooperative research laboratory may well develop under these conditions, supported by all or by a large number of firms in the industry, and undertaking research likely to be applicable to the technology of the industry. The motivation for these cooperative laboratories is only in part the high cost of research. More importantly, these laboratories are motivated by the fact that most of the firms will gain from the results of relatively basic research in certain fields whether or not they pay for it; hence little research will be undertaken in the absence of cooperation.

The preceding argument has been focused on external economies that open a gap between marginal private and marginal social benefit from basic research. Two other factors, working in the same direction, must be mentioned, if not discussed. Firstly, the long lag that very often occurs between the initiation of a basic-research project and the creation of something of marketable value may cause firms much concerned with short-run survival, little concerned with profits many years from now, to place less value on basic-research projects than does society, even in the absence of external economies. This is not to say that all firms have a greater time-discount factor than does society as a whole, but it can be argued that many firms do. Secondly, the very large variance of the profit probability distribution from a basic-research project will tend to cause a risk-avoiding firm, without the economic resources to spread the risk by running a number of basic-research projects at once, to value a basic-research project at significantly less than its expected profitability and hence, even in the absence of external economies, at less than its social value.

Is current social policy adequate?

It seems clear that, were the field of basic research left exclusively to private firms operating independently of each other and selling in competitive markets, profit incentives would not draw so large a quantity of resources to basic research as is socially desirable. But in fact basic research has not been the exclusive domain of private firms. Government and other non-profit institutions (principally universities) together spend more on basic research, and undertake more basic research in their own laboratories, than does industry. Since we are presently supporting collectively such a large share (more than half) of basic research, is it not possible that total basic-research expenditure (the sum of private and public efforts) equals or exceeds the social optimum? This is a tricky theoretical question. However, if basic research can be considered as a homogeneous commodity, like potato chips, and hence the public can be assumed to be indifferent between the research results produced in government or in industry laboratories; if the marginal cost of research output is assumed to be no greater in non-profit laboratories than in profit-oriented laboratories, and if industry laboratories are assumed to operate where marginal revenue equals marginal cost, then the fact that industry laboratories do basic research *at all* is itself evidence that we should increase our expenditure on basic research.

Public support of basic research has primarily been in the form of contracts let with private firms and in the establishment and support of a large number of non-profit laboratories. Save for the effects of tax laws (which apply to all business cost-incurring activities), public policy has not acted to *shift* the marginal cost curve of the basic-research industry. Public policy has resulted in shifts *along* the curve. Nor has public policy acted to drive marginal social utility to marginal private utility. External economies still exist at the margin. Clearly then, if industry laboratories are in profit-maximizing equilibrium, society would benefit from an increase in basic-research expenditure in industry laboratories, holding research efforts elsewhere constant, for the marginal social benefit of basic research in private laboratories exceeds marginal cost to the firm, which under our assumptions still equals alternative cost. But perhaps non-profit laboratories are spending too much on basic research – are operating beyond the point at which marginal cost equals marginal social benefit – and therefore it is desirable to reduce research expenditure in this sector. Given our assumptions, this cannot be. For, if marginal cost is no greater in non-profit laboratories than in industry laboratories, and society cannot distinguish between the fruits of research undertaken in the two kinds of laboratory – that is, if marginal social benefit is the same in the public and the private sector – and if it is socially desirable that expenditure on basic research be increased in industry laboratories, then it is also socially desirable that research expenditure be increased in non-profit laboratories. For if marginal social benefit exceeds marginal cost in industry laboratories, so does it in non-profit laboratories.

The assumptions on which the preceding argument is based rest but shakily on fact. Basic research certainly is not a homogeneous commodity. The types of knowledge generated, say, in an air-force project on high-speed gas flows, a Du Pont project on high polymer chemistry, or a Harvard project on solid-state physics are not perfectly substitutable. The knowledge generated will certainly be different, and in a reasonably predictable way. And, once the non-homogeneity of basic research is admitted, the concept of relative marginal cost becomes fuzzy. Thus one cannot make an airtight statement, based on welfare economics, that we are not spending as much on basic scientific research as we should. But I believe that the evidence certainly points in that direction.

Some policy implications

Though the profit motive may stimulate private industry to spend an amount on applied research reasonably close to the amount that is socially desirable, it is clear from the preceding analysis that under our present economic structure the social benefits of basic research are not adequately reflected in opportunities for private profit. Indeed, there is a basic contradiction between the conditions necessary for efficient basic research – few or no constraints on the direction of research with full and free dissemination of research results – and full appropriation of the gains from sponsoring basic research in a competitive economy.

This is not to say that some firms could not profitably increase their basic-research effort. Some may presently be operating well to the left of their maximum profit point. But to the extent that we want our economy to remain competitive and want efficient use of basic-research funds, the laboratories of colleges, universities and other non-profit institutions must perform a large share of our basic research if we are to put as much of our resources into basic research as we should. Although several laboratories of private industry have made significant contributions in the field of basic science, these contributions have been few and far between. If we advocate that basic research be increasingly undertaken by business and if we believe that business should be motivated by profit, we must accept the growth of large firms with a wide technological base, with virtual monopolies in several markets. If we do not want such an economic structure, then only to the extent that we think it desirable that private firms look to motives other than profit can we argue that industry laboratories should perform a significantly enlarged share of our basic research. In either case we undermine many of the economic arguments for a free-enterprise economy. If we want to maintain our enterprise economy, basic research must be a matter of conscious social policy.

This is not the place to suggest a menu of policies – the bill of fare offered in the National Science Foundation booklet on basic research lists some of the actions that might be considered. However, it does seem appropriate to suggest that public policy on basic research should recognize the following points:

1. The problem of getting enough resources to flow into basic research is basically the classical external-economy problem.[3] External economies result from two facts: firstly, that research results often are of little value to the firm that sponsors the

3. The external economy aspect of basic research reacts back through the price system to undervalue pure scientists relative to engineers.

research, though of great value to another firm, and, secondly, that research results often cannot be quickly patented. It therefore seems desirable to encourage the further growth of a 'basic-research industry', a group of institutions that benefit from the results of almost any basic-research project they undertake. University laboratories should certainly continue to be a major part of this industry. However, an increasingly important role should probably be played by industry-oriented laboratories not owned by specific industries but doing research on contract for a diversified set of clients. Such laboratories would usually have at least one client who could benefit from almost any research breakthrough.

2. The incentives generated in a profit economy for firms to keep research findings secret produce results that are, in a static sense, economically inefficient. The use of existing knowledge by one firm in no way reduces the ability of another firm to use that same knowledge, though the incentive to do so may be reduced. The marginal social cost of using knowledge that already exists is zero. For maximum static economic efficiency, knowledge should be administered as a common pool, with free access to all who can use the knowledge. But, if scientific knowledge is thus administered, the incentives of private firms to create new knowledge will be reduced. This is another case in which static efficiency and dynamic efficiency may conflict. It is socially desirable that as much of our basic research effort as possible be undertaken in institutions interested in the quick publication of research results if marginal costs are comparable. In the absence of incentives to private firms to publish research results quickly (such incentives might be legislated) a dollar spent on basic research in a university laboratory is worth more to society than a dollar spent in an industry laboratory, again, if productivity is comparable.

3. If society places the brunt of the basic-research burden on universities, funds must be provided for this purpose. The current Department of Defense policies of letting huge applied research projects with universities should either be reconsidered or complemented with other policies designed to prevent the increased applied-research burden from drawing university

facilities and scientists away from basic research. This is not to say that universities cannot effectively undertake applied research. Rather it is to say that their comparative advantage lies in basic research.

References

NATIONAL SCIENCE FOUNDATION (1956), *Science and Engineering in American Industry*, Washington, D.C.

NATIONAL SCIENCE FOUNDATION (1957a), *Basic Research: A National Resource*, Washington, D.C.

NATIONAL SCIENCE FOUNDATION (1957b), *Growth of Scientific Research in Industry, 1945–60*, Washington, D.C.

NATIONAL SCIENCE FOUNDATION (1957c), *Federal Funds for Science: The Federal Research and Development Budget Fiscal Years 1956, 1957 and 1958*, Washington, D.C.

8 K. Arrow

Economic Welfare and the Allocation of Resources for Invention[1]

K. Arrow, 'Economic welfare and the allocation of resources for invention', in *The Rate and Direction of Inventive Activity*, Princeton University Press, 1962, pp. 609–25.

Invention is here interpreted broadly as the production of knowledge. From the viewpoint of welfare economics, the determination of optimal resource allocation for invention will depend on the technological characteristics of the invention process and the nature of the market for knowledge.

The classic question of welfare economics will be asked here: to what extent does perfect competition lead to an optimal allocation of resources? We know from years of patient refinement that competition insures the achievement of a Pareto optimum under certain hypotheses. The model usually assumes among other things, that (a) the utility functions of consumers and the transformation functions of producers are well-defined functions of the commodities in the economic system, and (b) the transformation functions do not display indivisibilities (more strictly, the transformation sets are convex). The second condition needs no comment. The first seems to be innocuous but in fact conceals two basic assumptions of the usual models. It prohibits uncertainty in the production relations and in the utility functions, and it requires that all the commodities relevant either to production or to the welfare of individuals be traded on the market. This will not be the case when a commodity for one reason or another cannot be made into private property.

We have then three of the classical reasons for the possible failure of perfect competition to achieve optimality in resource allocation: indivisibilities, inappropriability and uncertainty. The first problem has been much studied in the literature under the

1. I have benefited greatly from the comments of my colleague, William Capron. I am also indebted to Richard R. Nelson, Edward Phelps and Sidney Winter of the Rand Corporation for their helpful discussion.

heading of marginal-cost pricing and the second under that of divergence between social and private benefit (or cost), but the theory of optimal allocation of resources under uncertainty has had much less attention. I will summarize what formal theory exists and then point to the critical notion of information, which arises only in the context of uncertainty. The economic characteristics of information as a commodity and, in particular, of invention as a process for the production of information are next examined. It is shown that all three of the reasons given above for a failure of the competitive system to achieve an optimal resource allocation hold in the case of invention. On theoretical grounds a number of considerations are adduced as to the likely biases in the misallocation and the implications for economic organization.[2]

Resource allocation under uncertainty

The role of the competitive system in allocating uncertainty seems to have received little systematic attention.[3] I will first sketch an ideal economy in which the allocation problem can be solved by competition and then indicate some of the devices in the real world which approximate this solution.

Suppose for simplicity that uncertainty occurs only in production relations. Producers have to make a decision on inputs at the present moment, but the outputs are not completely predictable from the inputs. We may formally describe the outputs as determined by the inputs and a 'state of nature' which is unknown to the producers. Let us define a 'commodity-option' as a commodity in the ordinary sense labeled with a state of nature. This definition is analogous to the differentiation of a given physical commodity according to date in capital theory or according to place in location theory. The production of a given commodity under uncertainty can then be described as the production of a vector of commodity-options.

This description can be most easily exemplified by reference to

2. For other analyses with similar points of view, see Nelson (1959, pp. 297–306) and Hitch (1958, p. 1297).

3. The first studies I am aware of are the papers of M. Allais and myself, both presented in 1952 to the Colloque International sur le Risque in Paris; see Allais (1953) and Arrow (1953). The theory has received a very elegant generalization by Debreu (1959, ch. 7).

agricultural production. The state of nature may be identified with the weather. Then, to any given set of inputs there corresponds a number of bushels of wheat if the rainfall is good and a different number if rainfall is bad. We can introduce intermediate conditions of rainfall in any number as alternative states of nature; we can increase the number of relevant variables which enter into the description of the state of nature, for example by adding temperature. By extension of this procedure, we can give a formal description of any kind of uncertainty in production.

Suppose – and this is the critical idealization of the economy – we have a market for all commodity-options. What is traded on each market are contracts in which the buyers pay an agreed sum and the sellers agree to deliver prescribed quantities of a given commodity *if* a certain state of nature prevails and nothing if that state of nature does not occur. For any given set of inputs, the firm knows its output under each state of nature and sells a corresponding quantity of commodity-options; its revenue is then completely determined. It may choose its inputs so as to maximize profits.

The income of consumers is derived from their sale of supplies, including labor, to firms and their receipt of profits, which are assumed completely distributed. They purchase commodity-options so as to maximize their expected utility given the budget restraint imposed by their incomes. An equilibrium is reached on all commodity-option markets, and this equilibrium has precisely the same Pareto-optimality properties as competitive equilibrium under certainty.

In particular, the markets for commodity-options in this ideal model serve the function of achieving an optimal allocation of risk bearing among the members of the economy. This allocation takes account of differences in both resources and tastes for risk bearing. Among other implications, risk bearing and production are separated economic functions. The use of inputs, including human talents, in their most productive mode is not inhibited by unwillingness or inability to bear risks by either firms or productive agents.

But the real economic system does not possess markets for commodity-options. To see what substitutes exist, let us first consider a model economy at the other extreme, in that no provisions for reallocating risk bearing exist. Each firm makes its input decisions; then outputs are produced as determined by the inputs and the state of nature. Prices are then set to clear the market. The prices that finally prevail will be a function of the state of nature.

The firm and its owners cannot relieve themselves of risk bearing in this model. Hence any unwillingness or inability to bear risks will give rise to a non-optimal allocation of resources, in that there will be discrimination against risky enterprises as compared with the optimum. A preference for risk might give rise to misallocation in the opposite direction, but the limitations of financial resources are likely to make under-investment in risky enterprises more likely than the opposite. The inability of individuals to buy protection against uncertainty similarly gives rise to a loss of welfare.

In fact, a number of institutional arrangements have arisen to mitigate the problem of assumption of risk. Suppose that each firm and individual in the economy could forecast perfectly what prices would be under each state of nature. Suppose further there were a lottery on the states of nature, so that before the state of nature is known any individual or firm may place bets. Then it can be seen that the effect from the viewpoint of any given individual or firm is the same as if there were markets for commodity-options of all types, since any commodity-option can be achieved by a combination of a bet on the appropriate state of nature and an intention to purchase or sell the commodity in question if the state of nature occurs.

References to lotteries and bets may smack of frivolity, but we need only think of insurance to appreciate that the shifting of risks through what are in effect bets on the state of nature is a highly significant phenomenon. If insurance were available against any conceivable event, it follows from the preceding discussion that optimal allocation would be achieved. Of course, insurance as customarily defined covers only a small range of events relevant to the economic world; much more important in shifting risks are securities, particularly common stocks and money. By shifting freely their proprietary interests among different firms, individuals can to a large extent bet on the different states of nature which favor firms differentially. This freedom to insure against many contingencies is enhanced by the alternatives of holding cash and going short.

Unfortunately, it is only too clear that the shifting of risks in the real world is incomplete. The great predominance of internal over external equity financing in industry is one illustration of the fact that securities do not completely fulfill their allocative role with respect to risks. There are a number of reasons why this should be so, but I will confine myself to one, of special significance with regard to invention. In insurance practice, reference is made to the moral factor as a limit to the possibilities of insurance. For example, a fire insurance policy cannot exceed in amount the value of the goods insured. From the purely actuarial standpoint, there is no reason for this limitation; the reason for the limit is that the insurance policy changes the incentives of the insured, in this case, creating an incentive for arson or at the very least for carelessness. The general principle is the difficulty of distinguishing between a state of nature and a decision by the insured. As a result, any insurance policy and in general any device for shifting risks can have the effect of dulling incentives. A fire insurance policy, even when limited in amount to the value of the goods covered, weakens the motivation for fire prevention. Thus, steps which improve the efficiency of the economy with respect to risk bearing may decrease its technical efficiency.

One device for mitigating the adverse incentive effects of insurance is co-insurance; the insurance extends only to part of the amount at risk for the insured. This device is used, for example, in coverage of medical risks. It clearly represents a compromise between incentive effects and allocation of risk bearing, sacrificing something in both directions.

Two exemplifications of the moral factor are of special relevance in regard to highly risky business activities, including invention. Success in such activities depends on an inextricable tangle of objective uncertainties and decisions of the entrepreneurs and is certainly uninsurable. On the other hand, such activities should be undertaken if the expected return exceeds the market rate of return, no matter what the variance is.[4] The

4. The validity of this statement depends on some unstated assumptions, but the point to be made is unaffected by minor qualifications.

existence of common stocks would seem to solve the allocation problem; any individual stockholder can reduce his risk by buying only a small part of the stock and diversifying his portfolio to achieve his own preferred risk level. But then again the actual managers no longer receive the full reward of their decisions; the shifting of risks is again accompanied by a weakening of incentives to efficiency. Substitute motivations, whether pecuniary, such as executive compensation and profit sharing, or non-pecuniary, such as prestige, may be found, but the dilemma of the moral factor can never be completely resolved.

A second example is the cost-plus contract in one of its various forms. When production costs on military items are highly uncertain, the military establishment will pay, not a fixed unit price, but the cost of production plus an amount which today is usually a fixed fee. Such a contract could be regarded as a combination of a fixed-price contract with an insurance against costs. The insurance premium could be regarded as the difference between the fixed price the government would be willing to pay and the fixed fee.

Cost-plus contracts are necessitated by the inability or unwillingness of firms to bear the risks. The government has superior risk bearing ability and so the burden is shifted to it. It is then enabled to buy from firms on the basis of their productive efficiency rather than their risk bearing ability, which may be only imperfectly correlated. But cost-plus contracts notoriously have their adverse allocative effects.[5]

This somewhat lengthy digression on the theory of risk bearing seemed necessitated by the paucity of literature on the subject. The main conclusions to be drawn are the following: (a) the economic system has devices for shifting risks, but they are limited and imperfect; hence, one would expect an under-investment in risky activities; (b) it is undoubtedly worthwhile to enlarge the variety of such devices, but the moral factor creates a limit to their potential.

5. These remarks are not intended as a complete evaluation of cost-plus contracts. In particular ,there are, to a certain extent, other incentives which mitigate the adverse effects on efficiency.

Information as a commodity

Uncertainty usually creates a still more subtle problem in resource allocation; information becomes a commodity. Suppose that in one part of the economic system an observation has been made whose outcome, if known, would affect anyone's estimates of the probabilities of the different states of nature. Such observations arise out of research but they also arise in the daily course of economic life as a by-product of other economic activities. An entrepreneur will automatically acquire a knowledge of demand and production conditions in his field which is available to others only with special effort. Information will frequently have an economic value, in the sense that anyone possessing the information can make greater profits than would otherwise be the case.

It might be expected that information will be traded in, and of course to a considerable extent this is the case, as is illustrated by the numerous economic institutions for transmission of information, such as newspapers. But in many instances, the problem of an optimal allocation is sharply raised. The cost of transmitting a given body of information is frequently very low. If it were zero, then optimal allocation would obviously call for unlimited distribution of the information without cost. In fact, a given piece of information is by definition an indivisible commodity, and the classical problems of allocation in the presence of indivisibilities appear here. The owner of the information should not extract the economic value which is there, if optimal allocation is to be achieved; but he is a monopolist, to some small extent, and will seek to take advantage of this fact.

In the absence of special legal protection, the owner cannot, however, simply sell information on the open market. Any one purchaser can destroy the monopoly, since he can reproduce the information at little or no cost. Thus the only effective monopoly would be the use of the information by the original possessor. This, however, will not only be socially inefficient, but also may not be of much use to the owner of the information either, since he may not be able to exploit it as effectively as others.

With suitable legal measures, information may become an appropriable commodity. Then the monopoly power can indeed be exerted. However, no amount of legal protection can make a thoroughly appropriable commodity of something so intangible as information. The very use of the information in any productive way is bound to reveal it, at least in part. Mobility of personnel among firms provides a way of spreading information. Legally imposed property rights can provide only a partial barrier, since there are obviously enormous difficulties in defining in any sharp way an item of information and differentiating it from other similar sounding items.

The demand for information also has uncomfortable properties. In the first place, the use of information is certainly subject to indivisibilities; the use of information about production possibilities, for example, need not depend on the rate of production. In the second place, there is a fundamental paradox in the determination of demand for information; its value for the purchaser is not known until he has the information, but then he has in effect acquired it without cost. Of course, if the seller can retain property rights in the use of the information, this would be no problem, but given incomplete appropriability, the potential buyer will base his decision to purchase information on less than optimal criteria. He may act, for example, on the average value of information in that class as revealed by past experience. If any particular item of information has differing values for different economic agents, this procedure will lead both to a non-optimal purchase of information at any given price and also to a non-optimal allocation of the information purchased.

It should be made clear that from the standpoint of efficiently distributing an existing stock of information, the difficulties of appropriation are an advantage, provided there are no costs of transmitting information, since then optimal allocation calls for free distribution. The chief point made here is the difficulty of creating a market for information if one should be desired for any reason.

It follows from the preceding discussion that costs of transmitting information create allocative difficulties which would be absent otherwise. Information should be transmitted at marginal cost, but then the demand difficulties raised above will exist. From the viewpoint of optimal allocation, the purchasing industry will be faced with the problems created by indivisibilities; and we still leave unsolved the problem of the purchaser's inability to judge in advance the value of the information he buys. There is a strong case for centralized decision making under these circumstances.

Invention as the production of information

The central economic fact about the processes of invention and research is that they are devoted to the production of information. By the very definition of information, invention must be a risky process, in that the output (information obtained) can never be predicted perfectly from the inputs. We can now apply the discussion of the preceding two sections.

Since it is a risky process, there is bound to be some discrimination against investment in inventive and research activities. In this field, especially, the moral factor will weigh heavily against any kind of insurance or equivalent form of risk bearing. Insurance against failure to develop a desired new product or process would surely very greatly weaken the incentives to succeed. The only way, within the private-enterprise system, to minimize this problem is the conduct of research by large corporations with many projects going on, each small in scale compared with the net revenue of the corporation. Then the corporation acts as its own insurance company. But clearly this is only an imperfect solution.

The deeper problems of misallocation arise from the nature of the product. As we have seen, information is a commodity with peculiar attributes, particularly embarrassing for the achievement of optimal allocation. In the first place, any information obtained, say a new method of production, should, from the welfare point of view, be available free of charge (apart from the cost of transmitting information). This insures optimal utilization of the information but of course provides no incentive for investment in research. In an ideal socialist economy, the reward for invention would be completely separated from any charge to the users of the information.[6] In a free-enterprise economy, inventive activity is supported by using the invention to create property rights; precisely to the extent that it is successful, there is an underutilization of the information. The property rights may be in the

6. This separation exists in the Soviet Union, according to N. M. Kaplan and R. H. Moorsteen of the RAND Corporation (verbal communication).

information itself, through patents and similar legal devices, or in the intangible assets of the firm if the information is retained by the firm and used only to increase its profits.

The first problem, then, is that in a free-enterprise economy the profitability of invention requires a non-optimal allocation of resources. But it may still be asked whether or not the allocation of resources to inventive activity is optimal. The discussion of the preceding section makes it clear that we would not expect this to be so; that, in fact, a downward bias in the amount of resources devoted to inventive activity is very likely. Whatever the price, the demand for information is less than optimal for two reasons: (a) since the price is positive and not at its optimal value of zero, the demand is bound to be below the optimal; (b) as seen before, at any given price, the very nature of information will lead to a lower demand than would be optimal.

As already remarked, the inventor will in any case have considerable difficulty in appropriating the information produced. Patent laws would have to be unimaginably complex and subtle to permit such appropriation on a large scale. Suppose, as the result of elaborate tests, some metal is discovered to have a desirable property, say resistance to high heat. Then of course every use of the metal for which this property is relevant would also use this information, and the user would be made to pay for it. But, even more, if another inventor is stimulated to examine chemically related metals for heat resistance, he is using the information already discovered and should pay for it in some measure; and any beneficiary of his discoveries should also pay. One would have to have elaborate distinctions of partial property rights of all degrees to make the system at all tolerable. In the interests of the possibility of enforcement, actual patent laws sharply restrict the range of appropriable information and thereby reduce the incentives to engage in inventive and research activities.

These last considerations bring into focus the interdependence of inventive activities, which reinforces the difficulties in achieving an optimal allocation of the results. Information is not only the product of inventive activity, it is also an input – in some sense, the major input apart from the talent of the inventor. The school of thought that emphasizes the determination of invention by the social climate as demonstrated by the simultaneity of inventions in effect emphasizes strongly the productive role of previous information in the creation of new information. While these interrelations do not create any new difficulties in principle, they intensify the previously established ones. To appropriate information for use as a basis for further research is much more difficult than to appropriate it for use in producing commodities; and the value of information for use in developing further information is much more conjectural than the value of its use in production and therefore much more likely to be underestimated. Consequently, if a price is charged for the information, the demand is even more likely to be sub-optimal.

Thus basic research, the output of which is only used as an informational input into other inventive activities, is especially unlikely to be rewarded. In fact, it is likely to be of commercial value to the firm undertaking it only if other firms are prevented from using the information obtained. But such restriction on the transmittal of information will reduce the efficiency of inventive activity in general and will therefore reduce its quantity also. We may put the matter in terms of sequential decision making. The *a priori* probability distribution of the true state of nature is relatively flat to begin with. On the other hand, the successive *a posteriori* distributions after more and more studies have been conducted are more and more sharply peaked or concentrated in a more limited range, and we therefore have better and better information for deciding what the next step in research shall be. This implies that, at the beginning, the preferences among alternative possible lines of investigation are much less sharply defined than they are apt to be later on and suggests, at least, the importance of having a wide variety of studies to begin with, the less promising being gradually eliminated as information is accumulated.[7] At each stage the decisions about the next step should be based on all available information. This would require an unrestricted flow of information among different projects

7. The importance of parallel research developments i n the case of uncertainty has been especially stressed by Klein (1958, p. 112 ff.) and Klein and Meckling (1958, pp. 352–63).

which is incompatible with the complete decentralization of an ideal free-enterprise system. When the production of information is important, the classic economic case in which the price system replaces the detailed spread of information is no longer completely applicable.

To sum up, we expect a free-enterprise economy to under-invest in invention and research (as compared with an ideal) because it is risky, because the product can be appropriated only to a limited extent, and because of increasing returns in use. This under-investment will be greater for more basic research. Further, to the extent that a firm succeeds in engrossing the economic value of its inventive activity, there will be an under-utilization of that information as compared with an ideal allocation.

Competition, monopoly, and the incentive to innovate

It may be useful to remark that an incentive to invent can exist even under perfect competition in the product markets though not, of course, in the 'market' for the information contained in the invention. This is especially clear in the case of a cost-reducing invention. Provided only that suitable royalty payments can be demanded, an inventor can profit without disturbing the competitive nature of the industry. The situation for a new product invention is not very different; by charging a suitable royalty to a competitive industry, the inventor can receive are turn equal to the monopoly profits.

I will examine here the incentives to invent for monopolistic and competitive markets, that is, I will compare the potential profits from an invention with the costs. The difficulty of appropriating the information will be ignored; the remaining problem is that of indivisibility in use, an inherent property of information. A competitive situation here will mean one in which the industry produces under competitive conditions, while the inventor can set an arbitrary royalty for the use of his invention. In the monopolistic situation, it will be assumed that only the monopoly itself can invent. Thus a monopoly is understood here to mean barriers to entry; a situation of temporary monopoly, due perhaps to a previous innovation, which does not prevent the entrance of new firms with innovations of their own, is to be regarded as more nearly competitive than monopolistic for the purpose of this analysis. It will be argued that the incentive to invent is less under monopolistic than under competitive conditions but even in the latter case it will be less than is socially desirable.

We will assume constant costs both before and after the invention, the unit costs being c before the invention and $c' < c$ afterward. The competitive price before invention will therefore be c. Let the corresponding demand be x_c. If r is the level of unit royalties, the competitive price after the invention will be $c' + r$, but this cannot of course be higher than c, since firms are always free to produce with the old methods.

It is assumed that both the demand and the marginal revenue curves are decreasing. Let $R(x)$ be the marginal revenue curve. Then the monopoly output before invention, x_m, would be defined by the equation

$$R(x_m) = c.$$

Similarly, the monopoly output after invention is defined by

$$R(x'_m) = c'.$$

Let the monopoly prices corresponding to outputs x_m and x'_m, respectively, be p_m and p'_m. Finally, let P and P' be the monopolist's profits before and after invention, respectively.

What is the optimal royalty level for the inventor in the competitive case? Let us suppose that he calculates p'_m, the optimal monopoly price which would obtain in the post-invention situation. If the cost reduction is sufficiently drastic that $p'_m < c$, then his most profitable policy is to set r so that the competitive price is p'_m, i.e. let

$$r = p'_m - c'.$$

In this case, the inventor's royalties are equal to the profits a monopolist would make under the same conditions, i.e. his incentive to invent will be P'.

Suppose, however, it turns out that $p'_m > c$. Since the sales price cannot exceed c, the inventor will set his royalties at

$$r = c - c'.$$

The competitive price will then be c, and the sales will remain at x_c. The inventor's incentive will then be $x_c(c-c')$.

The monopolist's incentive, on the other hand, is clearly $P'-P$. In the first of the two cases cited, the monopolist's incentive is obviously less than the inventor's incentive under competition, which is P', not $P'-P$. The pre-invention monopoly power acts as a strong disincentive to further innovation.

The analysis is slightly more complicated in the second case. The monopolist's incentive, $P'-P$, is the change in revenue less the change in total cost of production, i.e.

$$P'-P = \int_{x_m}^{x'_m} R(x)\, dx - c' x'_m + c x_m.$$

Since the marginal revenue $R(x)$ is diminishing, it must always be less than $R(x_m) = c$ as x increases from x_m to x'_m, so that

$$\int_{x_m}^{x'_m} R(x)\, dx < c(x'_m - x_m),$$

and $P'-P < c(x'_m - x_m) - c' x'_m + c x_m = (c-c') x'_m.$

In the case being considered, the post-invention monopoly price, p'_m, is greater than c. Hence, with a declining demand curve, $x'_m < x_c$. The above inequality shows that the monopolist's incentive is always less than the cost reduction on the post-invention monopoly output, which in this case is, in turn, less than the competitive output (both before and after invention). Since the inventor's incentive under competition is the cost reduction on the competitive output, it will again always exceed the monopolist's incentive.

It can be shown that, if we consider differing values of c', the difference between the two incentives increases as c' decreases, reaching its maximum of P (pre-invention monopoly profits) for c' sufficiently large for the first case to hold. The ratio of the incentive under competition to that under monopoly, on the other hand, though always greater than 1, decreases steadily with c'. For c' very close to c (i.e. very minor inventions), the ratio of the two incentives is approximately x_c/x_m, i.e. the ratio of monopoly to competitive output.[8]

The only ground for arguing that monopoly may create superior incentives to invent is that appropriability may be greater under monopoly than under competition. Whatever differences may exist in this direction must, of course, still be offset against the monopolist's disincentive created by his pre-invention monopoly profits.

The incentive to invent in competitive circumstances may also be compared with the social benefit. It is necessary to distinguish between the realized social benefit and the potential social benefit, the latter being the benefit which would accrue under ideal conditions, which, in this case, means the sale of the product at post-invention cost, c'. Clearly, the potential social benefit always exceeds the realized social benefit. I will show that the realized social benefit, in turn, always equals or exceeds the competitive incentive to invent and, *a fortiori*, the monopolist's incentive.

Consider again the two cases discussed above. If the invention is sufficiently cost reducing so that $p'_m < c$, then there is a con-

8. To sketch the proof of these statements quickly, note that, as c' varies, P is a constant. Hence from the formula for $P' - P$, we see that
$$d(P'-P)/dc' = dP'/dc' = R(x'_m)(dx'_m/dc') - c'(dx'_m/dc') - x'_m = -x'_m,$$
since $R(x'_m) = c'$. Let $F(c')$ be the difference between the incentives to invent under competitive and under monopolistic conditions. In the case where $p'_m \leqslant c$, this difference is the constant P. Otherwise
$$F(c') = x_c(c-c') - (P'-P),$$
so that $dF/dc' = x'_m - x_c$.
For the case considered, we must have $x'_m < x_c$, as seen in the text. Hence, $dF/dc' \leqslant 0$, so that $F(c')$ increases as c' decreases.
Let $G(c')$ be the ratio of the incentive under competition to that under monopoly, If $p'_m \leqslant c$, then
$$G(c') = P'/(P'-P),$$
which clearly decreases as c' decreases. For $p'_m > c$, we have
$$G(c') = x_c(c-c')/(P'-P).$$
Then $dG/dc' = \{-(P'-P)x_c + x_c(c-c')x'_m\}/(P'-P)^2$.
Because of the upper bound for $P'-P$ established in the text, the numerator must be positive; the ratio decreases as c' decreases.
Finally, if we consider c' very close to c, $G(c')$ will be approximately equal to the ratio of the derivatives of the numerator and denominator (L'Hopital's rule), which is, x_c/x'_m, and which approaches x_c/x_m as c' approaches c.

sumers' benefit, due to the lowering of price, which has not been appropriated by the inventor. If not, then the price is unchanged, so that the consumers' position is unchanged, and all benefits do go to the inventor. Since by assumption all the producers are making zero profits both before and after the invention, we see that the inventor obtains the entire realized social benefit of moderately cost reducing inventions but not of more radical inventions. Tentatively, this suggests a bias against major inventions, in the sense that an invention, part of whose costs could be paid for by lump-sum payments by consumers without making them worse off than before, may not be profitable at the maximum royalty payments that can be extracted by the inventor.

Alternative forms of economic organization in invention

The previous discussion leads to the conclusion that for optimal allocation to invention it would be necessary for the government or some other agency not governed by profit-and-loss criteria to finance research and invention. In fact, of course, this has always happened to a certain extent. The bulk of basic research has been carried on outside the industrial system, in universities, in the government and by private individuals. One must recognize here the importance of non-pecuniary incentives, both on the part of the investigators and on the part of the private individuals and governments that have supported research organizations and universities. In the latter, the complementarity between teaching and research is, from the point of view of the economy, something of a lucky accident. Research in some more applied fields, such as agriculture, medicine and aeronautics, has consistently been regarded as an appropriate subject for government participation, and its role has been of great importance.

If the government and other non-profit institutions are to compensate for the under-allocation of resources to invention by private enterprise, two problems arise: how shall the amount of resources devoted to invention be determined, and how shall efficiency in their use be encouraged? These problems arise whenever the government finds it necessary to engage in economic activities because indivisibilities prevent the private economy from performing adequately (highways, bridges, reclamation projects, for example), but the determination of the relative magnitudes is even more difficult here. Formally, of course, resources should be devoted to invention until the expected marginal social benefit there equals the marginal social benefit in alternative uses, but in view of the presence of uncertainty, such calculations are even more difficult and tenuous than those for public works. Probably all that could be hoped for is the estimation of future rates of return from those in the past, with investment in invention being increased or decreased accordingly as some average rate of return over the past exceeded or fell short of the general rate of return. The difficulties of even *ex post* calculation of rates of return are formidable though possibly not insuperable.[9]

The problem of efficiency in the use of funds devoted to research is one that has been faced internally by firms in dealing with their own research departments. The rapid growth of military research and development has led to a large-scale development of contractual relations between producers and a buyer of invention and research. The problems encountered in assuring efficiency here are the same as those that would be met if the government were to enter upon the financing of invention and research in civilian fields. The form of economic relation is very different from that in the usual markets. Payment is independent of product; it is governed by costs, though the net reward (the fixed fee) is independent of both. This arrangement seems to fly in the face of the principles for encouraging efficiency, and doubtless it does lead to abuses, but closer examination shows both mitigating factors and some explanation of its inevitability. In the first place, the awarding of new contracts will depend in part on past performance, so that incentives for efficiency are not completely lacking. In the second place, the relation between the two parties to the contract is something closer than a purely market relation. It is more like the sale of professional services, where the seller contracts to supply not so much a specific result as his best judgement. (The demand for such services also arises from uncertainty and the value of information.) In the third place, payment by results would involve great risks for the inventor, risks against which, as we have seen, he could hedge only in part.

9. For an encouraging study of this type, see Griliches (1958).

There is clear need for further study of alternative methods of compensation. For example, some part of the contractual payment might depend on the degree of success in invention. But a more serious problem is the decision as to which contracts to let. One would need to examine the motivation underlying government decision making in this area. Hitch (1958) has argued that there are biases in governmental allocation, particularly against risky invention processes, and an excessive centralization, though the latter could be remedied by better policies.

One can go further. There is really no need for the firm to be the fundamental unit of organization in invention; there is plenty of reason to suppose that individual talents count for a good deal more than the firm as an organization. If provision is made for the rental of necessary equipment, a much wider variety of research contracts with individuals as well as firms and with varying modes of payment, including incentives, could be arranged. Still other forms of organization, such as research institutes financed by industries, the government and private philanthropy, could be made to play an even livelier role than they now do.

References

ALLAIS, M. (1953), 'Généralisation des théories de l'équilibre économique général et du rendement social au cas du risque', *Économètrie*, vol. 40, Colloques Internationaux du Centre National de la Recherche Scientifique; reprinted in *Econometrica*, 1953, vol. 21, pp. 269–90.

ARROW, K. J. (1953), 'Rôle des valeurs boursières pour la répartition la meilleure des risques', *Économètrie*, vol. 40, Colloques Internationaux du Centre National de la Recherche Scientifique.

DEBREU, G. (1959), *Theory of Values*, Wiley.

GRILICHES, Z. (1958), 'Research costs and social returns: hybrid corn and related innovations', *J. polit. Econ.*, October, pp. 419–31. [Reprinted as Reading 9 in this selection.]

HITCH, C. J. (1958), 'The character of research and development in a competitive economy', Rand Corporation, May.

KLEIN, B. H. (1958), 'A radical proposal for R and D', *Fortune*, May.

KLEIN, B. H., and MECKLING, W. H. (1958), 'Application of operations research to development decisions', *Op. Res.*, vol. 6, no. 3, May–June.

NELSON, R. R. (1959), 'The simple economics of basic scientific research', *J. polit. Econ.*, June, pp. 297–306. [Reprinted as Reading 7 of this selection.]

9 Z. Griliches

Research Costs and Social Returns: Hybrid Corn and
Related Innovations[1]

Z. Griliches, 'Research costs and social returns: hybrid corn
and related innovations', *Journal of Political Economy*, October 1958,
pp. 419–31.

Introduction and summary

Both private and public expenditures on 'research and development' have grown very rapidly in the last decade. Quantitatively,
however, we know very little about the results of these investments. We have some idea of how much we have spent but very
little of what we got in return. We know almost nothing about
the realized rate of return on these investments, though we feel
intuitively that it must have been quite high. This article presents
a first step toward answering some of these questions. However,
all that is attempted here is to estimate the realized social rate of
return, as of 1955, on public and private funds invested in hybrid-
corn research, one of the outstanding technological successes of
the century. The calculated rate of return is an *estimate*, subject
to a wide margin of error, but it should provide us with an order
of magnitude for the 'true' social rate of return on expenditures
on hybrid-corn research. Actually, I believe that my estimate is
biased downward, for, whenever I had to choose among alternative assumptions, I chose the assumption that led to the lowest
estimate. This estimate will not tell us the global rate of return
on research expenditures, but even a modest step in that direction
may be of some use.

The following procedure is used to arrive at the estimate:
firstly, private and public research expenditures on hybrid corn,

1. This article is an outgrowth of a larger study of the economics of
hybrid corn. See my article (Griliches, 1957). I am indebted to A. C.
Harberger, Martin J. Bailey, Lester G. Telser and T. W. Schultz for
valuable comments and to the National Science Foundation and the
Social Science Research Council for financial support.

1910–55, are estimated on the basis of a mail survey and other data. Then the annual gross social returns are estimated on the assumption that they are approximately equal to the value of the resulting increase in corn production plus a price-change adjustment. The additional cost of producing hybrid seed is subtracted from these gross returns to arrive at an annual flow of net social returns. Using first a 5 and then a 10 per cent rate of interest, I bring all costs and returns forward to 1955, when the books are closed on this development and a rate of return is computed. Research costs are expressed as a capital sum, and returns are converted into a perpetual flow. The estimated perpetual flow of returns is divided by the cumulated research expenditures to arrive at a rate of return that will equalize the present value of the flow of returns with the cumulated value of research expenditures. This procedure leads to the estimate that *at least* 700 per cent per year was being earned, as of 1955, on the average dollar invested in hybrid-corn research.

Since this is not the only way in which a rate of return could be computed from these data, some alternative ways of defining and estimating the social rate of return are explored briefly. Comparisons are also made with estimates of returns in some other areas of technological change. Finally, I discuss the limitations of the procedure used and the implications of the results. In particular, I shall emphasize that almost no normative conclusions can be drawn from these few estimates.

Research expenditures

Inbred lines and hybrids have been developed by state agricultural experiment stations, the United States Department of Agriculture (USDA), and private seed companies. The distinction between the first two developing agencies is mainly in the source of funds. Except for funds spent on research and co-ordinating activities at Beltsville, Maryland, most of the USDA funds were spent on cooperative corn-breeding research at various experiment stations.

A mail inquiry to ascertain expenditures on hybrid-corn research was sent to all the agricultural experiment stations, and usable data were obtained from twenty of them. The twenty states represented by these replies include most of the important corn states in the country. Expenditures of non-responding stations were estimated by setting the expenditures of each of them equal to the expenditures of a 'similar' station.[2]

Data on USDA expenditures on 'corn-production research: agronomic phases' beginning with 1931 were obtained from the Agricultural Research Service and extrapolated back to 1910. They overestimate the USDA contribution substantially, because they include various other aspects of corn research besides hybrid corn. Moreover, some of the USDA funds have already been counted in the expenditures of agricultural experiment stations.

The research expenditures of one of the major private seed companies for the years 1925–55 were extrapolated back to 1911 and divided by that firm's estimated share of the total market for hybrid seed corn to arrive at an estimate of the research expenditures of the 'private' segment of the industry.[3]

The figures for 1955 may be used as an example of the numbers involved. I estimate that in 1955 the USDA spent about $300,000 on hybrid-corn research, the experiment stations about $650,000, and the private companies about $1,900,000.[4]

The historical research expenditure data, deflated by the Consumers Price Index (1955 = 100), are reproduced in column 1 of Table 1. In view of all the assumptions made, these figures should be taken with several grains of salt, the dosage increasing as one goes back into the past. In particular, for the years 1910–25, the figures are little more than guesses.

2. The pairing was made on the basis of geographic proximity and general information about the industry. For example, Indiana expenditures were assumed to equal the reported Illinois expenditures; Oklahoma's to equal Kentucky's, and so forth. These pairings probably overestimate the total expenditures on hybrid-corn research.

3. This will again overestimate expenditures, because 'public' hybrids make up 25–30 per cent of the total market, and research expenditures on these have already been counted once.

4. In 1951 M. T. Jenkins, of the USDA, estimated the total annual expenditures on hybrid-corn research as follows: USDA, $220,000; states, $600,000; and private industry, $1,100,000. My own independent estimate for 1951 is: USDA, $190,000; states, $550,000; private industry, $1,300,000. The two totals are $1,920,000 and $2,040,000, respectively. The agreement is very close, considering how arbitrary some of my assumptions are (see Jenkins, 1951, pp. 42–5).

Table 1 Hybrid Corn: Estimated Research Expenditures and Net Social Returns, 1910–55 (Millions of 1955 Dollars)

Year	Total research expenditures (*private and public*)	Net social returns*	Year	Total research expenditures (*private and public*)	Net social returns*
1910	0·008		1934	0·564	1·1
1911	0·011		1935	0·593	2·9
1912	0·010		1936	0·661	8·3
1913	0·016		1937	0·664	21·2
1914	0·022		1938	0·721	39·9
1915	0·032		1939	0·846	60·3
1916	0·039		1940	1·090	81·7
1917	0·039		1941	1·100	105·3
1918	0·039		1942	1·070	124·3
1919	0·044		1943	1·390	140·4
1920	0·052		1944	1·590	158·7
1921	0·068		1945	1·600	172·6
1922	0·092		1946	1·820	184·7
1923	0·105		1947	1·660	194·3
1924	0·124		1948	1·660	203·7
1925	0·139		1949	1·840	209·8
1926	0·149		1950	2·060	209·0
1927	0·185		1951	2·110	218·7
1928	0·210		1952	2·180	226·7
1929	0·285		1953	2·030	232·1
1930	0·325		1954	2·270	234·2
1931	0·395		1955	2·790	239·1
1932	0·495		Annually after		
1933	0·584	0·3	1955	3·000	248·0

* Net of seed production cost but not net of research expenditures. Net social returns are zero before 1933.

For the purpose of estimating the rate of return on these expenditures, I assume that the public sector will continue to invest in hybrid-corn research at an annual rate of $1 million, and the private sector at an annual rate of $2 million. No incremental returns, however, will be ascribed to these expenditures. I assume them to be 'maintenance' expenditures in face of a malevolent nature.

Cost of additional resources devoted to production of hybrid seed

I assume that the price of hybrid seed, approximately $11 per bushel in 1955, measures adequately the value of resources devoted to its production. If there were no hybrid corn, farmers would use mainly home-produced open-pollinated seed, which I value at $1·50 per bushel.[5] The quantity of hybrid seed used annually was estimated by multiplying the reported corn acreage planted with hybrid seed by the average seeding rate of corn. Multiplying the result by $9·50, the difference between the price of hybrid and non-hybrid seed, and subtracting $2 million research expenditures, I get $90 million as my estimate of the additional resources currently devoted each year to hybrid-seed production.[6] Using the average 1939–48 corn acreage (90 million), the 1951 seeding rate (7·5 pounds per acre), and the percentage planted with hybrid seed, I computed the additional cost of hybrid seed for the years 1933–55 and subtracted this from the subsequent estimate of gross returns to arrive at a net return figure.[7]

The value of hybrid corn to society

As everyone knows, hybrid corn increased corn yields. The figure most often quoted for this increase is 20 per cent. For my purpose, I assume that the superiority of hybrid over open-pollinated varieties is 15 per cent, the lower figure in most estimated ranges.[8] The value of this increase to 'society' will be measured

5. This is somewhat higher than the market price of corn because of the better quality of seed corn and the labor that would go into its selection. Since open-pollinated seed is now quoted at about $3·00 to $4·00 a bushel, this assumption also contributes to an overall overestimate of cost.

6. This result is reached as follows: 80 million acres × 90 per cent in hybrids × (8·6/64) average seeding rate × $9·50 = $92 million. Subtracting $2 million research expenditures, we get $90 million (source: *Agriculture Statistics, 1955*).

7. Throughout the period these computations use the average corn acreage planted in 1939–48; they disregard annual fluctuations in total corn acreage and seeding rates. For any year before 1956, the extra cost of seed equals the percentage planted with hybrid seed times $98 million (90 million × [7·5/64] × $9·50 − $2 million [research expenditures] = $98 million).

8. For example: 'Plant breeders conservatively estimate increase in yields of 15 to 20 per cent from using hybrid seed under field conditions. They expect about the same relative increases in both low- and high-yielding areas' (USDA, 1940, p. 7).

by the loss in total corn production that would have resulted if there were no hybrid corn. This hypothetical loss will be valued at the estimated equilibrium price of corn plus a price-change adjustment, a procedure equivalent to computing the loss in 'consumer surplus' that would occur if hybrid corn were to 'disappear'.

The amount of this loss will depend on our assumptions about the relevant demand-and-supply elasticities. As will be seen from the formulas presented below, these elasticities have only a second-order effect, and hence different reasonable assumptions about them will affect the results very little. I assumed that the price elasticity of the demand for corn is approximately -0.5.[9] Since we know much less about the supply elasticity of corn, I shall first explore the consequences of two different extreme assumptions about it.

Let us assume, first, that in the long run the supply of corn is infinitely elastic; that is, we face long-run constant costs. The 'disappearance' of hybrid corn would shift the supply curve upward by the percentage reduction in the yield of corn. The 'loss' to society, in this case, is the total area under the demand curve between the new and the old supply curves. In Figure 1 this area is $P_1 P_2 P_2' P_1''$. This area can be interpreted as the increase in the total cost of producing the quantity Q_2 in the new situation, the rectangle $P_1 P_2 P_2' P_1'$, plus the loss in consumer surplus caused by the rise in price, the triangle $P_1' P_2' P_1''$. A linear approximation of this area is given by the formula

$$\text{Loss } 1 = k P_1 Q_1 (1 - \tfrac{1}{2}kn),$$

where k is the percentage change in yield (marginal cost and average cost), P_1 and Q_1 are, respectively, the previous equilibrium price of corn and quantity of corn produced, and n is the absolute value of the price elasticity of the demand for corn.

Alternatively, it could be assumed that the elasticity of the supply curve is zero. In this case, the loss is measured by the area $Q_2 P_2' P_1'' Q_1$ in Figure 2. Instead of assuming that the supply curve shifts upward, we now assume that it shifts k per cent to the left. The rectangle $Q_2 P_1' P_1'' Q_1$ measures the loss in corn

9. This figure is based on a USDA demand analysis (see Foote, Klein and Clough, 1952).

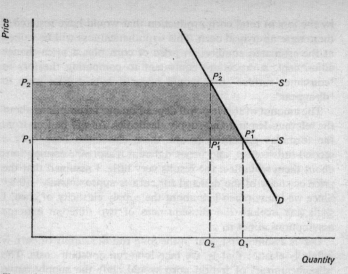

Figure 1

production at the old price P_1. The triangle $P_1'P_2'P_1''$ can be viewed as the additional loss in consumer surplus or as an adjustment for the increase in price from P_1 to P_2. The total loss is now given by the formula

$$\text{Loss 2} = kP_1Q_1(1+\tfrac{1}{2}kn).^{10}$$

10. A correction. I am indebted to T. D. Wallace of North Carolina State University and R. C. Lindberg of Purdue University for pointing out an error in the formula for Loss 2. The correct formula is

$kP_1Q_1\{1+(\tfrac{1}{2})k/n\}$

and not

$kP_1Q_1\{1+(\tfrac{1}{2})kn\}$

as in the text. This changes the estimated ratio of Loss 2 to Loss 1 from 1·07 to 1·13, which is still a very minor difference.

In addition, however, they have raised the question whether the difference between these Losses is due merely to different assumptions about the supply elasticity. Would not the area defined as $P_1P_2P_2'P_1'$ be a more relevant measure than the area $Q_2P_2'P_1'Q_1$ in Figure 2 which I actually used? The ratio of Loss 2 so defined to Loss 1 would be on the order of the reciprocal of the elasticity of demand, and thus larger than 1 for all of the relatively inelastic demand situations considered in the paper.

It is true that, in defining these Losses, I mentioned consumer surplus,

It is easily seen that the second assumption leads to a higher estimate of the loss. It can be also shown that the two estimates bracket estimates implied by assuming other intermediate supply elasticities. The ratio of the loss under assumption 2 to the loss under assumption 1 is $(1+\frac{1}{2}kn)/(1-\frac{1}{2}kn)$. In our case, this ratio is approximately $1\cdot07$.[11] The difference between these two extreme assumptions implies only a 7 per cent difference in the final estimate of the total loss. Because this difference is so small and because I am striving for a lower-limit estimate, I have chosen the first assumption – that of an infinitely elastic long-run supply of corn.

To calculate the loss, we must assume an equilibrium price of corn. I shall use $1·00 per bushel in 1955 dollars as a minimal estimate of the value of corn to society. The current price of corn, about $1·25, is affected by the existence of price-support programs and probably overestimates the social value of corn.[12]

---

and that as defined Loss 2 does not take all of the consumer loss into account. To that extent the original text is in error. But the actual definition of Loss 2 as used in the paper is the more sensible of the two. It is simply the gain (or loss) in output due to hybrid corn valued at average (pre- and post-hybrid) prices. The objection to the alternative suggestion can be illustrated by considering the case of an infinitely elastic demand function. The suggested definition would indicate no social gain from hybrid corn, which is clearly wrong, whereas the definition I used would still value the increase in output at the constant price.

In general, refinements in valuing the social gain are probably not worth the confusion they may create. As far as the substantive issues are concerned, the above reservations only reinforce the conclusion that, if anything, the estimated social gains are on the low side.

11. Assuming $k = 0·13$, i.e. 15/115, and $n = 0·5$, the ratio is
$(1 + 0·5 \times 0·13 \times 0·5)/(1 - 0·5 \times 0·13 \times 0·5) = 1·07$.

12. An approximate formula for determining the price of corn in the absence of price supports is given by Nerlove (1956, p. 497):

$$dp/p_0 = (dq/q)/(n+e),$$

where p_0 is the equilibrium price, n and e are the demand-and-supply elasticities, and dq is the quantity placed under loan. In recent years about 7 per cent of the annual corn crop, in the average, has been placed under loan with the Commodity Credit Corporation. The assumptions, $n = 0·5$, $e = 0·2$, imply that the current price is about 10 per cent above the equilibrium price. The current price is about $1·25 per bushel, which implies an 'equilibrium' price of corn of about $1·13. But this estimate does not take into account the impact on corn prices of the elimination of price supports

Because not all corn acreage was or is planted with hybrids, I multiply the percentage shift k by h, the percentage of all corn acres planted with hybrids [loss $= hkPQ(1 - \frac{1}{2}hkn)$]. This procedure disregards the fact that the acres first planted to hybrids

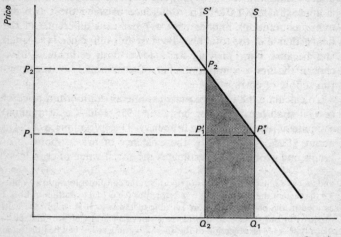

Figure 2

were higher-yielding acres than those planted later, and hence the procedure underestimates total returns.

In estimating past returns from hybrid corn, I ignore annual fluctuations in prices and production, basing my computations on the average 1937–48 level of production of 2900 million bushels[13] and a real price of corn of $1·00 per bushel in 1955

on all other agricultural commodities. Taking this into account, the equilibrium price would be closer to $1·00 per bushel. In any case, it is unlikely to be lower than $1·00.

13. This assumption was made to simplify the calculations. In the first part of the period, production was below this figure, and, since I use a relatively high rate of interest, this will result in an overestimate of returns. But the percentage planted with hybrids was also low then, while it was much higher during the period when production exceeded its average, and this will result in an underestimate of returns. On balance, the second effect should outweigh the first by a fair margin.

dollars. On the returns side, only the percentage planted with hybrid seed varies over time. To calculate the annual flow of future returns, I assume that the average 1943–52 level of production – approximately 3000 million bushels annually – will continue and that the percentage planted with hybrid seed will stabilize at 90. Both these assumptions are conservative and will result in an underestimate of returns.

Assuming that k, the relative shift in the supply curve, is 0·13 (15/115); that PQ is $3000 million; that n, the demand elasticity, is 0·5; and that h, the current and future fraction of all corn acres planted with hybrid seed, is 0·9, we can calculate the current and expected annual flow of gross social returns from hybrid corn as follows:

$$0.9 \times 0.13 \times \$3000 \text{ million} \ (1 - \tfrac{1}{2} \times 0.9 \times 0.13 \times 0.5)$$
$$= 0.117 \times \$3000 \text{ million} \ (1 - 0.029)$$
$$= 0.117 \times 0.971 \times \$3000 \text{ million} = \$341 \text{ million}.$$

Subtracting the projected annual cost of hybrid-seed production and research – $93 million – we get $248 million as the current and expected annual flow of net social returns.

Similarly, returns for the past years, beginning with 1933, are calculated by multiplying 2900 million 1955 dollars, the average total value of corn production, by the percentage of total corn acreage planted with hybrid seed in each year. Subtracting the estimated past costs of hybrid-seed production, we get the net social returns for the years 1933–55 shown in column 2 of Table 1.[14]

Calculation of a rate of return

Table 1 presents estimates of costs and returns. There are several ways in which these figures could be summarized and a rate of return calculated. My procedure is as follows: consider the development closed as of 1955. Future expenditures will not

14. These are calculated from the following formula:
$$h \times \{0.13 \times \$2.900 \text{ million} \ (1 - \tfrac{1}{2} \times 0.9 \times 0.13 \times 0.5) -$$
$$- 90 \text{ million} \times (7.5/64) \times \$9.50 - \$2 \text{ million}\} = h \times \$268 \text{ million}.$$

This procedure is an approximation, since h should also have entered into the second part of the first term of the formula, the 'triangle'. I neglect this. Because h is less than 1 and because that part is always subtracted, this procedure again underestimates total returns.

increase returns, nor will there be an expansion of hybrid-corn acreage. Standing in 1955, I cumulate and bring forward to 1955 all past costs and returns at a reasonable *external* rate of interest. To explore the impact of two quite different rates of interest, I perform the calculations twice, using first a 5 per cent and then a 10 per cent rate of interest. Past research costs are cumulated and expressed as a capital sum. Past returns are cumulated to 1955, and a 5 or 10 per cent rate of return on these cumulated returns is projected into the future. The estimated flow of future net returns is added to the flow from past returns to arrive at a perpetual flow of net social returns from hybrid corn. This flow, divided by the cumulated research expenditures, gives us our estimate of the realized perpetual rate of return.

Table 2 presents the calculations that lead to my estimate of approximately $7·00 as the annual return in perpetuity, as of 1955, for every dollar that has been invested in hybrid-corn research. Actually, even if we ignore all past returns completely, the figure is still very high – approximately $4·00 annually (using the 5 per cent interest rate) for every research dollar.

Table 2 Rate of Return on Hybrid-Corn Research Expenditures as of 1955 (Millions of Dollars)

	$r = 0·05$	$r = 0·10$
1. Net cumulated past returns	4405	6542
2. Past returns expressed as an annual flow	220	654
3. Annual future gross returns	341	341
4. Annual additional cost of production and research	93	93
5. Total net annual returns, (2)+(3)−(4)	468	902
6. Cumulated past research expenditures	63	131
7. Rate of return 100×(5)/(6)	743	689

This way of calculating a 'rate of return' is not really different from a benefit–cost ratio calculation. It may be useful to bring out explicitly the relationship between these two concepts. The preceding rate of return is defined as follows:

$$r = 100(PR \times k + AFR)/RC,$$

where PR = cumulated past returns,
k = the external rate of interest used to cumulate or discount returns,

AFR = annual future returns,

and RC = cumulated research costs.

A benefit–cost ratio from these same data would be

$$B/C = (PR + AFR/k)/RC.$$

Hence $r = 100k(B/C)$, and we can translate our calculation into a benefit–cost ratio, and vice versa. Using 5 and 10 per cent as the external rates of interest, the benefit–cost ratios for hybrid-corn research expenditures are 150 and 70, respectively.[15] When we recall that most Bureau of Reclamation watershed projects have *ex post* benefit–cost ratios of 1 or less, this does imply a certain misallocation of public resources.[16]

These calculations use an external rate of interest to bring all sums forward to 1955. It is reasonable to assume that the marginal productivity of capital in alternative investments is between 5 and 10 per cent and to use these rates as conversion factors for funds expended or earned at different dates. Alternatively, however, one could calculate an *internal* rate of return – that rate of interest which will equate the flow of costs and the flow of returns over time.[17] Such a rate has to be calculated using an iterative procedure, changing the rate used until the cumulated costs are equal to the discounted returns. The two procedures will give different answers when the time shape of costs differs markedly from the time shape of returns, as in our case. The internal rate of return on hybrid-corn research expenditures is between 35 and 40 per cent.[18] My objection to this particular procedure is that it

15. If before cumulating we were to subtract the research costs from net returns annually (that is, have in our denominator only research expenditures before 1934, the year when net returns began to exceed research costs), the benefit–cost ratios would be substantially higher (about 700 and 200, respectively), and so would also the rate of return as defined in the text.

16. For an evaluation of public investments in watershed projects see Renshaw (1957).

17. I am indebted to Martin J. Bailey for suggesting this alternative way of calculating the rate of return.

18. That is, 40 per cent is too high and 35 is too low. The iterative procedure was not carried further.

values a dollar spent in 1910 at $2300 in 1933. This does not seem very sensible to me. I prefer to value the 1910 dollar at a reasonable rate of return on some alternative social investment. Also, this procedure gives tremendous weight to the early expenditures, which are subject to the largest error of measurement. Actually, however, the two estimates are not very far apart. The estimate using an *external* rate of interest says that a dollar invested in hybrid-corn research earned 10 cents annually until 1955 and $7·00 annually thereafter. The *internal* rate estimate says that the dollar earned 40 cents annually throughout the whole period. If the average delay between investment and fruition is about ten years, then the two figures represent different ways of saying the same thing.[19]

Limitations

The estimate of 700 per cent is probably too low. At almost every point at which there was a choice of assumptions to be made, I have purposely chosen those that would result in a lower estimate. This is an attempt to arrive at an estimated lower limit of the social rate of return from hybrid corn.

Both the public and the private research expenditures are probably overstated substantially. In fact, the expenditure estimates supplied by some experiment stations are obviously too high. Of the USDA expenditures, perhaps less than half were devoted to hybrids. I did leave out all expenses incurred before 1910, but these in total could not have been more than a few thousand dollars. This should remind us, however, that the estimated rate of return is mainly a rate of return on applied rather than basic research. The basic idea of hybrid corn was developed between 1905 and 1920, with the help of very little money. The rate of return on this basic research, if it could be calculated, would be much higher. However, the idea had to be translated to commercial reality, and separate adaptable hybrids had to be developed for different areas. These are the activities reflected in my estimate of research cost.

19. An annual flow of 40 cents discounted at 40 per cent has a present value of $1·00. An annual flow of $7·00, discounted at 40 per cent, will also have a present value of $1·00 if there is a lag of approximately ten years between the date of investment and the date at which the perpetual flow of returns begins.

The returns, on the other hand, are understated. The assumed price of corn of $1·00 per bushel in 1955 dollars and the assumption of a 15 per cent superiority of hybrids are both conservative and probably result in an underestimate of the real returns to society. I have also assumed that all past research has already borne all its fruit and that all future research on hybrids will result in no benefit whatsoever. Nor has credit been given for the impact of hybrid corn on other fields: the research on hybrid poultry and hybrid sorghum which it stimulated or the reduction of farmer resistance to new technology as a result of the spectacular success of hybrid corn.

Hence, as far as costs and returns from hybrid corn *per se* are concerned, the estimate is too low. One troublesome problem, however, remains to haunt us. Does it really make sense to calculate the rate of return on a successful 'oil well'? What is the point of calculating the rate of return on one of the outstanding technological successes of the century? Obviously, it will be high. What we would like to have is an estimate that would also include the cost of all the 'dry holes' that were drilled before hybrid corn was struck.

The estimate does include the cost of all the 'dry holes' in hybrid-corn research itself. Hybrid corn was not a unique invention – it was an invention of a method of inventing. Many different combinations were tried before the right ones were found. One major seed company annually tests approximately fifteen hundred different combinations of inbred lines. Of these, at most three or four prove to be successful. The cost of the unsuccessful experiments is included in my estimate. What is excluded is investment in various other areas of agricultural research which has not borne fruit.

The problem of dry holes, however, can be reduced *ad absurdum*. What is the relevant segment for which an aggregate rate of return is to be computed? Is it really reasonable to ascribe the cost of unsuccessful gold exploration to the oil industry? If one takes this kind of reasoning too seriously, there is only one rate that has any meaning – the rate of return on research for the economy as a whole.

Nevertheless, the rate of return on a successful innovation may be of some interest. In particular, it may be useful, *ex ante*, to break down the probable rate of return into two components: the rate of return if the development turns out to be a success, and the probability that it will be a success. The approach outlined here is a way of estimating the first component. An estimate of the probability of success, however, must be made on the basis of data other than those presented in this article.

Returns in some other areas

Schultz (1953, pp. 114–22) has provided us with estimates of costs and returns of research for United States agriculture as a whole. His data can be used to estimate the rate of return on agricultural research as a whole. Schultz gives an upper- and a lower-limit estimate of how much more input it would have taken to produce the 1950 output with 1940 techniques and inputs. His upper estimate is that it would have taken 18·5 per cent more input; his lower estimate is 3·7 per cent. Let us use Schultz's figures to estimate the perpetual gross annual returns, beginning with 1951, from the agricultural research that bore fruit between 1940 and 1950. Using the lower estimate, there would be a loss of 3·7 per cent in output if the new technology were to disappear. Taking the total annual value of farm output as $30 billion ($32 billion cash receipts from farm marketings in 1951, minus approximately 10 per cent to allow for the impact of the support programs), and assuming a price elasticity of demand for agricultural products of -0.25 and an infinite supply elasticity, we get

$$k(1 - \tfrac{1}{2}kn) \times \$30 \text{ billion}$$

$$= 0.037 \times (1 - 0.5 \times 0.037 \times 0.25) \times \$30 \text{ billion}$$

$$= 0.037 \times 0.995 \times \$30 \text{ billion} = \$1110 \text{ million.}$$

Using the upper limit estimate of 18·5 per cent saving in inputs, we get

$$0.185(1 - 0.5 \times 0.185 \times 0.25) \times \$30 \text{ billion} = \$5430 \text{ million,}$$

as an upper-limit estimate of the gross annual social return from the technical change that occurred between 1940 and 1950.

Schultz also provides an estimate of total public expenditures on agricultural research for the years 1937–51. I assume that all these expenditures were used to produce the increase in output in 1951. This leaves out the contributions developed from funds spent before 1937, but, on the other hand, it disregards the possible returns after 1951 from the 1937–51 expenditures. On balance, we will probably overestimate the funds spent on the 1940–50 improvement in technology.

I will assume that total private agricultural research expenditures were of about the same magnitude as the public expenditures. This is approximately half the corresponding ratio for hybrid corn but is probably an overestimate for agriculture as a whole.[20]

Multiplying the 1937–51 public expenditures by 2, deflating them by the Consumers Price Index, and cumulating them at the rate of 5 per cent, I get the figure of 3180 million 1951 dollars as my estimate of total cumulated research expenditures in agriculture. Comparing this with the two estimated limits of the annual social returns of $1110 and $5430 million, I get a lower limit of 35 and an upper limit of 171 per cent as estimates of the annual rate of return per dollar spent on agricultural research. These are substantially lower than the estimated returns from hybrid corn but are comparable to estimates made by Ewell (1955, pp. 298–304) for the economy as a whole (100–200 per cent per year per dollar spent on 'research and development') and to figures quoted by major industrial companies on their returns on research.

Of course, these estimates are quite consistent with the estimated returns on hybrid corn, if the probability of success in research on innovations like hybrid corn is on the order of one-tenth or one-twentieth. Nevertheless, all these figures indicate that, in spite of the large growth in research expenditures during this century, the social returns to this activity are still very high.

20. Mighell (1955, p. 130) has estimated that recent annual expenditures by industry for research on agricultural products and on machinery and materials used in agriculture were in excess of $140 million. At the same time the USDA and state agricultural experiment stations spent about $118 million annually on research. However, only one-third of the industry research was in aid of farm production, mainly in machinery and chemicals: the rest was used in product research, while four-fifths of public expenditures were for research in aid of farm production.

Hybrid sorghum

The approach previously outlined can be used to estimate the probable rate of return on a new development: for example, hybrid sorghum. The development of hybrid sorghum began seriously only after the Second World War but has gained momentum rapidly since. Hybrid sorghum is now being introduced commercially. Very little was planted in 1956, but substantial amounts were planted in 1957. The experimental data to date suggest that the superiority of hybrid sorghum over previous seed may be even greater than that of hybrid corn (USDA, 1956).

The 1956 value of the grain sorghum crop was approximately $232 million. Assuming a 15 per cent superiority of hybrids, a demand elasticity of −1·0, and an infinitely elastic supply curve, the estimated gross social returns from sorghum hybrids would be about $37 million annually.[21] Assuming that the extra cost of hybrid seed will be about $7 million annually, the net annual returns would be about $30 million.[22] I have no official data on research expenditures on hybrid sorghum. The head of one of the major seed companies has estimated that to date all public and private expenditures on hybrid sorghum total approximately $1 million and that current expenditures are at an annual rate of $300,000. Doubling his estimate of past expenditures and projecting into the future an annual research expenditure rate of $500,000, I get $10 or $13 million, depending on the rate of interest used, as my 'estimate' of cumulated hybrid-sorghum research expenditures in 1967. I choose 1967 as the reckoning date on the assumption that it will take ten years for hybrid sorghum to capture most of the sorghum seed market. While the projected rate of development is faster than that of hybrid

21. This figure is derived as follows:

$0·15 (1 + \frac{1}{2} \times 0·15 \times 1·0) \times \$232$ million $= 0·16 \times \$232 = \$37$ million.

The figure would have been approximately twice as large if I had used the value of the 1957 crop − $493 million − as my base.

22. The additional cost of hybrid seed, $7 million, is estimated from the following data: ten million acres sown to sorghum; $10 difference between the average prices per hundredweight of hybrid and open-pollinated grain sorghum seed (*Agricultural Prices*, April, 1958, p. 42); and an average seeding rate of seven pounds per acre.

corn in the United States as a whole, it is equivalent to the rate of acceptance of hybrid corn in Iowa. It is reasonable to assume that hybrid sorghum will spread much faster than hybrid corn did. Sorghum production is more localized than corn production, almost all sorghum is grown for commercial purposes, and hybrid sorghum will probably encounter much less resistance than hybrid corn encountered. I also assume that the use of hybrid sorghum in the United States will follow the same time path that the use of hybrid corn followed in Iowa.[23] This assumption allows me to estimate the social returns during the 'transition period', 1957–67.

Table 3 outlines the calculation of the estimated rate of return on research expenditures on hybrid sorghum, which is approximately 400 per cent per annum. While this is somewhat lower than the estimated rate of return on expenditures on hybrid-corn research, it is still very high indeed.

Table 3 Estimated Costs and Returns of Hybrid-Sorghum Research as of 1967 (Millions of Dollars)

	$r = 0.05$	$r = 0.10$
Cumulated social net returns, 1957–66	155	171
Value in 1967 of returns beyond 1967	590	295
Total value of net returns in 1967	745	466
Cumulated research expenditures	9·4	13
Benefit–cost ratio	79	36
'Rate of return' in per cent per annum	395	360

Some implications

One might have expected, on *a priori* grounds, that the rate of return on expenditures on hybrid-sorghum research would be higher than the return on hybrid-corn research. The cost of hybrid sorghum has been and will be lower than the cost of hybrid corn both because of the cumulated experience in hybrid-corn breeding and because sorghum-growing is much more localized than corn-growing. Therefore, adaptable hybrids will

23. The percentage of all corn planted with hybrid seed in Iowa followed the following time path by years: 0·02, 0·06, 0·14, 0·31, 0·52, 0·73, 0·90, 0·99, and 1·00.

have to be developed for a much smaller portion of the United States. Nevertheless, the estimated returns from hybrid sorghum are lower than those from hybrid corn. Why? To some extent this lower rate may be a result of overestimating research costs for hybrid sorghum, but the principal explanation is that the total value of the sorghum crop is substantially smaller than that of corn; sorghum is a relatively unimportant crop. It has recently been suggested that we should redirect our research efforts away from the major commodities that are in 'surplus' and away from commodities with low elasticities of demand, where technical improvements result in reduced total returns to farmers in the long run. However, if we assume, in the absence of any other information, that technological change operates somewhat like a percentage increase in yield, and that the cost of achieving a given percentage boost in yield is the same for different crops or at least independent of their price elasticities and relative 'importance', then the highest *social* returns per research dollar are to be found in the important, low-elasticity commodities. For it can be easily shown that the absolute social gain from a given percentage increase in yield will vary proportionately with the total value of the crop and that the impact of different demand-and-supply elasticities is of a second order of magnitude. The latter affect only the 'triangle', or the magnitude of the price-change adjustment. Among all the different factors, the total value of the crop is by far the most important determinant of the absolute social gain from a given percentage increase in yield.

No matter how we calculate them, there is little doubt that the overall social returns on publicly supported technological research have been very high. It is not clear, however, whether or not this fact has any normative implications. I am afraid it has very few. More knowledge than is now at hand is required for prescription.

It is clear that we have not succeeded in equalizing the returns on different kinds of public investments. The returns from technological research have been much higher than the returns from reclamation and watershed projects. But should we have had more technological research? Surely, we have yet to reach the optimal level of expenditures on research, but this can only be a hunch. Should the public support agricultural research? My analysis illustrates and quantifies one of the major arguments for *public* investments in this area – the divergence between the social and private rates of return. Almost none of the calculated social returns from hybrid corn were appropriated by the hybrid-seed industry or by corn producers. They were passed on to consumers in the form of lower prices and higher output. Entry into the hybrid-seed industry was easy, and in the long run no 'abnormal' profits were made there. By valuing the extra cost of seed production at the market price, I have counted as a cost whatever profit was made in this area by private producers, and the resulting estimate of returns consists almost entirely of social rather than private returns. These social returns were diffused widely among consumers of corn and corn products. Given the difficulty of patenting most of the valuable ideas in this area, the short life of a patent, and the general precariousness of a monopoly position in the long run, the incentive for private investment was very much smaller than that implied by the social rate of return.

While a divergence between social and private rates of return is a necessary reason for public intervention, it is not, by itself, a sufficient reason. We must ask not only whether social returns are higher than private – this is also true of many private investments – but also whether the private rate of return is too low, relative to returns on alternative private investments, to induce the *right* amount of investment at the *right* time. The social returns from nylon were probably many times higher than DuPont profits, but the latter were high enough to induce the development of nylon without a public subsidy, although, perhaps, not soon enough. To establish a case for public investment one must show that, in an area where social returns are high, private returns, because of the nature of the invention or of the relevant institutions, are not high enough relative to other private alternatives. This was undoubtedly true of hybrid corn, and it is probably true of many other areas of agricultural research and basic research in general. But it is not universally true. Hence a high social rate of return is not an unequivocal signal for public investment.

In this paper I have estimated that the rate of return on public investments in one of the most *successful* ventures of the past has been very high. This may give support to our intuitive feeling that the returns to such ventures in general have been quite high and to our feeling that 'research is a good thing'. But that does not mean that we should spend any amount of money on anything called 'research'. The moral is that, though very difficult, some sort of cost-and-returns calculation is possible and should be made. Conceptually, the decisions made by an administrator of research funds are among the most difficult economic decisions to make and to evaluate, but basically they are not very different from any other type of entrepreneurial decision.

References

EWELL, R. H. (1955), 'Role of research in economic growth', *Chemical and Engineering News*, vol. 33.

FOOTE, R. J., KLEIN, J. W., and CLOUGH, M. (1952), *The Demand and Price Structure for Corn and Total Feed Concentrates*, USDA, Technical Bull. 1061, October.

GRILICHES, Z. (1957), 'Hybrid corn: an exploration in the economics of technological change', *Econometrica*, October.

JENKINS, M. T. (1951), 'Corn-breeding research – whither bound?', *Proceedings of the Sixth Annual Hybrid Corn Industry Research Conference*, American Seed Trade Association, November.

MIGHELL, R. (1955), *American Agriculture*, Wiley.

NERLOVE, M. (1956), 'Estimates of the elasticities of supply of selected agricultural commodities', *J. farm Econ.*, vol. 38, May.

RENSHAW, E. F. (1957), *Toward Responsible Government*, Idyia Press.

SCHULTZ, T. W. (1953), *The Economic Organization of Agriculture*, McGraw-Hill.

UNITED STATES DEPARTMENT OF AGRICULTURE (1940), *Technology on the Farm*, Washington, D.C.

UNITED STATES DEPARTMENT OF AGRICULTURE (1956), *Sorghum Hybrids*, ARS Special Report 22–6, May.

10 W. Fellner

Two Propositions in the Theory of Induced Innovations

W. Fellner, 'Two propositions in the theory of induced innovations',
Economic Journal, June 1961, pp. 305-8.

This note is intended to establish a presumption for the existence
of an adjustment mechanism which in market economies directs
inventive activity into more or less labour-saving (less or more
capital-saving) channels, according as one or the other factor of
production is getting relatively scarce on a macroeconomic level.
On the conventional static equilibrium assumptions for firms
which are very small in relation to the economy, it would be
inconsistent to assume the existence of such a mechanism. On
these assumptions macroeconomic resource-scarcities express
themselves to the individual firms exclusively in the ruling factor
prices, none of which is either 'high' or 'low' in relation to the
marginal productivity of the resource. Consequently, the firm is
not interested in whether any given product-raising or cost-
saving effect is achieved by raising primarily the marginal
productivity of the one or of the other factor of production.
However, the point to be argued here is that this negative conclusion must be qualified significantly – indeed, loses its validity
for pronounced scarcities – if we make the assumptions slightly
more realistic.

The writer has presented a more detailed analysis of the problem at a conference on inventive activity, but while the note that
follows here is less complete in several respects than the conference paper in question, it is more specific about some propositions whose relevance now seems greater to the writer than it
did earlier.[1]

1. The more detailed analysis, which was recently presented at a conference in Minneapolis, will be published in a volume of the National
Bureau of Economic Research. Both propositions to be developed in this
note below – the proposition concerning the 'learning process' and the

In Figures 1 and 2 the simplifying assumption is made that labour and capital are the only factors of production. Each technology is defined by isoquants. Economies of scale are disregarded, since their effects on the preference for labour-saving versus capital-saving inventions is unpredictable (hence these economies may be viewed as having a random influence). The relative factor prices, which must be known before the iso-cost functions can be drawn, may be interpreted as 'the real wage-rate' and 'the interest rate', that is, as the money price of hiring workers, and as the cost of borrowing capital (or of renting capital goods of given money value), where the price of the final output is assumed as given. Sir Roy Harrod called my attention to the fact that, contrary to frequent practice, it is wrong to take it *generally* for granted that in models of this sort a change in money wage-rates relative to interest rates will lead to a change in the capital–labour ratio; yet I shall explain in a footnote why I believe that for my specific purpose this result may indeed be taken for granted.[2]

---

proposition concerning monopsonistic imperfections – are elaborations on hypotheses expressed in (Fellner, 1956). Professor H. A. J. Green has rightly pointed out that the presentation of these propositions in Fellner (1956) needed a more detailed explanation (see his excellent article (Green, 1960, pp. 57–73)). The presentation of the first proposition (see Figure 1, below) has greatly benefited from discussions I had with Mr Richard R. Nelson of the RAND Corporation.

2. The difficulty which in analysis of this sort needs to be watched is the following. If money wage-rates change (say, rise) *but the rate of interest does not change* (say, does not decline), then this does not *necessarily* provide an inducement for using methods of greater capital-intensity. The result may simply be that all prices, including those of the capital goods, rise in the same proportion as the money wage-rate, and hence the price of real 'capital disposal', as well as the real wage-rate, stay unchanged. However, the reasoning in the present note will always imply assumptions on which it is legitimate to conclude that a change (say, a rise) in money wage-rates relative to interest rates does make it profitable to use methods of greater capital-intensity. This is because if the supply of capital rises relative to that of labour *along given production functions*, then interest rates do decline (and, of course, real as well as money wage-rates rise). If, on the other hand, the rise of the supply of capital relative to that of labour is accompanied by *technological progress*, then the rate of interest need not decline; yet even in this case unrealistic assumptions would have to be made with respect to extremely labour-saving innovations to arrive at the conclusion that real

To save space I deduced the two central propositions of the argument in the legends appearing under the two graphs. In the text of this note I shall attempt to express the meaning of these two propositions briefly in more general terms.

The meaning of the first proposition (see Figure 1) is that in some cases a preference may develop for inventions which are particularly factor-saving in the resource that is getting scarcer, because a *learning process* may induce atomistic firms to behave as if they were big enough to notice that *macroeconomically* the factors of production are *not* in infinitely elastic supply. To be more specific, in Western economies, in which the supply of capital is expected to rise in a higher proportion than the supply of labour, a labour-saving invention which at present is inferior to a capital-saving invention may in due time become superior to the latter; and this is a macroeconomic fact which the atomistic firm can *learn* by watching trends in factor prices. However, firms will, of course, give preference to a now inferior labour-saving invention over a now superior capital-saving invention only if they wish to make a substantial allowance for the possibility that no important further invention will be available for some time. This is because for any given period the superiority or inferiority of an invention to the atomistic firm does not depend on whether the invention is relatively labour-saving or capital-saving.

The meaning of the second proposition (see Figure 2) is that distortions or 'market imperfections' of certain kinds may call forth market imperfections of a different kind which counteract, or in some cases wholly neutralize, the initial distortion. If, for example, the macroeconomic capital–labour ratio is rising rapidly, and not enough labour-saving innovations are introduced, then after a sufficient decline of the profit rate this is apt to lead to a distortion which shows in the form of Keynesian unemployment. This distortion results from imperfections – money-wage rigid-

---

wage-rates will *not* rise. It seems quite appropriate to assume that in such circumstances a rise in money wage-rates expresses at the same time a rise in *real* wage-rates (say, with a constant price level); and if with a constant price level money and real wage-rates rise relative to interest rates, then this does make it profitable to use more capital-intensive methods of production.

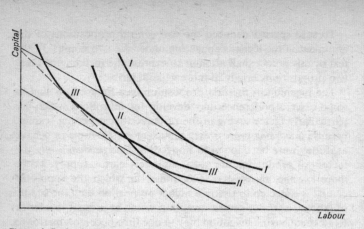

Figure 1 Perfectly competitive factor hiring

All three isoquants of the firm relate to the identical quantity of output.
The fully drawn straight lines are the iso-cost lines as of 'now'. Isoquant *I*
describes the old technology. The firm must choose between making an
effort to develop (acquire) 'innovation' *II* or *III*. Given the factor–price
ratios expressed by the present iso-cost lines, technology *II* is superior
to *III*. But if (as is the case in Western economies) a gradual increase in
real wage-rates relative to interest rates is anticipated (see the broken
iso-cost), then the firm expects that *III will in the future* become
superior to *II*. Thus, whatever the length of the relevant periods is for
which investment decisions are made, *it may be preferable to establish
III rather than II*, provided that it is necessary to make a substantial
allowance for the possibility that no further invention and innovation
will become available during the subsequent periods. The argument
gains in strength if *II* and *III* are nearly equivalent at present factor
prices. Note that *III* is more 'relatively labour-saving' than *II*. This is
true on microeconomic grounds because for any given factor–price ratio
the firm uses *III* with a smaller labour–capital ratio than *II*. Viewed
macroeconomically, *III* is more relatively labour-saving than *II* because if,
given the macroeconomic quantities of input, many firms use *III*, the
wage rate is lower relative to the interest rate than if many firms use
II. Hence the argument here points to a labour-saving bias in
circumstances where real wage rates are expected to rise relative to
interest rates. As here drawn, innovation *II* could be defined as 'neutral'
and *III* as 'labour-saving', but what matters for the argument is merely
that *III* should be more labour-saving than *II* and that they should
intersect.

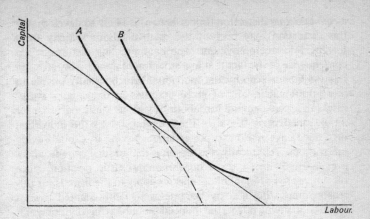

Figure 2 Monopsonistic imperfection

Legend: Isoquants *A* and *B* relate to the same output. Technology *A* is more relatively labour-saving than *B* (see legend under Figure 1). The slope of the fully drawn iso-cost line expresses the factor–price ratios in the event of perfectly competitive factor hiring. In this case *A* and *B* appear equally desirable to the firm. Hence both processes will be developed and used in the economy. However, if a 'monopsonistic' imperfection raises the wage-rate relative to the rate of interest whenever the firm increases its labour input, *A* becomes superior to *B*. The monopsonistic imperfection is here illustrated with the aid of an iso-cost line, which in the neighbourhood of the ordinate coincides with the fully-drawn line, but further down continues along the 'broken' path. Firms are here regarded as seeking (at a cost) innovation *A* or innovation *B*.

ities – such as prevent the Pigou–Patinkin process from going into effect. However, Keynesian unemployment presupposes a low marginal efficiency of capital, and hence it may be interpreted as developing after an initial period in which labour is getting increasingly *scarce* relative to capital.[3] During this initial period

3. The type of unemployment with which the Keynesian theory is concerned has the 'paradoxical' feature that it results indirectly from a labour supply which is *too* small to prevent the marginal efficiency of capital from falling to a critical level. It is this initial shortage of factors co-operating with capital which then turns into an excess supply of labour (because in the absence of the Pigou–Patinkin effect investment becomes insufficient to match savings at full employment). The other type of unemployment, which is usually attributed to the insufficiency of the capital stock (either

a *counteracting* distortion (imperfection) is likely to develop, and this distortion may prevent the relative labour scarcity from turning into unemployment. The counteracting distortion is a consequence of the fact that in a situation of appreciably growing relative labour scarcity the individual firm is usually unable to hire additional labour of given qualities at unchanging wage-rates (an excess demand for labour of various kinds may exist for an extended period) and in these circumstances the individual firm does have reason to prefer labour-saving to capital-saving innovations. Technically, the reason is that in such circumstances even a small firm is in a quasi-monopsonistic position, since resources are not in infinitely elastic supply to the firm. If for this reason innovating activity becomes sufficiently slanted toward the labour-saving effect, the induced labour-saving slant may put an end to the relative labour scarcity and the phase of Keynesian unemployment may thus be indefinitely postponed.

Similarly, if innovations become *too* labour-saving, and if for this or any other reason a relative shortage of *capital* develops,[4] the individual firm finds itself in a quasi-monopsonistic position in the capital market, and innovating activity may become directed into more capital-saving channels.

There is reason to assume that in those countries in which a genuine innovating process has originated, the character of innovating activity adjusted reasonably well to basic resource positions. I would like to suggest that this is unlikely to have been entirely the result of an historical accident.

References

FELLNER, W. (1956), *Trends and Cycles in Economic Activity*, Henry Holt.
GREEN, H. A. J. (1960), 'Growth models, capital and stability', *Econ. J.*, March.

---

because the labour supply is rising at a higher rate than the capital stock or because innovations become extremely labour-saving) does *not* have this 'paradoxical' feature. This second type of unemployment may be explained directly with reference to the fact that given the size of the capital stock, the marginal productivity of the fully employed labour force would be smaller than the ruling wage-rate.

4. Given a floor to real wage rates, such a shortage of capital can create the second type of unemployment discussed in the preceding footnote.

Part Three **The Diffusion of New Technology**

Popular discussions of changes in technology are often centered upon the actions of single individuals to whom history textbooks frequently assign the sole credit for an invention. But although the availability of a name and a date may simplify the arduous task of writing elementary histories, they add very little to our appreciation of the economic consequences of an invention. This is so because the economic impact of an invention is felt only to the extent that a superior technique or product is actually introduced and adopted by business firms. It now seems clear that an explanation of the diffusion of inventions is necessarily a major component in any deeper understanding of the process of economic growth. For this reason, studies of diffusion have recently become increasingly popular both among economists and economic historians.

The papers in this section, in spite of the diversity of their subject content, share the view that the observed changes, both spatial and chronological, in the distribution of particular innovations, can be largely accounted for by economic variables. Griliches's study of the diffusion of hybrid corn shows that the behaviour of both farmers and hybrid-seed producers were firmly grounded in expectations of profit. The time lag in the first introduction of hybrid seed into a particular region, the rate at which farmers shifted to the new seed once it became available, and the extent to which the new seeds replaced the old open-pollinated varieties, all turned upon questions of profitability. In areas where the profits to be realized from the shift were large and unambiguous, the transition was exceedingly rapid. In Iowa, where the hybrid corn was particularly well-suited – and the profitability of the switch was accordingly high – it took only four years for farmers to shift from 10 to 90 per cent of their acreage in hybrid corn.

The paper by David on the mechanization of reaping comes to grips with the intriguing historical fact that, although the reaper had been invented in the early 1830s, it experienced for some time only a slow and limited diffusion. In the mid-1850s, however, fully two decades after its introduction, mid-western farmers discarded their old, labour-intensive techniques of cutting grain and adopted the reaper on a large scale. By use of a threshold function pertaining to farm size, and by focusing upon the manner in which the rising relative cost of harvest labour lowered that threshold size, David accounts for the rapid introduction of the reaper on family-sized farms in the American mid-west during the 1850s.

Temin's paper also addresses itself to a long-standing, troublesome problem of historical interpretation. The use of coke in making pig-iron was one of the decisive steps in Britain's early industrialization. Yet the United States, which by a widely-accepted tradition was highly receptive to the adoption of new techniques throughout the nineteenth century, lagged several decades behind the British in the introduction of coke into blast furnaces. Temin shows that the timing of the diffusion of this new ironmaking technique in the United States revolved around unique supply considerations: specifically (a) the availability of the rich anthracite deposits of eastern Pennsylvania which postponed the introduction of coke and (b) the later discovery of high-quality coking coal in western Pennsylvania which eventually accelerated the introduction.

Finally, Mansfield's paper casts a wide net and studies the speed with which twelve important innovations spread from firm to firm in four industries – bituminous coal, iron and steel, brewing, and railroads. His empirical results demonstrate both the general overall slowness of the diffusion process as well as the wide variation in the speed at which particular techniques are adopted. His model also suggests, *inter alia*, that the rate of imitation was a direct function of the profitability of a given innovation and a decreasing function of the size of the investment required for its installation.

11 Z. Griliches

Hybrid Corn and the Economics of Innovation

Z. Griliches, 'Hybrid corn and the economics of innovation', *Science*, 29 July 1960, pp. 275–80.

The idea that a cross between plants that are genetically unlike can produce a plant of greater vigor and yield than either of the parental lines dates back to Darwin and earlier. Serious research on hybrid corn, however, did not begin until the first years of this century, and the first application of research results on a substantial commercial scale was not begun until the early 1930s. During the last twenty-five years, the change from open pollination to hybrid seeds has spread rapidly through the Corn Belt, and from the Corn Belt to the rest of the nation. The pattern of diffusion of hybrid corn, however, has been characterized by marked geographic differences. As shown in Figure 1, some regions began to use hybrid corn much earlier than others, and some regions, once the shift began, made the transition much more rapidly than others. For example, Iowa farmers began planting hybrid corn earlier than did Alabama farmers, and Iowa farmers increased their acreage in hybrid corn from 10 to 90 per cent more rapidly than did Alabama farmers.

Although the explanation of area differences in the pattern of diffusion of hybrid corn constitutes the main contribution of the study reported here,[1] it is worth drawing attention first to the striking similarity in the general pattern of diffusion of hybrid seed in the various areas. Almost everywhere the development followed an S-shaped growth curve. As illustrated in Figure 1, the rate of change is slow at first, accelerating until it reaches its peak, at approximately the mid-point of development, and then

1. A more detailed and technical account of this study can be found in Griliches (1957); see also Griliches (1958), a study supported by the Social Science Research Council and the National Science Foundation.

slowing down again as the development approaches its final level.[2]

Interestingly enough, this pattern of development also applies to increases in the use of farm equipment – combines, corn-pickers, pickup balers and field forage harvesters, as illustrated in Figure 2. Similar patterns occur in the use of new drugs by

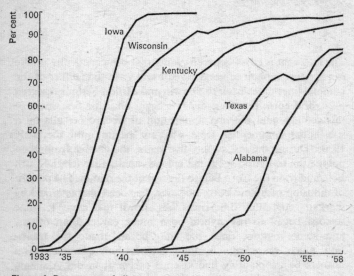

Figure 1 Percentage of all corn acreage planted to hybrid seed (*Agricultural Statistics*)

doctors and in the diffusion of other new items or ideas (see Coleman, Katz and Menzel, 1957). Thus, the data on hybrid corn and other technical changes in US agriculture support the general finding that the pattern of technical change is S-shaped.

Although the finding that technical change follows this pattern is not very surprising or new (see, e.g. Lotka, 1925, and Kuznets, 1930), it is very useful. It allows us to summarize large bodies of data on the basis of three major characteristics (parameters) of a

2. In Griliches (1957) I show that the data fit the logistic growth curve very well. Unpublished data by small sub-divisions – county and crop-reporting districts – give essentially the same picture, though the development is somewhat more irregular in the marginal corn areas.

diffusion pattern: the date of beginning (origin), relative speed (slope), and final level (ceiling). The interesting question then is, given this general S-shape, what determines the differences among areas in the origin, slope and ceiling? Why were some areas ahead of others in first using hybrid corn? Why did hybrid

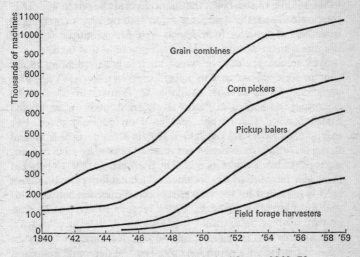

Figure 2 Machines in use on farms in the United States, 1940–59. Note the resemblance to Figure 1 (*US Dept. Agri. Statist. Bull.*, no. 233, 1959)

corn spread faster in some areas than in others? Why did some areas reach higher levels of equilibrium than others?

Date of availability

Although the *idea* of breeding hybrid corn as we know it today goes back at least to 1918, to D. F. Jones and the double cross, the dates at which superior hybrids actually became available in different areas varied widely. Hybrid corn was not a once-and-for-all innovation that could be adopted everywhere. Rather, it was an invention of a new method of innovating, a method of developing superior strains of corn for specific localities.[3] The

3. 'Hybrid corn is the product of a controlled, systematic crossing of specially selected parental strains called "inbred lines". These inbred lines

actual process of developing superior hybrids had to be carried out separately for each locality. It is important to remember this fact before one blames, for example, the southern farmers for being slow to plant hybrid corn. Although superior hybrids became available in the Corn Belt in the early 1930s, it was only in the middle of the 1940s that good hybrids began to appear in the South. Thus, the date for a given area on which commercial quantities of superior hybrid seed were first produced is one of the major determinants of the development in that area.

We can take the date on which an area began planting 10 per cent of its corn acreage to hybrid corn as the date on which superior hybrids became available to farmers in commercial quantities. As shown in Figure 3, different areas in the United States reached the 10 per cent level on different dates. For example, this level was reached in 1936 in some parts of Iowa and in northern Illinois, but was not reached until after 1948 in some parts of Alabama and Georgia. The usefulness of the 10 per cent level as a measure of the commercial availability of hybrid corn seed is indicated by the very close correspondence between this and an alternative measure. From records of state yield tests and from other publications it is possible to determine in what year hybrids first outyielded open-pollinated varieties by a substantial margin in a given locality. The '10 per cent' definition used in this study has a 0·93 Spearman rank correlation coefficient with the 'technical' definition.

Area differences in date of first planting of hybrid corn can be explained in terms of differences in date of availability of hybrid corn seed. Area differences in date of availability, in turn, can

---

are developed by inbreeding, or self-pollinating, for a period of four or more years. Accompanying inbreeding is a rigid selection for the elimination of those inbreds carrying poor heredity and which, for one reason or another, fail to meet the established standards' (Neal and Strommer, 1948, p. 4).

'The [inbred lines] are of little value in themselves for they are inferior to open-pollinated varieties in vigor and yield. When two unrelated inbreds are crossed, however, the vigor is restored. *Some* of these hybrids prove to be markedly superior to the original varieties. The development of hybrid corn, therefore, is a complicated process of continued self-pollination accompanied by selection of the most vigorous and otherwise desirable plants. These superior lines are then used in making hybrids' (Jugenheimer, 1939, pp. 3–4).

be explained, in part, in terms of some simple economic factors. Innovators among the seed producers first entered those areas where the expected profits from the commercial production of hybrid corn seed were largest. They entered the 'good' areas ahead of the 'poor' ones. It is no accident that, though the major innovation occurred in Connecticut, commercial development began in the heart of the Corn Belt where the potential market – farmers who buy and plant corn seed – was largest. The profits that seed producers can expect to make in a given region depend upon the size of the market for corn seed in that region and the cost of entry in that region.

Market density and cost of entry

The close correlation for an area between the date of availability of hybrid corn seed and the market for corn seed may be seen by comparing Figure 3 with two reasonable measures of the market. The first measure is the density of corn acreage in 1949, shown in Figure 4; the second measure is the density of corn pickers in use on farms, shown in Figure 5 (the second measure is the better index of the 'goodness' of a corn area and provides the best simple outline of the Corn Belt). This correlation is also demonstrated by plotting the date of entry of hybrid-seed producers into an area, as measured by the date on which farmers devoted 10 per cent of their corn acreage to hybrid corn, against the average market density of the area. As shown in Figure 6 the lower the market density, the later the date of entry into a given area. The rank correlation is high (0·7), and even higher (0·9) if the Southeast is excluded from the computations. The Southeast is a special case. It was entered later because of the relative lateness of the research contributions by the region's experiment stations and obstacles put in the way of private seed companies in that area. Moreover, when one gets down to a certain low level it does not really pay to discriminate between areas on the basis of their relative market densities, and other variables become more significant.

Deviations in the correlation between the spread of hybrid corn and the distribution of the market can be explained by the cost of entry factor. Cost of entry depends, among other things, on how different the area is from those already entered, and on whether

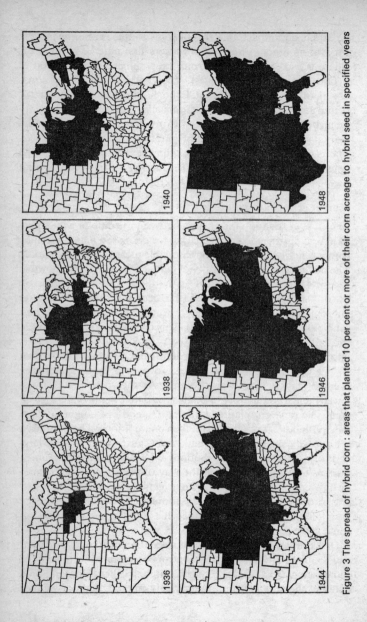

Figure 3 The spread of hybrid corn : areas that planted 10 per cent or more of their corn acreage to hybrid seed in specified years

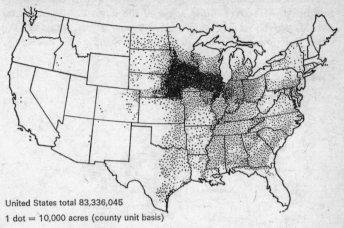

Corn for all purposes
acreage, 1949

United States total 83,336,045
1 dot = 10,000 acres (county unit basis)

Figure 4 The market for hybrid seed. Corn acreage in 1949

experiment stations have already developed inbred lines and
whole hybrids adaptable to the area. Study of Figures 3, 4 and 5
shows first that the spread was much faster latitudinally than it
was longitudinally. The reason, in part, is that an important
factor determining the range of adaptability of a particular
hybrid is the length of the growing season. To a large extent
this is a function of latitude, and as one moves east or west the
chances that the same hybrid will be adaptable to new areas are
much higher than they are if one moves north or south. Nevertheless, the movement north seems to have been faster than the
movement south. This is partly because of the larger markets
in the north but is also a reflection of the special contributions of
the Minnesota and Wisconsin agricultural experiment stations.
They entered hybrid corn research very early in the game and
contributed a great deal more than one would have expected from
them just on the basis of the relative importance of corn in their
states. Similarly, the contributions of Texas, Louisiana and Florida
stations came earlier and were relatively larger than those of the
other stations in the South, which produced little of importance

till the middle 1940s. This would explain to some extent why hybrid corn moved into the Southwest before it did into the Southeast. Moreover, quite a few of the Corn Belt inbreds and hybrids proved adaptable in the Southwest. It was more like the Corn Belt than was the Southeast, and it did not suffer to the

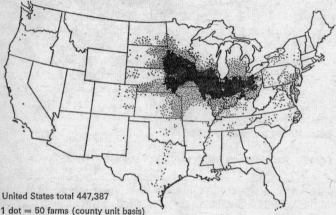

Corn pickers
number of farms reporting,1 April 1950

United States total 447,387
1 dot = 50 farms (county unit basis)

Figure 5 Corn pickers on farms, 1950. This is a better index than corn acreage of the 'goodness' of an area with respect to corn growing and provides the best single outline of the Corn Belt

same extent from insect and disease problems that corn-breeders in the Southeast had to deal with.[4]

Since cost factors in areas that are close together are likely to be similar, we may assume that entry into the neighborhood of an area makes entry into the area itself more likely, and thus we may use the earliest date of entry into any of the immediately

4. One can approximate a measure of the difference between the Corn Belt and some other areas with respect to the kind of hybrids they take by computing from lists of recommended hybrids in various areas and their pedigrees the percentage of all inbred lines accounted for by Corn Belt inbred lines. Such an index of 'Cornbeltlines' had a correlation of 0·8 with the 'date of origin' on the state level and of 0·7 with that on the crop-reporting-district level.

neighboring areas as a proxy variable for the cost of entry into the given area. This variable and the measure of market density, taken together, explain to a large extent the variability in the dates on which hybrid corn was introduced in different parts of the country and support the statement that the innovators were influenced by considerations of profit, entering those areas first where the expected profits from innovation were highest.

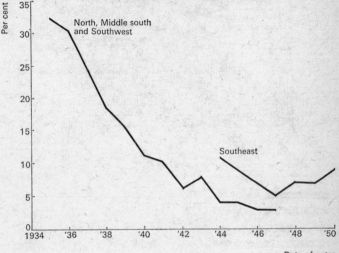

Figure 6 Average market density by date of entry: corn acres as a percentage of land in farms for crop-reporting districts reaching 10 per cent use of hybrid seed corn in a specified year

Rates of acceptance

The rate at which farmers in a region accepted hybrids, once hybrid corn became available, also varied from area to area. As shown in Figure 7, this rate was highest in Iowa and the surrounding area and lowest in some of the areas of the Southeast and the Mississippi Delta states. The differences in the rates of acceptance are largely demand phenomena, not a result of different supply conditions. After the first few years, in most of the places and most of the time, the supply of seed was not the limiting factor.

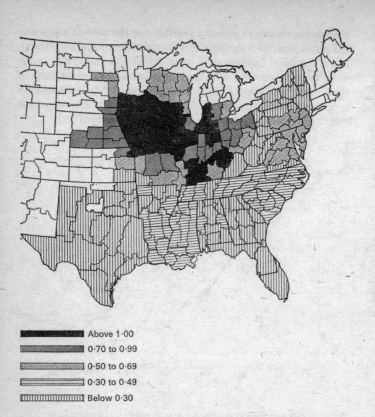

Above 1·00
0·70 to 0·99
0·50 to 0·69
0·30 to 0·49
Below 0·30

Figure 7 Estimated rate of acceptance of hybrid corn by farmers in different parts of the country. A value of 1·00 means that it took four years for an area to increase its percentage of corn acres planted to hybrid seed from 12 to 88 per cent. A value of 0.5 implies that it took twice as long, eight years, to accomplish the same change

The rate of acceptance is taken to be the relative speed of the diffusion process – that is, it is the slope coefficient of the S-shaped curve.[5] The measure is such that a value of 1·00 means that it takes four years for the acreage devoted to hybrid corn to

5. More exactly, the measure is the slope coefficient of the logistic growth curve as adjusted for ceiling differences. For an explanation of this measure see Griliches (1957).

rise from 12 to 88 per cent, while a value of 0·5 implies that it would take eight years, or twice as long, for this same rise to occur.

The rate at which farmers accept a new technique depends, among other things, on the magnitude of the profit to be realized from the change-over. This hypothesis is based, firstly, on the general observation that the larger the stimulus the faster the reaction to it, and, secondly, on the fact that in an uncertain environment it takes a shorter time to find out that there is a difference, if that difference is large.[6] Farmers doubted that this new hybrid corn was any good, and it took them some time to become convinced of its superiority. Individual farmers followed a development pattern of their own in shifting from open-pollinated to hybrid seed in planting their corn acreage (see Figure 8). Almost no farmer planted 100 per cent of his corn acreage to hybrid seed the first time he tried it.[7]

Yield per acre and acres per farm

The rate at which farmers shifted to hybrid corn depends, among other things, upon the profitability of such a shift. This in turn depends upon the absolute superiority of hybrids in corn yield in bushels per acre, and on the average number of acres per farm planted to corn.

It is widely accepted that hybrids out-yielded open-pollinated varieties by approximately 15 to 20 per cent and that this *percentage* superiority did not vary much between different areas.[8] Since similar percentage increases in yield imply different absolute increases in bushels per acre in areas where the previous yields were different, a good measure of the absolute superiority

6. This is analogous to sequential sampling. The 'average sample number' – that is, the expected length of the experiment – will depend, among other things, inversely on the difference between the means of the two populations being sampled and directly on their variance.

7. That this is rational behavior in the face of uncertainty is intuitively obvious but difficult to prove. See Simon (1957).

8. Of course, hybrids differed from open-pollinated varieties not only in yield but also in improved stand, in uniformity, and in resistance to disease. However, there are no good quantitative measures of improvement in factors other than yield, and besides, most of the other improvements were correlated with the increases in yield.

of hybrids over open-pollinated varieties is given by the long-run level of corn yields in various areas. The distribution of corn yields in the United States, as shown in Figure 9, shows strikingly close correlation with the distribution of rates of acceptance of

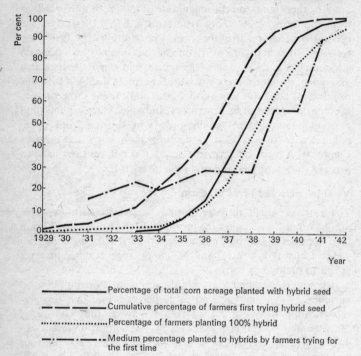

Figure 8 The acceptance of hybrid corn in Iowa. Each farmer planted only a small fraction of his corn acreage to hybrid seed on his first trial. Only very late in the spread process did the 'first timers' become bolder (B. Ryan, *Rural Sociol.*, vol. 13, p. 273, 1948; *Agricultural Statistics*)

hybrid corn, as shown in Figure 7; the higher the yield, the higher the rate of acceptance.

Where the rate of acceptance fails to correlate with the yield per acre, the failure can be explained by taking into account the difference in the average number of acres of corn per farm (corn acreage per farm increasing as one moves from east to west), since what is important is not only the profitability per acre but also the profitability per farm. A large fraction of the variability between areas in the rate of acceptance of hybrid corn by farmers can be explained with the help of these two 'profitability' variables.[9]

Equilibrium level

In an analysis of the use of hybrid corn in this country, one must consider, finally, differences in the equilibrium level reached – that is, differences in the fraction of the acreage which is ultimately devoted to hybrid seed. As shown in Figure 10, different levels were found in different areas of the country. By 1959, close to 100 per cent of the corn acreage in most of the Corn Belt and in its northern and eastern fringes was planted to hybrid seed. Substantially lower percentages were found only in the western fringes of the Corn Belt and in the deep South. In the South, the level is still changing, moving towards an equilibrium level of approximately 70 to 80 per cent of the corn acreage planted to hybrid seed. The western parts of Nebraska, South Dakota and Kansas have already reached their equilibrium level of approximately 30 to 60 per cent. These are areas of very low and very variable yields, where the use of hybrid seed is unprofitable except on the better land or on land under irrigation.

Differences in the equilibrium level are explained by differences in the *average* profit to be realized from the shift to hybrid seed. In an area of high average profit no farmer faces a loss from the shift. In areas of low average profit a substantial proportion of the farmers face the possibility of having no return or even of sustaining a loss on their investment. The ceiling, or the fraction of the corn acreage that will ultimately be planted to hybrid seed, is not unique or constant. It will change with the introduction of better hybrids, with improvements in the market for corn, and with large changes in corn acreage. Nevertheless, for almost all of the areas except the very marginal ones, a constant ceiling

9. Similar results were obtained from unpublished Agricultural Marketing Service data on the actual difference in the yield of hybrid and of open-pollinated varieties as a measure of the superiority of hybrids. Both the fit and the coefficients were very similar.

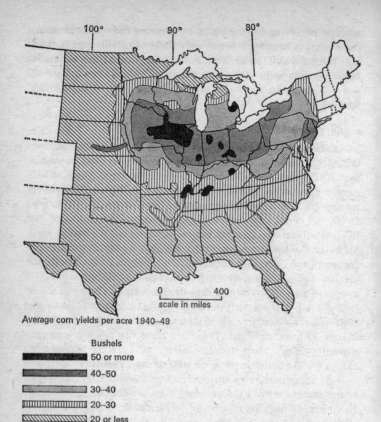

Average corn yields per acre 1940–49

Bushels

████████ 50 or more

▨▨▨▨▨▨ 40–50

▦▦▦▦▦▦ 30–40

|||||||||||| 20–30

▧▧▧▧▧▧ 20 or less

Figure 9 The profitability of hybrid corn, as measured by average corn yields per acre, 1940–49. Since the superiority of hybrids was a constant percentage at different yield levels, differences in the *absolute* superiority of hybrids between areas are indicated by differences in the long-run levels of corn yields in the various areas. The period 1940–49 is neither strictly a pre- nor a post-hybrid period, but the differences in the levels of hybrid seed use in different areas during this period are unlikely to affect the relative ranking of the various regions with respect to their long-run corn-yield potentials (from Grotewold, 1955)

fits the data well. Variation in these ceilings across the country can be explained in good part by the same two measures of profitability: the average absolute superiority of hybrids and the average number of acres per farm planted to corn.

Relation to studies by sociologists

It may prove useful to relate the results of this study to earlier work by sociologists in this area (see, e.g., Ryan and Gross, 1950 and 1955). In previous analyses of similar data it was mainly individual behavior that was investigated – that is, who are the

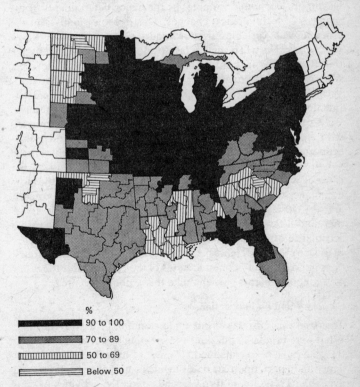

%
90 to 100
70 to 89
50 to 69
Below 50

Figure 10 Hybrid corn today, 1959. Approximate percentage of total corn acreage planted to hybrid seed (State Agricultural Statisticians' release, and unpublished *Agricultural Marketing Service* data)

first and who are the last to adopt hybrid corn? – and an attempt was made to explain such behavior on the basis of differences in personality, education, economic status and social environment. An attempt to use some of these variables (for example, level-of-living indexes) in explaining differences between states in the rate of acceptance of hybrid corn proved unsuccessful in this study.

It is my belief that in the long run, and when the country is taken as a whole, many of these variables either do not vary enough to be significant or tend to cancel themselves out, leaving the economic variables as the major determinants of the patterns of technological changes. This does not imply that the 'sociological' variables are not important if one wants to know which *individual* will be first or last to adopt a particular technique, only that these factors do not vary much from area to area. Moreover, the distinction between 'economic' and 'sociological' variables is partly semantic, and a very difficult one to make in practice. Some of the variables used in this study – for example, yield of corn and corn acres per farm – are closely correlated with such variables as education, level-of-living and socio-economic status. It is very difficult to discriminate between the validity of the assertion that hybrids were accepted slowly because it was a 'poor corn area' and the assertion that the slow acceptance was due to 'poor people'. Poor people and poor corn are very closely correlated in the United States. Nevertheless, one may find a few areas where this is not so. Obviously, the slow acceptance of hybrids on the western fringes of the Corn Belt – in western Kansas, Nebraska, South Dakota and North Dakota – does not reflect low economic status of the people but is the result of 'economic factors' which make this a poor corn area.

Summary and an implication

This study has increased our understanding of a body of data. What were originally puzzling and seemingly peculiar patterns in the data have been explained. The use of hybrid seed in an area depends, in part, upon the date at which superior hybrids become available. This date, in turn, depends upon the activities of seed producers guided by their expectations of profits, and upon the contributions of the various experimental stations. Thus, the

South was late in getting hybrids because the market for seed was substantially poorer there than in other areas and because southern experiment stations produced few hybrids of importance until the middle 1940s. The use of hybrid seed in an area also depends upon the rate at which hybrids are accepted by farmers. This rate, in turn, depends upon the profit farmers expect to realize from the shift to hybrids. Thus, farmers in the Corn Belt accepted hybrids at a faster rate than farmers in the South because the absolute magnitude of profit was higher in the Corn Belt than in the South. Similarly, the fraction of acreage ultimately planted to hybrid seed depends upon expectations of profits to be realized from the change and the distribution of these expectations around their mean.

When uncertainty and the fact that the spread of knowledge is not instantaneous are taken into account, it appears that American seed producers and American farmers have behaved, on the whole, in a fashion consistent with the idea of profit maximization. Where the evidence appears to indicate the contrary, I predict that a closer examination of the relevant economic variables will show that the change was not as profitable as it appeared to be.[10]

This study of hybrid corn has at least one interesting implication. Hybrid corn was an innovation which was more profitable in the 'good' areas than in the 'poor' areas. This, probably, is also a characteristic of many other innovations. Obviously, tractors contribute more on large than on small farms, and so forth. Hence, there may be a tendency for technological change to accentuate regional disparities in levels of income and rates of growth. Moreover, this tendency is reinforced by the economics of the innovation process, which results in the new techniques being supplied to the 'good' areas before they are supplied to the 'poorer' areas, and also in the more rapid acceptance of these

10. That these findings are not restricted to hybrid corn has been confirmed by a recent study of the spread of a series of industrial innovations (diesel locomotives, continuous mining machines, and so on) within particular industries. It was found there that (a) the logistic growth curve summarized the data well, and (b) most of the variability in the rate of acceptance of different innovations can be explained on the basis of the profit to be realized from the innovation and the size of the required initial investment. See Mansfield (1959).

techniques in the 'good' areas. A lag of this sort can by itself cause long-run regional differences in levels of income. The kinds of inventions we get, and the process by which they are distributed, may lead to aggravation of the already serious problem of regional differentials in levels of income and growth.

References

COLEMAN, J., KATZ, E., and MENZEL, H. (1957), 'The diffusion of an innovation among physicians', *Sociometry*, December.

GRILICHES, Z. (1957), 'Hybrid corn: an exploration in the economics of technological change', *Econometrica*, October.

GRILICHES, Z. (1958), 'Research costs and social returns: hybrid corn and related innovations', *J. polit. Econ.*, October. [Reprinted as Reading 9 in this selection.]

GROTEWOLD, A. (1955), *Regional Changes in Corn Production in the US from 1909 to 1949*, Dept of Geography University of Chicago.

JUGENHEIMER, R. W. (1939), 'Hybrid corn in Kansas', *Kansas Agri. Expt. Stn. Circ.*, no. 196, February.

KUZNETS, S. (1930), *Secular Movements in Production and Prices*, Houghton Mifflin.

LOTKA, A. (1925), *Elements of Physical Biology*, Dover.

MANSFIELD, E. (1959), 'Technical change and the rate of imitation', *Econometrica*, October, pp. 741–66. [Reprinted as Reading 14 of this selection.]

NEAL, N. P., and STROMMER, A. M. (1948), 'Wisconsin corn hybrids', *Wisconsin University Agri. Expt. Stn. Bull.*, no. 476, February.

RYAN, B., and GROSS, N. (1950), 'Acceptance and diffusion of hybrid corn seed in two Iowa communities', *Iowa Agri. Expt. Stn. Res. Bull.*, no. 372, January.

RYAN, B., and GROSS, N. (1955), 'How farm people accept new ideas', *Iowa State College Special Report*, no. 15, November.

SIMON, H. A. (1957), *Models of Man*, Wiley.

12 P. David

The Mechanization of Reaping in the Ante-Bellum Midwest[1]

P. David, 'The mechanization of reaping in the ante-bellum midwest',
H. Rosovsky (ed.), *Industrialization in Two Systems: Essays in Honor
of Alexander Gerschenkron*, Wiley, 1966, pp. 3–28.

The widespread adoption of reaping machines by Midwestern
farmers during the years immediately preceding the Civil War
provides a striking instance of the way that the United States'
nineteenth-century industrial development was bound up with
concurrent transformations occurring in the country's agricultural sector. On the record of historical experience, as Alexander Gerschenkron has cogently observed, 'the hope that industry
in a very backward country can unfold from its agriculture is
hardly realistic' (1962, p. 215). Indeed, even when one considers
countries that are not very backward it is unusual for agricultural
activities to escape an uncomplimentary evaluation of their
efficacy in creating inducements for the growth and continuing
proliferation of industrial pursuits. As Albert Hirschman puts it,
'agriculture certainly stands convicted on the count of its lack of
direct stimulus to the setting up of new activities through linkage
effects: the superiority of manufacturing in this respect is crush-
ing' (1959, pp. 109–110).[2] But having conceded that much

1. I wish to acknowledge my gratitude to Peter Temin for stimulating
criticism and helpful suggestions offered when this paper was first being
drafted. The present version has benefited from the comments of my
colleagues in the Economics Department, the members of the Graduate
Seminar in Economic History at Stanford University, 1964–5, and many
participants in the Purdue Conference on Quantitative and Theoretical
Research Methods in Economic History (4–6 February 1965). My debts on
this account are so numerous that those who hold them must perforce
remain anonymous. Errors or deficiencies that have survived all this counsel
are assuredly mine alone.
2. On the now familiar concepts of 'forward' and 'backward' linkages
between a sector of the economy (or an industry) and other sectors (or
industries) that buy its output and supply it with inputs, respectively,
see Hirschman (1958,) ch. 6.

regarding the general state of the world, the student of economic development in nineteenth-century America is compelled to stress the anomalous character of his subject, to insist that in a resource-abundant setting, highly market-oriented, vigorously expanding and technologically innovative agriculture did provide crucial support for the process of industrialization.

Such support in the form of sufficiently large demands for manufactures and supplies of raw material suitable for industrial processing would, undoubtedly, have been less readily forthcoming from a small or economically backward agrarian community. It is precisely in this regard that United States industrialization may be seen as having diverged most markedly from the historical experience of continental European countries, where backward agriculture militated against gradual industrial growth, and the successful pattern of modernization of the economy tended to be characterized by an initial disengagement of manufacturing from the agrarian environment (Gerschenkron, 1962, pp. 107–108, 125–6, 215, 354).

However, to treat the generation of demand for manufactures during the process of industrialization as taking place within a framework of static, pre-existing intersectoral relations, summarized by a set of input-output coefficients, does not prove to be an entirely satisfactory way of looking at the connections between the character of agriculture and the growth of industrial activities in the United States. Adherence to such an approach leads one, *inter alia*, to gloss over the problems of accounting for alterations in the structure of intersectoral dependences, although those alterations often constitute a vital aspect of the process of industrialization. It is not wholly surprising that pursuit of a static 'linkage' approach has tended to promote the misleading notion that the expansion and modernization of the agrarian sector constituted a temporal pre-condition for rapid industrial development in the United States (Rostow, 1960, pp. 17–18, 25–6), whereas in many crucial respects it is far more useful to regard the two processes as having gone hand in hand. As a small contribution to the study of the interrelationship between agricultural development and industrialization in the American setting, this essay ventures to inquire into the way that – with the adoption of mechanized reaping – an important element was added to the set of linkages joining these two sectors of the mid-nineteenth century economy.

The spread of manufacturing from the eastern seaboard into the transmontane region of the United States during the 1850s derived significant impetus from the rise of a new demand for farm equipment in the states of the Old Northwest Territory. That impetus was at least partially reflected by the important position which activities supplying agricultural investment goods came to occupy in the early structure of Midwestern industry. In the still predominantly agrarian American economy of the time it is not unexpected that a substantial segment of the total income generated by industrial activities was directly attributable to the manufacture of durable producers' goods specifically identified with the farmer's needs – leaving aside the lumber and related building materials flowing into construction of farm dwellings, barns, sheds and fences. If, in addition to value added in the production of agricultural implements and machinery in 1859–60, one were also to include half the value added by the manufacture of wagons and carts, saddles and harnesses, and the variety of items turned out by blacksmiths' shops, the resulting aggregate would represent over 4 per cent of the value added by the nation's entire industrial sector. That is, rather more than the proportion contributed by the manufacture of machine shop and foundry products, which at the date in question ranked as the country's seventh largest industry in terms of current value added (US Eighth Census, 1860, pp. 733–42). However, on the eve of the Civil War the production of agricultural implements and machinery *alone* generated just as large a proportion of total industrial value added in the preponderantly agrarian Western states; in Illinois, this single branch of manufacturing accounted for fully 8 per cent of the total value added by the state's industries in 1859–60 (US Eighth Census, 1860, pp. ccxcii, 725, 729).[3]

To appreciate the importance of the position that the agricultural implements and machinery industry assumed in the

3. The 'Western' states here are: Ohio, Indiana, Michigan, Illinois, Wisconsin, Minnesota, Iowa, Missouri and Kansas. The share of the agricultural implements industry in aggregate value added by US manufacturing was 1·4 per cent, according to the Census of 1860.

structure of Illinois' early manufacturing sector, it must be realized that at the time there was no single branch of industry which in the nation as a whole contributed so large a portion of aggregate value added in manufacturing. Cotton goods production, America's largest industry in 1859–60, accounted for only 6·6 per cent of the national aggregate.

When one looks at a rapidly developing center of industrial activity in the Midwest such as was Chicago during the 1850s, the manufacture of agricultural implements and machinery is found to have had still greater relative importance as a generator of income. The growth of agricultural commodity-processing industries, especially meat-packing in Chicago during the latter half of the century suggests that the Garden City's meteoric rise to the status of second manufacturing center in the nation by 1890 might be taken as a demonstration of the strength of *forward* linkages from commercial agriculture. It is not an object of the present essay to assess the validity of that impression. Nevertheless, it should be remarked that during Chicago's first major spurt of industrial development, a movement which saw manufacturing employment in the city rise from less than 2000 in 1850 to approximately 10,600 by 1856, the forward-linked processing industries were less significant to the industrial life of the city than was an activity based on *backward* linkage from agriculture. The branch of manufacturing in question was the farm implements and machinery industry: in 1856 it accounted for 10·8 per cent of total value added by Chicago's industrial sector, compared with 6·3 per cent contributed by the principal processing industries, meat-packing, flour- and grist-milling, and distilling, combined (David, n.d., Appendix C, Table C-2, Appendix A-III, ch. 3).[4]

Among the salient characteristics of the agricultural scene in the ante-bellum Midwest, two appear as having been crucial to

4. Estimates of value added in Chicago industries are based on local census statistics for gross value of product (in 1856) and the ratios of value added to gross product in the corresponding industries reported by the US Eighth Census (1860) for Cook County, Illinois. In 1859–60, according to the latter source, meat-packing, milling and distilling together accounted for 8·7 per cent of manufacturing value added in Cook County, compared with 7·9 per cent contributed by the agricultural implements and machinery industry.

the emergence during the 1850s of a substantial regional manufacturing sector bound by demand-links reaching backward from commercial agriculture. First, the settlement of the region and the extension of its agricultural capacity during that decade proceeded with great rapidity, encouraged by favorable terms of trade and improvements in transportation facilities providing interior farmers with access to distant markets in the deficit foodstuff areas to the east. Between the Seventh and Eighth Censuses of Agriculture over a quarter of a million farming units came into existence, and about nineteen million acres of improved farm land were added in Illinois, Indiana, Michigan, Iowa and Wisconsin. This represented a rate of increase in the number of farms of 7 per cent per annum, and a 9 per cent annual rate of expansion in improved acreage (*US Eighth Census*, 1860, p. 222).

Secondly, agricultural practice in this region of recent settlement was not the static crystallization of long experience typical of stable agrarian societies. Far from being a closed issue, choices among alternative production techniques were rapidly being altered and Western farming was thereby being carried in the direction of greater capital-intensity and higher labor productivity. On the eve of the Civil War this burgeoning farm community was in the midst of a hectic process of transition from hand methods to machine methods of production, from the use of rudimentary implements to reliance on increasingly sophisticated machinery. Among the items of farm equipment being introduced on a large scale in the Midwest during the 1850s were steel breakers and plows, seed drills and seed boxes, reapers and mowers, threshers and grain separators (Rogin, 1931, pp. 33–4, 47, 72–80, 165–6, 196, 201). An editorial pronouncement appearing in the *Scientific American* during 1857 suggests the extremes to which the mechanization of farming had proceeded:

. . . every farmer who has a hundred acres of land should have at least the following: a combined reaper and mower, a horse rake, a seed planter, and mower; a thresher and grain cleaner, a portable grist mill, a cornsheller, a horse power, three harrows, a roller, two cultivators and three plows (Danhof, 1951, p. 150).

The importance that the newly introduced reaping and mowing

machines (especially the former) had assumed among the products of the agricultural implements and machinery industry of the Midwest by the end of the 1850s provides some indication of the direct impact of the shift to more capital-intensive farming techniques upon the expansion of an agrarian market for industrial products.[5] According to the Census of 1860, reapers and mowers accounted for 42 per cent of the gross value of output of all agricultural implements and machinery in Illinois and for 78 per cent of the gross value of output of the corresponding industrial group in Chicago. A few years earlier, in 1856, when the Midwestern boom was still in full swing, reaper and mower production in Chicago had dominated that center's farm equipment output-mix to an even greater degree (*US Eighth Census* 1860, Table 3, pp. 11, 86).[6]

Despite the fact that the history of commercial production of mechanical reaping machines in the United States stretched back without interruption to the early 1830s, this industry was one that only began to flourish in the 1850s. From 1833, the date of the first sale of Obed Hussey's reaping machine, to the closing year of that decade, a total of forty-five such machines had been purchased by American farmers. At the end of the 1846 harvest season Cyrus H. McCormick determined to abandon his efforts

5. The term 'direct impact' is used here with two considerations in mind. First, this neglects the indirect (input-output) effects of expanded reaper and mower production on the production of intermediate inputs used by the industry, e.g. pig iron, bar iron, malleable castings, cast steel, brass castings, sheet zinc, leather, oils, paint, turpentine, physical input coefficients for each of which are available (David, ch. 3). Secondly, we here neglect the favourable indirect impact on the growth of agricultural demand in general, arising from the fact that substitution of machinery for labor on the farms raised labor productivity and facilitated faster expansion of agriculture during this, and subsequent periods. The latter point is discussed further below.

6. The enumeration of establishment output given in the (Chicago) *Daily Democratic Press*, 'Review of 1856', shows that 5860 reapers and mowers contributed 87 per cent of the gross value of all agricultural implements and machinery produced in Chicago in that year. (Separate mowers, in contrast to reapers and combined reaper-mowers, accounted for less than 32 per cent of the value of reaper and mower production in Chicago in 1856.) The balance of Chicago's production for 1856 consisted of a miscellany of 541 threshers, 200 separators and horse powers, 1000 plows, and an unknown number of corn-shellers and cob-crushers.

of the previous six years at manufacturing his reaping machine on the family farm in Rockbridge County, Virginia, and set about transferring the center of his activities to a more promising location, Chicago. The known previous sales of all reaping machines at that time aggregated to a mere 793, but by 1850 some 3373 reapers in all had been produced and marketed in the United States since 1833. A scant eight years later it was reckoned that roughly 73,200 reapers had been sold since 1845, fully 69,700 of them since 1850. And most of that increase had resulted from the burst of production enjoyed by the industry during the five years following 1853![7]

The major portion of this production had taken place in the interior of the country, and it is apparent that in the absence of farmers' readiness to substitute machinery for labor during the 1850s, an equally rapid pace of agricultural expansion – had such in fact been feasible – would have provided a considerably weaker set of demand stimuli for concurrent industrial development in the region. The latter facet of the late ante-bellum agrarian scene must, therefore, be the prime focus of our interest; it cannot be taken as given, but must be explained. That should not, however, be regarded as a dismissal of the first-mentioned aspect of Midwestern agricultural development in this period. As shall be seen when we come to grips with the problem of explaining the movement of mechanization, the speed of agricultural expansion and the substitution of machines for farm labor were intimately connected developments between which causal influences flowed in both directions.

7. See Rogin, 1931, pp. 72–8, for the record of reaper production before 1860. The figures given above are cumulated from yearly sales data, save for the estimate of 73,200 reapers sold between 1845 and 1858. The latter figure can be traced to pro testimony in the litigation over extension of the McCormick Patent of 1845, not the original 1834 Patent. At that time it was asserted to represent the number of machines sold that had made use of principles patented by McCormick; the claim was sweepingly inclusive, as it covered the 23,000 machines sold by McCormick (directly and under licence) and all other machines sold since 1845 (Hutchinson, 1930, p. 470). Leo Rogin (1931, pp. 78–9) erroneously accepts this figure as an estimate of the stock of reaping machines in operation on farms west of the Alleghenies in 1858, evidently following a misinterpretation perpetrated by *Country Gentleman*, vol. 13 (1859), pp. 259–60, the proximate source cited by Rogin for the number in question.

In view of the consequences for agricultural and industrial development that followed from the mechanization of reaping during the 1850s, it might be supposed that this episode in the modernization of American farming and the formation of backward linkages between the enterprises of field and factory would have been thoroughly explored by economic historians. To be sure, virtually all the standard accounts of the development of agriculture in the United States up to 1860 mention the introduction of the machines that Obed Hussey and Cyrus H. McCormick had invented in the 1830s. Yet, the literature remains surprisingly vague about the specific technical and economic considerations touching the adoption of these devices by American farmers. We have called attention to the fact that although the twenty years prior to 1853 had witnessed a slow, limited diffusion of the new technique, the first major wave of popular acceptance of the reaper was concentrated in the mid-1850s. Thus, the intriguing question to which an answer must be given is: why only at that time were large numbers of farmers suddenly led to abandon an old, labor-intensive method of cutting their grain, and to switch to the use of a machine that had been available since its invention two decades earlier?

In this inquiry, the impact of the mechanization of small grain harvesting upon US agriculture is not the prime subject of concern.[8] Nevertheless, it would hardly be possible to account for the upsurge of demand for reaping machinery without considering the economic implications of the new harvesting technology and the specific circumstances surrounding its introduction. The traditional story of the ante-bellum adoption of mechanical reaping, a version to be found in any number of places (Bidwell and Falconer, 1925, pp. 281–94; Danhof, 1951, pp. 144–6; Gates, 1960, pp. 258–88; Robertson, 1964, pp. 257–8), is set out along the following lines. During the first half of the nineteenth century arable land was abundant in the United States, but the amount of small grains (especially the amount of wheat) that an indi-

8. See William N. Parker (4–5 September 1963), for a recent quantitative study which attributes to mechanization the major part of the increase of labor productivity in US small grain production during the nineteenth century. Wheat, oats and rye are the small grain crops considered in the present paper.

vidual farmer could raise was limited by the acreage that could be harvested soon after the ripening of the crop. Labor was scarce, and harvest labor notably dear as well as unreliable in supply. Compared with the method of harvesting using the grain cradle – an improvement on the sickle that had come into quite general use even in the transmontane wheat regions by the middle of the century – the new mechanical reapers effected a saving in labor. When Midwestern farmers were led to increase production as a result of the rise in wheat prices during the 1850s (a rise augmented by the impact of the Crimean War upon world grain markets), the demand for reaping machines rose, and their adoption went forward at an accelerated pace. The movement thus initiated received renewed impetus from the extreme shortage of agricultural manpower occasioned by the Civil War. By saving labor, and therein relaxing the constraint on cultivated acreage imposed by hand methods of harvesting, the introduction of the reaper made possible the rapid expansion of small grain production that occurred during the latter half of the nineteenth century.

This account may vary in some details from any particular historian's version of the events in question, but it contains all the generally accepted elements of the story. It specifically follows the historiographic tradition of ascribing to the rise in wheat prices during the 1850s a causal role in bringing about the transition from cradling to mechanical reaping prior to the outbreak of the Civil War.[9] Upon a moment's reflection, the latter is seen

9. Although it is generally accepted that the Civil War provided an important stimulus to the widespread use of agricultural machinery in Northern agriculture, the extensive use made of reapers and mowers in the West during the 1850s is now regarded as a well established point (Rogin, 1931, p. 79; Fite, 1910, p. 7), Rogin estimates that about 70 per cent of the wheat harvested west of the Alleghenies was cut by mechanical reapers on the eve of the War. There are grounds for questioning the validity of that figure, although they are not such as to lead us to doubt that the reaper had won general acceptance, if not universal adoption, in Midwestern agriculture by 1860.

Rogin's estimate is based on his acceptance of a figure (73,200), giving the number of reaping machines sold in the US between 1845 and 1858 as an estimate of *the stock of reapers in operation on Western farms* in 1859. Since the average life of a reaper was roughly ten years (see Appendix A, section 2b), a calculation of the stock of machines net of replacements (but gross of depreciation, i.e. the gross stock, assuming reapers simply fell apart after

to be the analytically unexpected aspect of this tale of a change in technology; it is far more usual for discussions of the choice of technique to be couched, implicitly or explicitly, in terms of the relative prices of the substitutable factors of production (grain cradlers and reaping machines in this instance) and to say nothing about the price of the commodity being produced.

Yet, precisely why this departure from the classical (or, properly speaking, neo-classical) treatment of the choice between labor-intensive and capital-intensive factor proportions is called for in the case of the adoption of the mechanical reaper, is not revealed by the statement. That it remains rather ambiguous about the lines of causation linking dear labor, high grain prices, expanded production acreage, and the spreading use of the reaper must, with all diffidence, be attributed to the ambiguities of the literature from which the statement itself has been constructed. See Bidwell and Falconer's classic work on Northern agriculture prior to the Civil War:

During the early fifties the reaper was gradually supplanting the cradle in the wheat fields of the country, but as yet the acreage in grain in the Western States was largely limited to the capacity of the cradle. . . .

ten years in the manner of the 'one horse shay') would require disregarding at least the sales of machines prior to 1848. This would merely lower the stock estimate from 73,200 to approximately 72,300 (see annual sales data given by Rogin, 1931, pp. 72–8). However, if one allows for continuous straight-line decay of the machine at the rate of 0·1 per annum, and takes the time pattern of known sales of machines by the McCormick Company during 1848–58 (from data in Hutchinson, 1930, p. 369, n. 60), as a representation of the time distribution of all reaper sales, it is found that the net stock figure for 1858 should be only 56 per cent of the cumulated number of machines sold during 1848–58, or the equivalent of 43,400 full-capacity machines. This probably understates the true net stock, since the time distribution of McCormick's production was less skewed towards the latter half of the period of 1848–58 than was the time distribution of aggregate reaper production in the country; 50,000 full-capacity machines might be accepted as not too high a figure for the net reaper stock in 1858.

But, it is necessary to compare both the net and the gross stock estimates with the *national* wheat harvest; rather than with the Western wheat harvest as Rogin does. Taking the average national wheat yield per acre as 11·4 bushels in 1859 (a figure computed by weighting regional yield estimates given in Table 5 of Parker, 1963), and following Rogin's procedure of comparing the wheat acreage harvested, as estimated from the Census of 1860 output figure and the yield per acre, with that acreage which could be cut

Moreover, in a large part of the West there was little incentive to produce large amounts of wheat on account of the lack of markets and low prices. Rising prices of wheat caused a 'boom' in agriculture from 1854 to 1857 and caused almost universal demand for reapers in the wheat-growing regions (1925, p. 293).

P. W. Gates's recent study of ante-bellum agriculture proceeds in much the same general vein:

With wheat well above the dollar mark from 1853 to 1858, Illinois, Wisconsin, Iowa and Minnesota farmers enjoyed real prosperity and were in a position to buy and pay for reapers. . . . Since the amount of wheat a man could sow was limited by the short period in which it had to be harvested and by the man-days of labor required to cut it, it can readily be seen how much the reaper expanded the possibilities of wheat growing (1960, p. 287).

Comparable passages of other contributions to American economic history could be examined without finding any clear views as to whether it became profitable for farmers to adopt the mechanical reaper only when they found it profitable to increase the acreage of wheat sown per farm, or whether it was the expansion of grain cultivation in Midwestern agriculture as a whole that led to a general substitution of machinery for labor in harvesting operations. The literature is no less ambiguous in the answers it offers to two closely related questions. Did the adoption of mechanical reapers make rapid expansion of grain cultivation in the Midwest possible by raising the scale on which it could be profitably grown (and harvested) by individual farming units? Or, was it simply that the widespread substitution of the reaper for the grain cradle alleviated the scarcity of agricultural

by the stock of reapers if each machine were worked at the normal seasonal capacity rate of one hundred acres, one obtains the following estimates as alternatives to that given by Rogin. At a maximum, if we accept the gross reaper stock figure, 48 per cent of the fifteen million acres of wheat harvested in the US in 1859 was cut by machines; at a minimum, if we accept the net stock figure, 33 per cent was thus harvested. Rogin's figure of 70 per cent seems somewhat exaggerated for the West, since even were the entire stock of reapers (implausibly) thought to have been located west of the Alleghenies, the above revisions of the reaper stock figure would suggest that the proportion of wheat acreage in the West cut by horse power at the end of the 1850s lay somewhere between 70 per cent and 50 per cent.

labor which otherwise would have restricted wheat production in the newly developing Western regions to appreciably lower levels?

There is no question that mechanical reaping effected a saving in harvest labor requirements; the evidence marshalled in Leo Rogin's path-breaking work, *The Introduction of Farm Machinery in its Relation to the Productivity of Labor*, is nothing if not conclusive on that point (Rogin, 1931, pp. 125–37, Appendix A). Yet, to the writer's knowledge, no systematic attempt has been made to compare the magnitude of the savings in wage costs with the capital costs of a reaper to Western farmers during the first decade of the machine's widespread adoption. It is therefore not surprising that the accounts cited fail to divulge whether (or not) the new harvesting technique proved more profitable than grain cradling under all plausible relative factor prices, or whether (or not) it was economically superior to cradling for all scales of farm operations.

These are, indeed, crucial questions. If the answers are in the affirmative, then the rate at which the reaper replaced the cradle in Western grain fields during the 1850s depended solely upon the capacity of the agricultural machinery industry; Bidwell and Falconer's assertion that 'Reapers were introduced as fast as they could be manufactured' (1925, p. 293), would be more than a mere figure of speech. It would be literally true and would carry the implication that the replacement of hand-harvesting methods would have occurred much earlier in American history were it not for technically unsolvable problems of designing and manufacturing a reaping machine. One would then have to find more convincing technical obstacles than are discussed in the authoritative works on the reaper to account for the lag between the first sale of Hussey's machine in 1833, the filing of the original McCormick patent in 1834, or even the first sale of McCormick's machine in 1840, and the eventual adoption of the innovation in the 1850s (Hutchinson, 1930, chs. 5–10; Rogin, 1931, pp. 72–5, 85–91).[10] If the mechanical reaping technique actually was superior to hand-harvesting with the cradle, regardless of relative factor prices or scale, this would also pose something of a problem for those writers who, like H. J. Habakkuk, regard the

10. It is not, however, necessary to depend upon a completely supply-determined explanation of this lag.

mechanization of agriculture in the United States as an 'obvious' illustration of the labor-saving bias of American technology fostered by nineteenth century conditions of relative labor scarcity (1962, pp. 100–102).

It is quite clear, however, that the traditional accounts of the introduction of the reaper do not entertain such notions. By placing emphasis on the effects of rising grain prices and the extension of wheat production, they imply that altered demand for agricultural products was of fundamental significance in determining the rapid rate at which the innovation supplanted hand methods of harvesting small grains in the Midwest during the 1850s. This line of explanation, taken broadly, would suggest that the sudden growth of the market for the reaper – coming nearly two decades after the machine first began to be sold – was a consequence of the specific conditions surrounding Midwestern agricultural development towards the close of the late ante-bellum era. Even had the rise of the market for farm machinery not provided significant impetus to the initial industrialization of the region, the implications of this hypothesis for our general view of the process of the diffusion of technology would make it important to try to formulate the economics of the traditional account in a fashion sufficiently precise to permit its re-examination in the light of pertinent evidence.

Suppose, for the moment, that it is justifiable to assert that the saving of labor achieved with the mechanical reaper was not so great as to render cradling an inferior technique in all relevant factor price situations. It may then be argued that mechanization of reaping spread through the agricultural sector as a result of an alteration in factor prices which accompanied the expansion of grain cultivation in the West.[11] In other words, the standard

11. It is sometimes argued that the rate of growth of an industry is an important determinant of the extent to which it adopts new techniques of production (see, e.g. Temin, 1966), because with equipment of given durability the rate of growth of the industry will affect the equilibrium age of the capital stock and, therefore, the extent to which the capital stock embodies the newest techniques. This line of reasoning, which would connect the rising demand for wheat and the rapid growth of the Midwestern agricultural sector in the 1850s with the adoption of the reaper on a large scale in that part of the country – even if mechanical reaping was an unambiguously superior technique – does not carry much force in the present

versions can be read as saying that the 'agricultural boom' set in motion by rising grain prices in the mid-1850s added to already existing pressures upon the available harvest labor force in the region, drove up the farm wage rate relative to the cost of harvesting machinery, and, thereby, created a situation in which it became profitable for farmers to substitute machinery for labor in harvesting small grain. This argument requires the not unreasonable assumption that the supply schedule of harvest labor facing the farm sector in the Midwest was less elastic than the supply schedule for agricultural machinery; otherwise, the outward shift of the demand schedules would not have resulted in the relative price of harvest labor being raised to a level at which continuing substitution of machines for cradlers would take place. Granting that assumption, the argument may be completed by noting that as the availability of the new method of harvesting rendered the demand schedule for labor more elastic than would otherwise have been the case, substitution itself tended to check the extent of the actual rise in relative wages caused by the expansion of aggregate grain production. In this manner, the use of the reaper throughout the grain regions held down the total cost of production, although it could not prevent some rise in costs, and made possible a large volume of total output at any given level of grain prices.

For this analysis, in which mechanization appears as a change 'imposed' upon grain farmers by the general expansion of midwestern agriculture, the relative inelasticity of the farm labor supply schedule is crucial. The greater the emphasis that is placed upon the role of related competitive demands for labor, such as regional railroad construction, to cite but one significant source, the less thoroughly tied to exogenous events (e.g. the

---

context. The reason is simply that the 'older' technique (i.e. cradling) of cutting grain was not embodied in any fixed capital on farms; since they had virtually no sunk costs connected with the old method, established farmers would not *on that account* find it more expensive to switch to mechanical reaping than would new entrants to the industry or farmers who were making significant increases in their productive capacity. In explaining the adoption of the reaper in Midwestern agriculture it therefore does not seem important to concentrate on any distinction between new and old farms in the industry.

Crimean War) affecting world grain prices is the explanation offered for the timing of the adoption of mechanical reaping.[12]

In the picture just presented, *the individual farmer's* desire to increase his acreage under wheat does not appear as influencing his decision to purchase a reaper and dispense with the services of cradlers. Nor can the personalization of the collective market process described be justified with any plausibility. Since there is no reason to suppose that the labor supply schedule facing the individual farmer was less elastic than the supply curve for agricultural machinery that confronted him, why should there be any connection between *the individual farmer's* decision to sow more wheat and his choice of the new reaping technique? Yet, the literature is replete with statements suggesting such a connection: 'When the wheat from an acre of land would sell for more than the price of the land, it was considered a safe investment to sow more land in wheat and buy a reaper' (Bidwell and Falconer, 1925, p. 293); '. . . Americans also had a very strong incentive to develop machines which would enable farmers to cultivate a larger area. The alternative was to leave land uncultivated' (Habakkuk, 1962, p. 101). If such statements represent something more than illustrations of the ease with which efforts to write economic history as the intended outcome of purposive individual actions, rather than their interplay in impersonal markets, can lead to what may be called 'fallacies of decomposition', their authors must have in mind a set of considerations influencing the introduction of mechanical reaping which is quite distinct from the process of market-imposed adjustments already set forth.

To put it most simply, these statements may be taken to imply either that there were significant economies of scale associated with the use of the reaper, or that diseconomies of scale existed in the use of labor for cradling grain that were not encountered with the mechanized technique in the range of farm size relevant to the ante-bellum Midwest.[13] Both situations would arise from

12. See David (n.d.), ch. 5, for discussion of the competing sources of demand in the Midwestern labor market during the 1850s, in which it is argued that the effects of urban construction and internal improvements activity in creating a relative scarcity of unskilled labor in the region should be accorded more importance than they usually receive.

13. It must be confessed that what Habakkuk has in mind in this connection is less than completely clear, and that some doubt remains whether

the presence of indivisibilities among the inputs of the micro-production function for harvesting small grains.

In the apparent absence of feasible co-operative arrangements for sharing the use of harvesting machinery among farms, at this time the reaping machine itself constituted an indivisible input for the farmer.[14] Since he typically had to purchase it, rather than rent it when it was needed, the relevant cost of using a reaper in harvesting was the average annual cost over the life of the machine. Within a particular season, however, the cost of a reaper per acre harvested would fall as the acreage was increased. It would continue falling to the point at which the cutting capacity of a single machine during the feasible harvest was reached. By contrast, given a perfectly elastic supply of labor and no diseconomies of scale in its use, the saving in wage costs obtained by substituting the mechanical reaper for cradlers would remain constant per acre harvested. It is possible, therefore, that below some level of acreage to be harvested – which we shall call the 'threshold' farm size – the total capital cost of a reaper (or of more than one reaper) exceeded the potential reduction in wage

the rationalization put forward in the text is appropriate. The last sentence of the quotation – i.e. 'The alternative was to leave land uncultivated' – does suggest that the American farmer had a passion for cultivating all available land without consideration of profitability and that a machine permitting greater acreage to be cultivated with a given amount of labor would be adopted by the farmer under any product and factor price conditions. The former part of this view is perhaps shared by other writers. Of the American farmer, Hutchinson (1930, pp. 50–51), says, '. . . environment compelled him to be acquisitive, and he was prone to add more acres to his freehold than he could well keep under cultivation.' Yet, to proceed along these lines is to turn the question of the choice of harvesting techniques into one more properly dealt with by psychologists and sociologists, whereas, as will be seen, a satisfactory explanation can be provided in traditional economic terms.

14. See Shannon (1945, pp. 329–48), for farm groups' efforts at co-operative manufacture and large-scale purchase of agricultural machinery after the Civil War. Such co-operative ventures did not emerge prior to the War, nor are they equivalent to the co-operative use of farm machinery. There is little evidence of commercial renting of reaping and mowing machines, or commercial grain harvesting by horsepower, such as developed in connection with the use of the large breaking-plows on the prairies during the 1850s. Neighbors may well have shared the use of a reaping machine owned by one farmer, compensating the owner on an informal basis, but

costs, making its adoption unprofitable in comparison with the method of cradling.

Exactly where the threshold point was located in the spectrum of farm acreage devoted to the small grains was determined by relative factor prices. The saving of labor achieved with the reaper being essentially technologically fixed per acre harvested, a doubling of the total yearly reaper cost relative to the money wage cost of harvesting an acre with the cradle would double the number of acres that would have to be harvested before the costs per acre would be the same with either technique.[15] While it is conceivable that so great a saving of labor was effected by the reaper that the costs per acre harvested by machine were lower for any finite total acreage, so long as both the money wage rate and the cost of a reaper were positive and finite, the existence of *significant* economies of scale associated with mechanical

some inquiry into contemporary accounts, farm newspapers, and such sources does not suggest that this was common practise. It certainly does not appear to have been as characteristic a feature of inter-farm relations as was the joint use of corn shellers, threshers and other equipment used in *post-harvest* tasks. The foregoing impressions are, perhaps, not sufficiently firmly established to demand a hypothesis which would account for the absence of commercial renting of early reapers and the lack of arrangements for sharing the use of jointly (or singly) owned machines. Nor are we able at this point to offer more than a possible line of explanation. The fact that the maximum cutting capacity of the early machines was not very large, especially when the time constraint on harvesting in any given locale is taken into consideration, coupled with the high costs of overland transport for a bulky machine weighing upwards of half a ton, would appear to have militated against operating a profitable itinerant commercial reaping enterprise during the ante-bellum era. The time constraint seems the crucial factor, since it was not present to the same degree either in prairie-breaking operations or in post-harvest tasks such as threshing and corn shelling; the former became established on a commercial basis, while sharing of equipment among neighbors was not unusual in the latter cases. Furthermore, as a consequence of the time constraint, the problems of deciding who was to have priority of use of a jointly owned reaper would have required the owners (users) to form some compensation arrangement, equivalent to an output pool or a profit pool. The economic, not to mention the political and sociological consequences of a reorganization of independent small farms into such pooling arrangements would have profoundly altered the course of agrarian history in the United States.

15. The relationship between the relative price of labor, *vis-à-vis* the reaper, and the threshold size is developed formally in Appendix A. Even

harvesting cannot be taken to have been a purely technical matter; relative wage rates must not have been so high that it was profitable for the farmer to adopt the new method at any level of grain production.

In principle, the existence of diseconomies of scale in the use of labor for harvesting grain with cradles would operate in the same manner as economies of scale associated with the mechanical reaper. Harvest workers required a certain amount of supervision to maintain their efficiency, and the addition of hands required to cut the grain on a larger acreage presumably taxed the farmer's supervisory capacities. The harvest had to be carried out in a limited number of days lest the ripe grain be lost through shattering or spoilage. It was therefore not feasible simply to employ the optimum number of hired hands that could be supervised at any one time for as long a period as it would take to cut the grain. Assuming that the average productivity of cradlers would begin to decline when the amount of supervision they received fell below some minimum, we may say that the manpower requirements per acre would have been greater for larger acreages. Consequently, savings in labor obtained by switching to mechanical reaping would tend to rise as the acreage to be harvested increased. Even if the capital costs of a reaper were constant per acre, this could define a threshold size beyond which farmers would find it advantageous to mechanize.

Although there is evidence of considerable contemporary dissatisfaction about the quality of hired help on farms during the first half of the nineteenth century, and notwithstanding the comments of American farmers on the need to supervise temporary help in order to keep them on the job and careful in their work,[16] it is difficult to gauge the extent to which the inferior quality of hired help failed to reflect itself in the general level of

without that derivation it is readily seen that if the labor cost per acre harvested by cradle is constant, threshold size must vary in direct proportion to the *relative* cost of capital.

16. See, e.g. Gates (1960, pp. 272, 275) and also the passage cited from Ruffin's essay on 'Management of wheat harvest' (1851) in Rogin (1931, p. 131). One of the state sales agents employed by the McCormicks in the 1850s took it as the object of his work to 'place the farmer beyond the power of a set of drinking Harvest Hands with which we have been greatly annoyed', (Hutchinson, 1930, pp. 355–6.)

farm wages. Moreover, among the complaints registered by farmers are those specifically citing the carelessness of hired hands with machinery. If there were diseconomies of scale in the use of harvest labor, we do not know that these were restricted to the employment of cradlers rather than labor in general, and that they did not simply place a limit on the scale of farm operations in the free states.[17] It therefore seems justifiable to restrict the discussion of scale effects to consideration of those which arose from spreading the fixed cost of harvesting machinery over large grain acreage.

Figure 1 depicts the hypothesized situation in terms of alternative long-run unit cost curves for hand methods and machine methods of grain harvesting on independent farms. The supposed existence of some fixed costs with either process causes both the hand-method cost curve, $C_H C_H$, and the machine-method curve, $C_M C_M$, to decline over a range of total harvested acreage, but, because of the *additional* fixed cost entailed by the reaper, the curve $C_M C_M$ goes on falling after the hand method unit costs begin rising. The rise in unit costs results from the limitation on the total supervisory capacity of the farmer, a restraint which eventually also causes the $C_M C_M$ curve to turn upward at a larger scale of operations. In the situation shown, the representative farm operating with the hand method of harvesting is taken to be in equilibrium at size S_0, with the market price of grain (per acre harvested, assuming constant yield per acre) equal to minimum units costs at P_0. Beyond S_0 lies S_T – the threshold size at which unit costs with the hand method are equal to those with the machine method – whose location is determined by the factors influencing the relative positions of the two curves.

The argument that the individual farmer's decision to adopt mechanical reaping was tied up with a simultaneous decision to expand his grain acreage, the latter being prompted by a rise in the relative price of grain, may now be quickly restated in terms of Figure 1. If there were no initial costs involved in increasing grain acreage under cultivation, with the market price at P_0 it would clearly pay to abandon the hand method of harvesting

17. Monographic studies of the effects of mechanization in agriculture, such as Rogin's, make no mention of increases in labor productivity associated with the reaper having been influenced by the size of the farm on which the machines were used.

and expand the representative farm from S_0 to S_M. However, the costs of acquiring, clearing and fencing new land, or simply of preparing land already held, were hardly insignificant even on the open prairies. If these costs, relative to the market price of grain, were sufficiently large to prevent expansion beyond S_T, they would have effectively blocked the concomitant adoption of the mechanical reaper (Bogue, 1963; Danhof, 1941; Gates, 1960, pp. 34, 186–8; Primack, 1962).[18] The significance of the rise in grain prices during the 1850s in bringing about widespread introduction of the reaper accordingly was, that by lowering the relative unit costs of expansion, higher prices induced the typical farmer to increase his grain acreage beyond the previous threshold size. In so doing, the farmer would take advantage of the $C_M C_M$ cost curve.

The presence of scale considerations in the choice between hand method and machine method of harvesting grain was not explicitly recognized in setting forth our earlier argument, in which the change in technique was depicted as an adjustment imposed by the relative inelasticity of the aggregate supply schedule for farm labor. That hypothesis may, nonetheless, be quite readily stated within the framework of the micro-analysis summarized by Figure 1. The contention is simply that because the threshold point will move inversely with changes in the price of labor relative to the cost of the reaper, the relative rise in farm wage rates (resulting from the collective response to higher grain prices in the mid-1850s) drove the threshold size downward to the point at which adoption of the reaper became profitable even on farms that had not enlarged their cultivated acreage. This is shown in Figure 1 as a downward shift in the position of the long-run machine-method cost curve relative to $C_H C_H$, lowering the threshold size (S_T) to the optimum acreage (S_0) established under the regime of the older harvesting technique.[19]

On formal grounds our two versions of the switch to mechanical

18. In terms of Figure 1 we can say that, spread over the total acreage to be harvested, the unit costs of expanding farm acreage beyond S_T was at least equal to the difference between P_0 and the $C_M C_M$ cost curve at S_M.

19. Note that the downward shift in $C_M C_M$ is equivalent to a rise in the market price of grain being accompanied by an upward shift in the $C_H C_H$ cost curve which is not matched by a rise in unit costs with the machine technique.

reaping thus turn out to be entirely compatible; they merely stress what may have been different aspects of a single story – one directing attention to forces working to push farms across the old threshold point, and the other emphasizing that an alteration in relative factor prices may have brought the threshold down to the vicinity of previously established farm size.

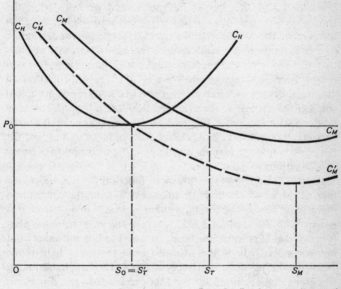

Small grain acreage harvested

Figure 1 Hypothetical long-run average-cost curves for harvesting

The empirical requirements of these two hypotheses are equally apparent. To credit either version we must at least have some evidence that at the beginning of the 1850s the threshold size for adoption of the reaper lay above the average small grain acreage on Midwestern farms, not only in the region as a whole, but in those areas especially devoted to these crops. Once that is established, further evidence of a substantial decline in the threshold size *as the result of an alteration in relative factor prices* during the decade would lend credence to the view that the adjustment in technique was imposed by the inelasticity of the labor supply, whereas acceptance of the pure individual farm-size adjustment view would hinge on a finding that over the course of the 1850s average grain acreage on Midwestern farms rose to the neighborhood of the threshold size that had existed at the opening of the period.

Appendix A provides a detailed discussion of the evidence assembled and the way it has been used in calculating the threshold acreages for adoption of mechanical reapers by grain farmers in the Western states during the period under study. The computations are readily made by linearizing the cost functions for the hand and machine methods of harvesting. It is sufficient here to consider the results of those calculations in conjunction with such information as is available regarding actual small grain acreage on the average Midwestern farm.[20] This may be done with the aid of Figure 2, which depicts the relationship between threshold size and relative costs of labor and capital for the basic hand-rake reaping machine on the assumption of linear cost functions.

Figure 2 also shows the threshold function for the self-raking type of reaper which, by mechanically delivering the cut grain to the ground beside the machine either in gavels or in swath, dispensed with the need for a man to sweep the grain from the platform of the hand-rake machine. However, since self-rakers did not win popularity in the Midwest until the latter half of the 1850s, well after the initial acceptance of the basic hand-raking machines manufactured by the McCormick Co., the discussion will focus on the pioneering hand-rake model.[21] Our conclusions,

20. It should be noted that although the traditional accounts consider the introduction of the reaper only in connection with the harvesting of wheat, the Midwest's principal cereal and leading commercial crop at the time, the discussion here has been phrased in terms of the small grains – wheat, oats and rye, all of which were harvested with the cradle and could be cut with the early reaping machines (Bidwell and Falconer, 1925, p. 353; Hutchinson, 1930, p. 310). This means that calculated threshold sizes for adoption of the reaper must be compared to estimates of average small grain acreage, rather than average wheat acreage alone (Appendix A, section 4).

21. The McCormicks did not manufacture a self-rake model until the introduction of the 'Advance' in the post Civil War period. See Appendix A for further discussion of the self-rakes and of the combined reaper-mowers which came into use in the Western states before the war.

therefore, relate to the influence of alterations in market forces on the adoption of the basic reaper during the 1850s, rather than to the role played by the continuing technical refinement and elaboration of the device.

At the daily wage rate paid grain cradlers during the harvest and the average delivered price of a McCormick hand-rake

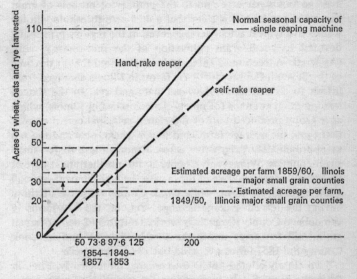

Ratio of delivered reaper price to harvest labor cost (cradlers) per man-day

Figure 2 Threshold functions for adoption of the reaper

reaper that prevailed in Illinois at the end of the 1840s and early in the 1850s, specifically in the period 1849–53, the purchase of a reaper was equivalent to the hire of 97·6 man-days labor with the cradle. From Figure 2 it is seen that these factor prices established a threshold level at 46·5 acres of grain. Where it was possible to hire cradlers on a monthly basis, instead of by the day, and therefore to pay the lower per diem wages implied in typical monthly agreements, the abandonment of cradling in favor of mechanical reaping would have reduced harvesting costs only on farms with more than seventy-four acres of grain to be cut.

Hiring all the labor required for the harvest on a monthly basis was, however, generally not worthwhile for the farmer, so the lower threshold level is more relevant to the problem at hand.

Although there are no direct statistics for the average acreage sown with wheat, oats and rye on Midwestern farms at the beginning of the 1850s, from the available data pertaining to average yields per acre and to the number of bushels of grain harvested per farm it is clear that a 46·5 acre threshold still lay well above typical actual acreage, even in the regions principally devoted to small grain cultivation at the mid-century mark (Appendix A, section 4, Table 2). It is estimated that at the time of the Seventh Census (1850) the farms in Illinois averaged from fifteen to sixteen acres of wheat, oats and rye. In the sixteen leading grain counties (of ninety-nine counties in Illinois) which as a group produced half of the state's principal cereal crop at that time, the average farm land under wheat, oats and rye ran to approximately twenty-five acres. Among these major small grain counties, Winnebago County in the northernmost part of the state is representative of those with the highest average small grain acreage per farm, while Cook County was one of those having the lowest average acreage. Yet, on the 919 farms in Winnebago County the average worked out to 37·2 acres of small grain, still ten acres under the threshold level, and in Cook County the 1857 farms averaged but 18·6 acres apiece.

Two closely related points thus emerge quite clearly. First, in the years immediately surrounding the initiation of reaper production in the Midwest (1847) and the establishment of the McCormick Factory at Chicago (1848), the combination of existing average farm size and prevailing factor prices militated against widespread adoption of the innovation. The admonitions appearing in Western agricultural journals during 1846 and 1847 against purchase of the new reaping machine by the farmer 'who has not at least fifty acres of grain', would appear in the light of the considerations presented here to have been quite sound advice (*Ohio Cultivator*, 1 October 1846; 15 April 1847; *Chicago Daily Journal*, 2, 24, 28, 30 July, 15 August 1846; abridged in Hutchinson, 1930, p. 234, n. 15).[22] Secondly, observations of the

22. Note that, as one might expect in view of the upward trend of relative farm wages from the trough of the depression of the 1840s, the 46·5 acre

sort made by a reliable contemporary witness, Lord Robert Russell, who travelled in the prairie country in 1853, that 'the cereals are nearly all cut by horsepower on the *larger farms* in the prairies' [italics added] (1857, p. 114, and in Rogin, 1931, p. 79),[23] become understandable simply on the grounds of the scale considerations affecting the comparative profitability of the reaper. It is not necessary to explain them by contending that the larger farms were run by men more receptive to the new methods of scientific farming or less restrained by the limitations of their financial resources and the imperfections of the capital market, however correct such assertions might prove to be.

An initial empirical foundation for the plausibility of both our hypotheses having thus been established, we turn now to consider the evidence relating to the character of the adjustments themselves. During the mid-1850s, as the aggregate labor supply constraint hypothesis suggests, the price paid for harvest labor in Illinois did rise more rapidly than the average delivered price of a hand-rake reaper; in the period 1854–7 a McCormick reaper cost the farmer, on average, the equivalent of only 73·8 man-days of hired cradlers' labor, compared with 97·6 man-days in the preceding period 1849–53. In consequence, as Figure 2 shows, the threshold point dropped from over forty-six to roughly thirty-five acres. By the middle of the decade, then, the average small grain acreage above which it paid the farmer to abandon cradling had fallen below the average acreage that had existed in a leading grain-producing area like Winnebago County, Illinois at the beginning of the decade, and lay only ten acres above the average on the 21,634 farms in the state's leading grain counties in 1849–50.

At the same time, there is some evidence pointing to a rise in the average grain acreage harvested per farm, such as is proposed by the individual farm-size adjustment version. Just how large

threshold estimated here to be appropriate for the period 1849–53 lies very close to, but slightly *below* the fifty-acre threshold level implied by the advice to farmers in the years 1846–7.

23. Rogin says he regards this statement as 'more in accord with the facts' than are claims that in the early 1850s cereals as a rule were cut by machinery, but he offers no further evidence or reasoning to support this judgment.

an increase in average acreage occurred during the 1850s in the specialized small grain regions of the Midwest is difficult at present to say, for the simple reason that the Census of Agriculture in 1860, unlike the previous Census, neglected to publish the statistics of the number of farms on a county-by-county basis. In Illinois as a whole, however, the number of acres of wheat, oats and rye harvested per farm is estimated to have been roughly 19 per cent higher in 1859–60 than it had been ten years earlier. Of course, it is possible that in the transition from cereal to corn and livestock production that was under way in the state during this decade, specialization in small grain cultivation became more concentrated and, therefore, that the increase in the typical farm acreage sown with those crops in the leading regions of their cultivation was considerably greater than 19 per cent. But, such evidence as can be brought to bear on the matter does not point in that direction. Broadly speaking, small grain cultivation was spatially no more concentrated in Illinois at the end of the 1850s than at the beginning, and wheat production became, if anything, geographically more dispersed.[24] It is therefore not wholly unreasonable to assume that small grain acreage per farm in the areas especially devoted to those crops increased at the same rate as it did in Illinois as a whole. On that basis one may conjecture that the number of acres under wheat, oats and rye on a typical farm in the state's leading small grain regions increased from twenty-five to thirty in the course of the 1850s.

The story of the adoption of the mechanical reaper in the years immediately before the Civil War should thus be told in terms of the effects of both an expansion of grain acreage sown on individual farms and the downward movement of the threshold size as a result of the rising relative cost of harvest labor. But, of the two types of adjustment taking place during the 1850s, the former must properly be accorded lesser emphasis. As Figure 2 reveals, on farms in the leading grain regions of Illinois the estimated

24. Whereas only sixteen counties made up the group producing half the total wheat crop, and nineteen counties accounted for half the total number of bushels of wheat, oats and rye harvested in Illinois in 1860, it took twenty-one counties to account for 50 per cent of the wheat and nineteen counties to account for 50 per cent of the three small grain crops harvested at the end of the decade (*US Eighth Census*, 1860, 'Agriculture', pp. 31–5).

increase in average small grain acreage was responsible for less than a third of the subsequent reduction of the gap existing between threshold size and average acreage at the opening of the decade. Moreover, among the Midwestern states experiencing rapid settlement during the 1850s, Illinois was singular in the magnitude of the expansion of its average farm size.[25] Elsewhere in the Midwest, the relative rise in farm wage rates is likely to have played a still greater role in bringing the basic reaping machine into general use during the decade preceding the Civil War.

Although the questions considered in the preceding pages are very specific, we have arrived at answers with rather broader implications. Historians of United States agriculture have maintained that during the nineteenth century the transfer of grain farming to new regions lying beyond the Appalachian barrier played a significant part in raising labor productivity in agriculture for the country as a whole. The connection between the spatial redistribution of grain production and the progress of farm mechanization figures prominently among the reasons that have been advanced to support this contention. Some writers suggest that inasmuch as heavier reliance was placed on the use of farm machinery in the states of the Old Northwest before the Civil War, and, similarly, in the Great Plains and Pacific Coast states during the last quarter of the century, the geographical transfer of agriculture into these areas was tantamount to a progressive shift of grain farming towards the relatively capital-intensive region of the technological spectrum.[26] But the

25. Between 1850 and 1860 improved total acreage per farm increased from 66·2 to 91·3 in Illinois, from 53·4 to 62·4 in Indiana, from 55·6 to 62·1 in Iowa, and from 51·8 to 54·0 in Wisconsin. Illinois average farm size was thus not only growing more rapidly than that in the surrounding states, but was initially larger (*US Eighth Census*, 1860, 'Agriculture', p. 222). On these grounds alone, Cyrus McCormick's decision to move his base of operations from Rockbridge County, Virginia in 1847 and embark upon manufacturing the reaper at Chicago during the following year proved extremely cogent.

26. As an indication of the extent of the geographical redistribution of grain farming prior to the Civil War, it may be noted that Pennsylvania, Ohio and New York were the leading wheat growing states in 1849–50, but, a decade later roughly half the total US wheat crop was raised in Ohio,

mechanism of this putative interaction between spatial and technological change has not been fully clarified, and as a result, important aspects of the interrelationship between the historical course of industrialization and the settlement of new regions in the United States remain only imperfectly perceived.

To make some headway in this direction it is necessary to distinguish two possible modes of interaction between spatial and technological changes in United States agriculture: one involves adjustments of production methods in response to alterations of relative prices that were associated, either causally or consequentially, with the geographical relocation of farming; the other turns on purely technological considerations through which regional location influenced choices among available alternative techniques. Now, the general statement that the conditions under which farmers located in the country's interior carried on grain production especially favored the spread of mechanization is sufficiently imprecise to embrace both interaction mechanisms, the influences of market conditions as well as those of technological factors peculiar to farming in the different regions. One may well ask whether such ambiguity is justified. Without establishing the dominance of purely technical considerations it would be unwarranted to suggest that shifts of small grain farming away from the Eastern seaboard automatically, in and of themselves, accounted for increases in the extent to which that branch of United States agriculture became mechanized.

In the case of reaping operations, it is certainly true that there were technical features of Midwestern farming which in contrast with those characteristic of the Eastern grain regions proved inherently more congenial to the general introduction of antebellum reaping machines. On the comparatively level, stone-free terrain of the Midwest, the cumbersome early models of the reaper were less difficult for a team to pull, less subject to malalignment

---

Indiana, Michigan, Illinois and Wisconsin – i.e. in the Old Northwest Territory (Edwards, 1940, p. 203). Among recent explicit treatments of this question, Parker's (1963) offers a set of calculations designed to gauge the impact of regional shifts in the pattern of production on the productivity of labor in small grain farming, and attributes much of the 'region effect' (on national productivity) to the interaction between location and the degree of mechanization.

and actual breakage; because the fields were unridged and crops typically were not so heavy as those on Eastern farms, the reapers cut the grain close to the ground more satisfactorily, and the knives of the simple cutting mechanism were not so given to repeated clogging.

Yet, despite the relatively favorable technical environment (and larger average small grain acreages on farms) in the Midwest, we have seen that the prevailing factor and product market conditions during the 1840s and early 1850s militated against extensive adoption of mechanical reaping equipment even in that region. Against such a background the fact that a large-scale transfer of small grain production to the Old Northwest Territory took place during the 1850s does not appear so crucial a consideration in explaining the sudden rise in the proportion of the total American wheat crop cut by horse power between 1850 and 1860.[27] Instead, it seems appropriate to emphasize that during the Midwestern development boom that marked the decade of the 1850s the price of labor – as well as the prices of small grains – rose relative to the price of reaping machines, and that the pressure on the region's labor supply reflected not only the expanded demand for farm workers, but also the demand for labor to build railroads and urban centers throughout the region – undertakings ultimately predicated on the current wave of new farm settlement and the expected growth of the Midwest's agricultural capacity. If one is to argue the case for the existence of an important causal relationship between the relocation of grain production and the widespread acceptance of mechanical reaping during the 1850s, the altered market environment, especially the new labor market conditions created directly and indirectly by the quickening growth of Midwestern agriculture, must be accorded greater recognition, and the purely technical considerations be given rather less weight than they usually receive in this connection.

There is, however, a sense in which the decline in the cost of

27. This proportion rose from a negligible level at the beginning of the decade to somewhere between 33 and 50 per cent by the close of the 1850s. At the latter date the proportion of the trans-Appalachian wheat crop cut by horse-power was appreciably higher than that for the nation as a whole. See footnote 9 of this Reading for the sources and some discussion of these estimates.

reaping machines relative to the farm labor wage rate may be held to have reflected the interaction of the technical factors favoring adoption of the early reapers in the Midwest with that region's emergence as the nation's granary during the 1850s. The rising share of the United States wheat crop being grown in the interior did mean that, *ceteris paribus*, a larger proportion of the national crop could be harvested by horse power without requiring the building of machines designed to function as well under the terrain and crop conditions of the older grain regions as the early reapers did on the prairies. For the country as a whole, as well as for the Midwest, this afforded economies of scale in the production of a simpler, more standardized line of reaping machines. It thereby contributed to maintaining a situation in which the long run aggregate supply schedule for harvesting machinery was more elastic than the farm labor supply. Thus it may be said, somewhat paradoxically, that the movement towards regional specialization in small grain farming directly made possible greater efficiency in manufacturing and thereby promoted the simultaneous advance of mechanized agriculture and industrial development in the ante-bellum Midwest.

Appendix A: Threshold farm size

The element of fixed cost present with the mechanical reaping process for harvesting grain makes it necessary to take account of the scale of harvesting operations in cost comparisons of hand and machine methods. One means of doing this would be to stipulate the acreage to be harvested and then proceed to ask how the profitability of mechanical reaping compared to cradling was affected by the prevailing level of factor prices. This appendix tackles essentially the same question by posing it in a slightly different way. The question can be put as follows: given alternative sets of factor prices, at what alternative scales of harvesting operations would it be a matter of indifference (on cost grounds) to the farmer of the 1850s whether he adopted the reaper or continued to rely on the cradle? The answer is to be found from a computation of that acreage, called the threshold size, at which the total costs of the two processes were just equal and beyond which abandoning the cradle would become profitable, other things remaining unchanged.

The form of the calculation

It is not, however, necessary to calculate the total cost of harvesting an acre of grain at different scales of operation, as depicted by the long-run unit cost curves of Figure 1. In the first place, as will be seen, the activities of the harvest other than the actual cutting of grain – raking, binding and shocking – may be omitted from consideration as not significantly influencing the choice between machine and hand reaping. Secondly, all that is required is the computation of the total *saving* of money wages effected by adoption of the mechanical reaper and the *additional* fixed capital charge that the farmer would incur in order to have the machine at his disposal during the harvest season. What must be known, therefore, is

L_s: the number of man-days of labor dispensed with by mechanizing the cutting operation, per acre harvested;
w: the money cost to the farmer of a man-day of harvest labor;
c: the fixed annual money cost of a reaper to the farmer.

From this information the threshold size, S_T, in acres to be harvested, can be determined, since, *ceteris paribus*, total costs of cutting the grain are the same for both processes when

$$c = \sum_{i=1}^{S_T} L_{si} w,$$ 1

where the index i designates the acre in the sequence of acres ($i = 1, ..., S_T$) harvested.

Actually, the problem can be further simplified. From the available information it appears justifiable to assume that within the range of normal daily cutting capacity of the reaper there were no economies or diseconomies of scale in the use of labor in cradling, and that the saving of labor effected by the mechanical reaper per acre harvested was a technically determined constant. In a word, the cost functions for the two processes may be taken as linear over the relevant range of operations. Consequently, we may replace equation 1 with the much simpler expression,

$$c = S_T L_s w.$$ 1a

The average annual capital cost, or effective rental rate on a piece of durable equipment may be reckoned as the sum of the annual depreciation of the equipment and the annual interest cost on the capital invested in it. One can think of the latter as an opportunity cost, since for half the year, on average, the owner's funds are tied up in the machine instead of being lodged elsewhere at interest. Alternatively, the interest cost is to be thought of as the actual charge made for a loan of the purchase price of the equipment. Strictly speaking, in calculating the interest cost of a mechanical reaper, allowance should be made for the fact that funds were locked up in these machines for periods longer than a year; it is known that Midwestern farmers purchased reapers on credit during the 1850s and paid them off only over an extended period of time (Hutchinson, 1930, pp. 368–9). Yet, within the range of accuracy we can hope to attain in the present calculations, the niceties of compound interest may be foregone and the interest cost computed on a simple basis. The half life of reapers was not, in any case, so long as to render this a serious omission.

An equivalently liberal attitude is warranted regarding the question of depreciation charges. Rather than play with formulas that attempt to take into account the actual time pattern of loss of value through wear and tear and obsolescence, straight-line depreciation over the physical life of the machine will be assumed.

As a result of the foregoing simplifications, the average annual gross money rental charge is quite straightforwardly given by

$$c = \{d+0\cdot5(r)\}\,C, \qquad\qquad 2$$

where $d \equiv$ the straight-line rate of depreciation,
  $r \equiv$ the annual rate of interest,
  $C \equiv$ the purchase price of a reaper.

Putting this together with equation 1a, we have the relation defining the threshold size in terms of the prices of harvest labor and reapers, the rate of interest, and the 'technical' coefficients L_s and d:

$$S_T = \left(\frac{d+0\cdot5r}{L_s}\right)\left(\frac{C}{w}\right). \qquad\qquad 3$$

From the form of equation **3** it is apparent that, given the rate of interest, the threshold for harvesting machines of specified durability and labor-saving characteristics is directly proportional to the relative price of the reaper (C/w). Thus, the threshold functions shown in Figure 2 of the text appear as positively sloped rays from the origin of a graph of acreages against the ratios of reaper prices to wage rates. In the following section we proceed first to consider the evidence that establishes the slope of the threshold function, and then to take up the question of the relevant range of variation of factor prices in the Midwest during the period preceding the Civil War.

Parameters and variables

1. *Labor savings*. The reduction in harvest labor requirements achieved by the introduction of the mechanical reaper is perhaps the single most widely cited instance of the improvement of agricultural technology in small grain production during the antebellum era. Various estimates of the magnitude of the saving in labor appear in the secondary literature and the economic history texts, but virtually all of them derive from the evidence and conclusions presented in Leo Rogin's pioneering study (Bidwell and Falconer, 1925, pp. 293–4; Danhof, 1951; Gates, 1960, p. 237; Robertson, 1964, p. 258). A crucial issue that arises in this connection is one that is frequently overlooked, namely, the amount of grain that the mechanical reaper could cut in a normal day's work. Notwithstanding the fact that during the 1850s the McCormick reapers were warranted to cut fifteen acres of wheat in a twelve-hour day and that frequently mentioned records set in reaper trials cite fifteen as the acreage cut in a day, Rogin's survey of the evidence leads him to conclude that ten to twelve acres per day was closer to normal practise even on the broad prairies of Illinois:

. . . the foregoing rate is the one most frequently mentioned in other contemporary accounts and may be taken to represent the average performance with the hand-rake reaper after it came into prominence, as well as with the self-rake which superseded it (1931, pp. 134–5).[28]

28. The Census of 1860 also maintained that 'a common reaper will cut from ten to twelve acres in a day of twelve hours'. (*US Eighth Census*, 1860.) Bidwell and Falconer (1925, pp. 293–4), who base their discussion on

Following Rogin, then, eleven acres per day may be taken as the average cutting rate with the reaper during its first decade of widespread use.

The significance of establishing the normal daily rate of harvesting with the reaper lies in the fact that the information available on the labor requirements of the hand method of cutting grain comes in the form of statements about the number of acres that it was common for a man working with a cradle to harvest in a day. In Rogin's judgement, 'there appears . . . to have been a norm of performance for the country as a whole which approximated two acres' per man day (1931, p. 128), despite regional and local variation in speed caused by differences in the heaviness of the grain. If an allowance of a 10 per cent difference in speed above the national average is made in recognition of the lighter yields in small grains typical in the Midwest (Parker, 1963, Tables 5, 14),[29] we are led to conclude that during the 1850s a mechanical reaper cutting eleven acres accomplished the work of five man-days' labor with cradles. Two men were required to operate the hand-rake reapers, whereas the self-rakers, first marketed commercially in 1854, needed but a driver. Therefore, in the cutting operation itself the use of the hand-rake reaper effected a saving of three man-days in cutting eleven acres, while the self-rakers saved four man-days.[30]

Although two other sources of economy are sometimes cited as associated with the mechanization of reaping, their inclusion does not seem warranted in the present connection. The reduction of losses due to the shattering of overripe grain by cradlers cannot be considered as anything more than a consequence of the saving in labor already mentioned: if this source of losses

the results of the Geneva, NY Trials of 1852, where mechanical reapers were matched in competition with cradlers cutting fifteen acres, point out that the latter 'was a large area for an early reaper to cut in one day'. See Hutchinson (1930, p. 336), for the McCormick warranties, and (p. 73) for a characteristic example of McCormick advertising giving the savings achieved with the reaper, based on an assumed daily cutting capacity of sixteen acres.

29. Wheat and oats yields were higher in the Northeast and Middle Atlantic states.

30. Rogin (1931, p. 135) gives the same savings in cutting ten acres a day, but in some places says ten to twelve acres per day.

had proved sufficiently costly in the era prior to the introduction of the reaper, farmers would have found it worthwhile to hire enough cradlers to ensure that their grain was harvested before it became especially prone to shattering.[31] The second point is somewhat more complex. Both the hand-rake and the self-rake models left the grain swept into gavels, or lying in swath, for the binders who followed behind the machine. Rogin (1931, p. 136) notes that binding behind machines was done at a faster rate than when the binders followed cradlers through the fields, and allows an additional saving of labor (amounting to two binders per day in harvesting ten to twelve acres) on this account. This leads him to state that the total saving in labor connected with the use of the reaper amounted to five man-days, instead of just three in the cutting operation itself. However, it is pointed out that the reduction in the labor requirements for binding that accompanied the mechanization of grain-cutting may be attributed to the increased pressure put on the binders to get the grain out of the way of the machine, and the consequent adoption of the system of 'binding stations' in place of the traditional practice of permitting binders to range over the field at their (comparative) leisure (Rogin, 1931, pp. 103, 136, n. 339). Since it would appear that the 'saving' of the hire of two binders per ten to twelve acres was achieved by having those employed behind the reaping machines work at a faster pace, the binders thus engaged would ask a higher daily wage, and, so long as there was work available for those who preferred the less demanding task of binding for cradlers, the differential wage would have to be paid. Thus, in terms of the cost of labor to the individual farmer, the greater efficiency exacted in binding by the

31. Note that there is no evidence of smaller losses from shattering resulting from machine cutting as contrasted with hand cutting of *equally* ripe grain. Yet the tradition has grown up in the secondary literature of treating the reduction of shattering losses owing to more rapid harvesting as distinct and supplementary to the reduction in labor requirements with the mechanical reaper. See e.g. Bidwell and Falconer (1925, p. 294) and Robertson (1964, p. 258), where specific mention is made of the reduction of output losses due to damage from the elements as a further benefit bestowed on the farmer by the reaper. This would be justified if, for purely technical reasons, expected post-harvest yields, as well as harvest labor requirements per acre were different with the two methods.

introduction of the reaper cannot be considered to have resulted in any further economies. Indeed, on this account the reaper may have entailed additional monetary costs for the farmer, since when the system of binding stations was employed the use of women and children was precluded by the pace of the work (Rogin, 1931, p. 103, n. 179).[32]

In summary, for a normal day's work cutting eleven acres, the saving in labor per acre (L_s) with the hand-rake reaper is esti-mated at 0·273 man-days hire of cradlers services, whereas with the self-raker introduced in the latter half of the 1850s the saving amounted to 0·364 man-days.

2. *Durability and the interest rate*. Information as to the average length of life of the reaping machines produced in the Midwest prior to the Civil War is very scanty. Yet those statements regard-ing durability that have come to light agree in placing the useful life of a reaper at close to ten years (Hutchinson, 1930, pp. 73, 311; Rogin, 1931, p. 95). This figure apparently assumed good care and normal use of the machine. Just what constituted 'good care', and how closely such a standard was approached by the notoriously casual practices of American farmers during the nine-teenth century in the maintenance and storage of equipment, is extremely difficult to say (Carman, 1934; Hutchinson, 1930, p. 365). What 'normal use' entailed is somewhat clearer; for, as a rule, the rate of ripening of the grain confined the feasible period for the harvest of small grain crops to roughly a two-week inter-val, with ten full working days (Rogin, 1931, p. 95). Thus, approx-imately 110 acres of wheat would represent the annual normal use cutting capacity of a reaper. On the assumption that the life of the machine would not be prolonged beyond ten years by utilization

32. Even if the reduction in manpower requirements for binding were due to purely technical considerations, e.g. the way the grain was delivered from the machines, it would be necessary to take account of the lower daily wage rates typically received by binders, compared to cradlers. In Ohio during the harvest of 1857 binders were paid roughly two-thirds the rate received by cradlers (*Ohio State Board of Agriculture Report*, 1857, pp. 181–2). With this as a basis for weighting the reductions in labor requirements by relative wage rates, Rogin's procedure should give a total saving in labor with the hand-rake machine equivalent to 4·33 cradler man-days, on ten to twelve acres, rather than the five (heterogenous) man-days saving frequently mentioned in the literature.

at rates below the capacity level imposed by custom, the available hours of daylight, and the length of the period within which the grain could be safely gathered, we shall take the annual straight-line depreciation rate as $d = 0.10$.

Selection of an appropriate rate of interest proves to be quite simple. It appears that the McCormick Co. charged farmers a standard rate of 6 per cent on the unpaid balance of their reaper notes throughout the 1850s (Hutchinson, 1930, pp. 337, 369, n. 31). Even if the farmer was able to secure a higher rate of return by placing his funds elsewhere than in a reaper, the 6 per cent rate is none the less appropriate from an opportunity cost viewpoint; so long as the McCormick Co. was willing to encourage sales by foregoing the opportunity to earn more than 6 per cent on its funds, the farmer who purchased a reaper would find it profitable to borrow at that rate and free such capital as he had for investment at higher rates of return.

Combining the 10 per cent depreciation rate and the simple interest cost, according to formula 2, we estimate the annual average rental rate of a reaper at 13 per cent of its purchase price.

3. *Threshold functions and relative factor prices.* The data assembled in the two preceding sections define a pair of functions relating threshold size to the relative price of reapers – one for the choice between cradling and hand-rake reaping machines, the other for the choice between cradling and self-rakers. With the parameter values substituted for d, r, and L_s in equation 3, the functions are found to be,

$$\text{hand-raker versus cradle:} \quad S_T = 0.4765 \frac{C_{HR}}{w},$$

$$\text{and self-raker versus cradle:} \quad S_T = 0.3572 \frac{C_{SR}}{w}.$$

Both functions are shown in Chart 1, but, since the self-raker was introduced in the Midwest only after 1854 and the reapers for which reliable price information is readily available are those of the hand-rake type manufactured by C. H. McCormick in Chicago, rather than the Wright–Atkins self-raking variety, the present discussion has been restricted to consideration of the threshold levels for adoption of the hand-rake machine.

We have also chosen to avoid the complications that would

arise in attempting to work out the threshold function for the basic McCormick reaper with the mower attachment (perfected in 1855) that enabled the machine to cut grasses as well as grain. The technical advance embodied in the reaper-mowers obviously contributed to the abandonment of the cradle by permitting the fixed cost of the machine, which was sold either with or without the mower attachment, to be spread over a greater total (grain and grass) acreage harvest. It thus became profitable to use the reaper on farms with small grain acreages too low to justify the expense of a machine that could not also be converted to cut grass crops (Hutchinson, 1930, pp. 309–16; Rogin, 1931, pp. 96–102). Similarly, since the advent of the self-raker did not drive the McCormick hand-rake machines out of the market, it must be inferred that the former were not less expensive than the hand-rake model on which McCormick continued to rely until after the Civil War. Instead, they must have been priced low enough to drop the threshold grain acreage at which their introduction as a replacement for the cradle was profitable to a level under the hand-raker threshold.[33]

The price of a reaper. From 1849 through the harvest of 1853 the price of the McCormick hand-rake reapers actually sold to farmers averaged 113 dollars plus freight charges, whereas the average unit price, also f.o.b. Chicago, for the period 1854–58 was 133 dollars.[34] Comparable average figures for the transport charges obviously present a greater problem, for such charges varied from place to place according to the distance from Chicago and the mode of transportation available.[35] An element

33. From the threshold functions given above, one would surmise that, at a maximum, the self-raking machine could have carried a delivered price as much as a third higher than that of the hand-rake reaper, and still have been competitive with the latter, i.e. $(C_{SR}/C_{HR}) = (0.4765/0.3572) = 0.33$.

34. Average unit prices, reflecting the proper mixture of cash and credit prices on actual transactions have been computed from financial data of the McCormick Co. presented in Hutchinson (1930, p. 369, n. 60). The figures given above agree fairly closely with the announced f o b prices of 115 dollars prior to 1854 and 130 dollars for the harvest of 1854 (p. 323).

35. Strictly speaking this was not true: the McCormick Co., when competing for sales with its rivals in the Eastern market – outside McCormick's home territory in the region west of Cincinnati – absorbed the total freight charges and sold reapers on the seaboard at Chicago prices on at least one occasion during the 1850s (Hutchinson, 1930, p. 324).

of arbitrariness is thus unavoidable in fixing upon any particular figure to represent the transport cost component of the total delivered price of a reaper. For the years before 1854 we shall take the eleven dollar freight charge paid by the farmers in Peoria, Illinois, to get their reapers by canal and steamboat from the McCormick Factory wharf in Chicago; in 1854, by which time Chicago's tributary railroad connections in the West totalled 814 miles of line, the printed McCormick sales forms stipulated a maximum charge of five dollars for freight, which we shall accept as representing the average transport cost during the period 1854–8 (Hutchinson, 1930, pp. 322–3).[36] Therefore, in the absence of any additional warehousing charges – sometimes imposed on the farmer by local sales agents – the delivered price of a hand-rake reaper averages out at roughly 124 dollars for 1849–53, and 138 dollars in years immediately thereafter. The estimate for 1849–53 is in close agreement with the figure of 125 dollars cited in 1851 by an Illinois periodical, the *Prairie Farmer*, as the amount the farmer would have to spend for a reaper (1851, vol. 11, p. 482).

The cost of labor. Obtaining appropriate figures for the average daily money wage paid grain cradlers during the harvest season poses a problem no less difficult than that of settling upon representative delivered prices of reaping machines during the 1850s. It must be noted, first, that daily wages for farm labor in the Midwest during this period appears to have been roughly 37 per cent higher than the per diem rate received by workers hired on a monthly basis, even after allowing for the money equivalent of board usually provided workers hired by the month (Lebergott, 1964, pp. 244–5). The mass of wage rate quotations for farm labor, however, reports monthly wages in addition to board and not the wage cost to the farmer. To get an idea of movements in the general level of daily wage rates for farm labor one must, therefore, rely on the movements in the level of daily rates for common labor during the decade. Finally, the existence of skill differentials within the structure of farm wages, often overlooked in studies that deal with problems of industrial skill differences

36. The Western tributary railroad mileage figure is for 1 January 1854. At the beginning of the previous year, Western tributary rail mileage amounted to only 157 miles (David, n.d., Appendix C, Table C-5).

and skill margins, must be taken into account. Cradlers tended to receive higher money wages than those who followed in their wake through the fields during the harvest; the differential was as much as 50 per cent over the general average daily wage paid farm workers during the harvest of 1857 in Ohio.[37]

To obtain estimates of cradlers' daily wage rates in Illinois we shall start with two quotations for the common day labor wage in the state: the rate of eighty-five cents per day reported as the state average by the Census of 1850 can be taken as roughly representative of rates prevailing in the period 1849–53, while from a number of contemporary sources it appears that $1·25 per day is an appropriate average rate for the 'boom' years 1854–7 (Debow, 1854, p. 164; David, n.d., p. 35; Gates, 1934, pp. 94–8, 1960, p. 278). Adjusting these figures on the assumptions that they were equal to the general level of money rates in the agricultural sector and that cradlers received a constant 50 per cent differential over the general farm labor rate, one arrives at $1·27 and $1·87 as estimates for 1849–53 and 1854–7, respectively. In view of the report at the end of the 1850s (after the collapse of the Midwestern construction boom) that Wisconsin farmers were still paying harvest workers wages ranging from one dollar to three dollars per day (Pierce, 1940, p. 156, n. 31) the estimate of $1·87 per day for the boom period does not appear too generous. If the level of the estimated rate for the latter period is accepted, that for the earlier years follows from the assumption that over short periods the whole structure of farm wages moved with the regional common labor wage rate. Since the most plausible objection that could be raised against this assumption is that the adoption of mechanical reaping may have prevented cradlers' wages from rising as rapidly as the general level of farm wages (and the common labor rate) it appears that $1·27 cannot be regarded as *too high* an estimate of the daily wage received by cradlers at the close of the 1840s. Actually, that estimate comes remarkably close to the figure of '$1·25 or more' per day mentioned in contemporary

37. See Ohio State Board of Agriculture, *Report* (1857) 181–2. At the end of the 1840s, Wm Marshall of Cordova, Illinois, testified that many farm laborers seeking work in the harvest did not known how to use a grain cradle. It is quite likely that they were recently arrived immigrants, probably Irish but possibly Germans (Hutchinson, 1930, p. 206, n. 8).

sources as being paid for 'extra laborers' taken on during the 1847 harvest in Illinois (*Prairie Farmer*, 1847, 1848).

As we have already noted, farmers paid considerably more for a day's work when labor was hired by the day than when a monthly agreement was made. With a 37 per cent differential required to compensate farm hands for the insecurity of day labor, the implied per diem wage for cradlers hired by the month would have been 80 cents instead of $1·27 for 1849–53, and $1·18 instead of $1·87 during 1854–7. Of course, given that differential, it would be cheaper for the farmer to hire a worker by the month only if his services were needed for seventeen or more out of the twenty-six working days in a month, that is, for a period considerably longer than the average ten working days within which small grain crops like wheat could be safely harvested on a single farm. Farmers undoubtedly hired some hands on that basis, and we shall therefore make use of the lower rates to compute the relative price of a reaper when labor was hired by the month. Nevertheless, it is more reasonable to suppose that the marginal workers replaced in the harvest by the mechanical reaper were those that would have been taken on at the higher daily wage rates just for the duration of the harvest.

4. *Estimates of threshold size and actual small grain acreage.* In Table 1 the delivered prices of reapers and the alternative wage rate data developed in the preceding section have been brought together to provide a set of estimates of the relative price of a hand-rake reaper for the early and middle years of the 1850s. Columns 3 and 6 of the table present the corresponding threshold acreages at which the costs of harvesting with cradles and with the reaper would have been exactly equal. The latter are computed from the threshold function for the hand-rake reaper given in Section 3 of the Appendix, and provide the basis for the discussion of Figure 2 in the text.

The acreage figures in Table 1 are discussed in conjunction with estimates of the actual average acreage sown with small grain crops on farms in Illinois during the 1850s, which are presented in summary form by Table 2. A few words about the sources of the latter figures are therefore in order.

The estimates for the harvest of 1849–50, in panel 1 of Table 2,

Table 1 Threshold Grain Acreage for Adoption of Hand-Rake Reapers in Illinois

| | Period: 1849–53 | | | Period: 1854–7 | | |
	1 w	2 C_{HR}/w	3 S_T	4 w	5 C_{HR}/w	6 S_T
Terms of hire of farm labor	cradlers, in current dollars	in man-days per reaper	acres harvested	cradlers, in current dollars	man-days per reaper	acres harvested
Day	1·27	97·6	46·5	1·87	73·8	35·1
Month	0·80	155·0	73·8	1·18	117·0	55·6

are based on the statistics given in the Census of Agriculture for 1850 on the volume of wheat, oats and rye harvested and the number of farms in each county of Illinois, as well as in the state as a whole (DeBow, 1854, pp. 21–7). From this information, the bushel harvest per farm of wheat, and of oats and rye (combined) were obtained and then converted into acreage estimates per farm for these crops by employing coefficients of the average grain yield per acre. For the harvest of 1859–60 only the estimate of the state average is given, since the Census of 1860 does

Table 2 Average Acreage per Farm in Illinois

Region	Number of farms	Wheat acreage per farm	Total wheat, oats and rye acreage per farm
Harvest of 1849–50			
State	76,208	11·0	15·4
Major Wheat Counties*	21,634	19·3	25·2
Winnebago County	919	30·4	37·2
Cook County	1857	11·4	18·6
Harvest of 1859–60			
State	143,310	14·6	18·3

* The group of leading wheat-growing counties which accounted for 50 per cent of the total wheat production in the state.

not supply statistics of the numbers of farms on a county basis (US Eighth Census, 1860, pp. 31–5).

Fortunately, the choice of appropriate average grain yield coefficients for Illinois at the beginning and end of the 1850s is immensely simplified by the availability of William Parker's recent exhaustive research on the problem of wheat and oat yields per acre in nineteenth-century America. In comparison with oats, the rye crop of the Western states was insignificant prior to the Civil War; application of the yield estimates for oats to the combined oats and rye output statistics – figures for the harvest of these crops not being given separately on a county basis by the Census of 1850 – should not, therefore, introduce serious error into the acreage estimates.[38] In constructing the acreage figures shown in Table 2, the yield of wheat in Illinois from the harvests of 1849–50 *and* 1859–60 was taken to have been 11·3 bushels per acre. The yield of oats we take to have been 30·5 bushels per acre. The following evidence, drawn from Parker's work, may be adduced in justification of these coefficients:

(a) Between 1849 and 1859 the average yields of wheat and oats in the region comprised of Ohio, Indiana, Missouri and Iowa were virtually unchanged; the wheat yield dropped from 12·4 to 12·1 bushels per acre, and the oats yield rose from 29·3 to 30·4 bushels per acre (Parker, 1963, Table 10, p. 32).

(b) From the US Department of Agriculture revised data on acreage yields in the period 1866–75 it is found that the wheat yield in Illinois was slightly below the typical yields in the Midwestern region, whereas the yield of oats was virtually the same as that for the region as a whole. The Illinois yield of wheat per acre was 11·3 bushels, compared with the mid-range figure of 12·2 bushels per acre for the group of states mentioned above, while the Illinois oats yield was 30·5 bushels per acre compared with the mid-range figure of 30·3 for the region (Parker, 1963, Table 1, p. 4).

38. In Illinois the number of bushels of rye harvested was a mere 0·8 per cent of the oats harvest of 1849–50, and even after rapid expansion during the decade the rye harvest of 1859–60 amounted to but 6 per cent of the volume of oats produced. See Bidwell and Falconer, 1925, pp. 350–56 for summary tables.

Since the Department of Agriculture regional yield data for 1866–75 agree so well with the regional data for 1849 and 1859, the USDA data on Illinois were accepted as appropriate for the decade of the 1850s.

The average yield coefficients employed for Illinois may be, if anything, somewhat low for the major grain regions of the state.[39] Consequently, if the derived acreage estimates do contain a bias, it is certainly not one which would favor our conclusion that a considerable gap existed between the threshold size for adoption of the hand-rake reaper and the actual average small grain acreage per farm in the early 1850s.

References

BIDWELL, P. W. and FALCONER, J. I. (1925), *History of Agriculture in the Northern United States*, Carnegie Institution of Washington.

BOGUE, A. G. (1963), 'Farming in the Prairie Peninsula 1830–1890', *J. econ. Hist.* vol. 23, pp. 3–29

CARMAN, H. J. (1934), 'English views of Middle Western agriculture 1850–1870', *Agric. Hist.*, vol. 8, pp. 3–19.

DANHOF, C. (1941), 'Farm-making costs and the safety valve 1850–1860', *J. polit. Econ.*, vol. 49, pp. 317–59.

DANHOF, C. (1951), 'Agriculture', *The Growth of the American Economy*, H. F. Williamson (ed.), 2nd edn, Prentice-Hall.

DAVID, P. A. (n.d.), *Factories at the Prairie's Edge: A Study of Industrialization in Chicago 1848–1893*.

DEBOW, J. D. B. (1854), *Compendium of the Seventh Census of the United States*, AOP Nicholson, Printer.

EDWARDS, E. E. (1940), 'American agriculture and the first 300 years', *Yearbook of Agriculture 1940*, Government Printing Office.

FITE, E. D. (1910), *Social and Industrial Conditions in the North during the Civil War*, Macmillan.

GATES, P. W. (1934), *The Illinois Central Railroad and its Colonization Work*, Harvard University Press.

GATES, P. W. (1960), *The Farmer's Age: Agriculture 1815–1860*, Holt, Rinehart & Winston.

GERSCHENKRON, A. (1962), *Economic Backwardness in Historical Perspective*, Harvard University Press.

HABAKKUK, H. J. (1962), *American and British Technology in the Nineteenth Century*, Cambridge University Press.

39. A sample of Illinois county estimates drawn from different dates in the period 1843–55 gives a median wheat yield of sixteen bushels per acre, and a median yield for oats of forty bushels per acre (Parker, 1963, Table 3, p. 8).

HIRSCHMAN, A. O. (1958), *The Strategy of Economic Development*, Yale University Press.

HUTCHINSON, W. T. (1930), *Cyrus Hall McCormick*, vol. 1, *Seed Time, 1809–1856*, Appleton-Century-Croft.

LEBERGOTT, S. (1964), *Manpower in Economic Growth. The United States Record since 1800*, McGraw-Hill.

OHIO STATE BOARD OF AGRICULTURE (1857), *Report*.

PARKER, W. N. (1963), 'Productivity change in the small grains', National Bureau of Economic Research Conference on Research in Income and Wealth, mimeographed.

PIERCE, B. L. (1940), *A History of Chicago*, vol. 2, *From Town to City 1848–1871*, Knopf.

Prairie Farmer (1847), October, pp. 374.

Prairie Farmer (1848), January, pp. 26–7.

PRIMACK, M. (1962), 'Land clearing under nineteenth century techniques: some preliminary calculations', *J. econ. Hist.*, vol. 21, pp. 484–97.

ROBERTSON, R. M. (1964), *History of the American Economy*, 2nd edn, Harcourt, Brace & World.

ROGIN, L. (1931), *The Introduction of Farm Machinery in its Relation to the Productivity of Labor in the Agriculture of the United States during the Nineteenth Century*, University of California Press.

ROSTOW, W. W. (1960), *The Stages of Economic Growth*, Cambridge University Press.

RUFFIN, E. (1851), 'Management of wheat harvest', *Amer. Farmer*, vol. 6, pp. 453–60.

RUSSELL, R. (1857), *North America, Its Agriculture and Climate*, Black.

SHANNON, F. A. (1945), *The Farmer's Last Frontier 1860–1897*, Holt, Rinehart & Winston.

TEMIN, P. (1966), 'The relative decline of the British iron and steel industry', in H. Rosovsky (ed.) *Industrialization in Two Systems: Essays in Honor of Alexander Gerschenkron*, Wiley.

UNITED STATES EIGHTH CENSUS (1860), US Government.

13 P. Temin

A New Look at Hunter's Hypothesis About the Ante-Bellum Iron Industry

P. Temin, 'A new look at Hunter's hypothesis about the ante-bellum iron industry', *American Economic Review Papers and Proceedings,* May 1964, pp. 344–51.

The process whereby innovations diffuse throughout an economy or throughout the world has interested many people at many times. One of the more intriguing examples of this process is the spread of the use of coke to make pig iron, which has interested investigators because of two seemingly incompatible properties. The use of coke in the blast furnace was one of the enabling innovations of the industrial revolution in Britain, and it played a similarly important role in the industrialization of many, if not all, other industrializing countries in the nineteenth century. Despite this importance, however, the spread of coke was not rapid; there was often a delay of as much as a half-a-century in the spread of this innovation from Britain to other countries. Coke was the fuel for well over 90 per cent of the blast furnaces in Great Britain before 1810 (Ashton, 1924, p. 99). But half-a-century later, at the start of the Civil War, the proportion of pig iron smelted with this fuel in the United States had barely attained the level of 10 per cent of the total (Swank, 1892, p. 376).

The example of America is particularly intriguing. The contrast between the long delay in adopting the new process and the subsequent rapid expansion of its use is striking (Swank, 1892, p. 376). In addition, the delay has been made to appear even more problematical by the recent book of Habakkuk (1962), which has emphasized the extent to which the United States led Britain technologically in certain areas before the Civil War. The problem before us is the reconciliation of the evidences of continuous progress in the United States with the neglect of this important innovation.

Many reasons have been given to explain the half-century delay in the adoption of coke in America. The strongest reasons given in the nineteenth century were that charcoal was much more plentiful in this country than in England and that American ironmasters were opposed to the use of coke due to ignorance, irrationality, or both (Swank, 1892, p. 366). Louis C. Hunter, in a classic article published in 1929, showed conclusively that these reasons were false and propounded a new hypothesis to explain the delay. This paper will build upon Hunter's masterful analysis and suggest some alterations in his hypothesis about the ante-bellum iron industry.

Hunter pointed out that the cost of charcoal was almost entirely the cost of labor incurred in preparing it, the cost of the original wood accounting for little or nothing in the price. The availability of wood in the United States consequently had little influence on this price, and charcoal – here as in England – was a more expensive fuel than coke. As for ignorance and irrationality, Hunter showed that the pig iron made with coke was inferior to the iron made with charcoal, which implies that the reluctance of the ironmasters to accept a new fuel in place of the old was an accurate, rational reaction to quality differences in the iron produced. These two observations by Hunter form the cornerstone of any modern discussion of the diffusion of coke smelting in America.

Hunter did not spell out the mechanism by which the low quality of pig iron made with coke restricted its production, but the obvious inference is that the low quality implied a low price which destroyed the profitability of the new process. Pig iron made with coke sold for a 20 per cent discount from the price of charcoal pig iron during the 1850s, a difference of six dollars, The cost of charcoal to make a ton of pig iron was approximately nine dollars, and I have endeavoured to show elsewhere that the reduction in the price of fuel attendant upon the use of coke was not sufficient to compensate for the reduced price of the product (Temin, 1964, ch. 3). The ante-bellum ironmaster was offered no inducement in the form of higher profits to use coke – a fact which is borne out in the not completely pleasant experiences of those ironmasters who tried. There were good reasons, in other words, why the technical progress in the ante-bellum era noted by Habakkuk for light manufacturing did not extend to all areas of the economy.

By the advent of the Civil War, however, a change was beginning to be evident, and the amount of pig iron made with coke began to rise faster than the total production of pig iron in America. The second part of Hunter's hypothesis endeavors to explain this development. It states that the character of the demand for iron was changing at the time of the Civil War, from a price-inelastic demand that was sensitive to changes in quality to one that was more price-elastic and less sensitive to changes in quality. The former pattern was characteristic of an agricultural demand in which small lots of iron were bought for use under a wide variety of conditions; the latter was the result of an industrial demand in which larger amounts of iron – relative to the total purchases of a firm or farm – were purchased for a specific purpose. The most important specific purpose was for the production of rails – a product that did not benefit from the use of high quality iron.

Two factors urge the reconsideration of this part of Hunter's hypothesis. As with the earlier part of the argument, Hunter did not outline a mechanism by which the change in demand caused a change in the method of production. Such a mechanism is considerably harder to specify than the one which translated a low quality of product into low profits because it involves a shift in demand curves, and even a shift in the elasticity of demand curves – always a difficult event to identify. In addition, recent work on the proportion of pig iron used for the production of rails has raised the possibility that the shift in demand noticed by Hunter was not large enough to effect the change in productive techniques under discussion (Fogel, 1963, ch. 5).

I would like to suggest an alternative to this part of Hunter's hypothesis. In the course of discriminating between my suggestion and the theory proposed by Hunter, both the mechanism involved and the relation of this mechanism to other work will be shown.

The increasing use of coke for a blast furnace fuel at the time of the Civil War may be seen as a reaction to changes in the cost and price structure of the American iron industry. Hunter's hypothesis is that the changes were in the demand for iron, that the demand for low-quality, low-cost iron rose relative to the demand for high-cost and quality iron. An alternative explanation is that the cost of making coke pig iron fell or that its quality improved, i.e. that the changes were on the supply side. The cost of making pig iron with coke does not seem to have changed with respect to the cost of using charcoal before the 1870s and the introduction of 'hard driving', and this explanation may be dismissed. Discrimination betweeen the two remaining alternatives – Hunter's hypothesis of changing demand and my suggestion of quality improvement in coke pig iron – is a harder task.

This task is facilitated by the existence of a third fuel – different from both charcoal and coke – suitable for use in the blast furnace and a third type of iron. The fuel was anthracite, found in eastern Pennsylvania and parts of Wales. Coke was used before anthracite in Britain, as the technical knowledge necessary for the ignition of the somewhat recalcitrant anthracite was not discovered until the late 1830s, but anthracite became the variety of mineral fuel used initially in America. Its production rose rapidly in the 1840s, and it accounted for over half of American pig iron production in 1860 (Swank, 1892, p. 376; Temin, 1964, Appendix A).

There are two possible reasons why anthracite was adopted before the Civil War and coke was not, corresponding to the two reasons proposed to explain why coke was adopted later. Anthracite is found east of the Alleghenies, while bituminous coal lies to the west. Demand may have differed in the two regions in such a fashion as to induce the production of a type of pig iron in the east that could not have been profitably produced in the west. Alternatively, the quality of the pig iron made with anthracite may have differed from that of coke pig iron, while the character of demand was unimportant. (The costs of making pig iron with the two different mineral fuels were approximately equal at this time. If there was a difference between them, it was that it was cheaper to produce pig iron with coke than with anthracite.)

As long as the commerce in heavy products such as pig iron was separated by the Allegheny Mountains into two markets, there was little evidence generated that would enable the modern observer to discriminate between these hypotheses. In 1852, the Pennsylvania Railroad completed its through line from Philadelphia to Pittsburgh, and the cost of transportation across the mountains was reduced. Pig iron made in eastern Pennsylvania with anthracite began to be sold in Pittsburgh in competition with the locally produced pig iron made with coke. If the character of demand was different in the two regions, we would expect pig iron made with anthracite to sell in Pittsburgh at approximately the price of pig iron made with coke, i.e. at a substantial discount from the price of pig iron made with charcoal. If, on the other hand, it was the nature of the varieties of pig iron that made a difference, we would expect to find an equally clear price differential existing between pig iron made with anthracite and with coke.

As it turns out, pig iron made with anthracite sold in Pittsburgh at the same price as charcoal pig iron (Hunter, 1928, pp. 393–433). Consumers in the west were willing to pay substantially more for pig iron made with anthracite than for pig iron made with coke in the ante-bellum era, and we conclude that pig iron made with anthracite differed significantly in its quality from pig iron made with coke. The reasons for this are clear. The main debilitating element in the coking coal available in the 1840s and 1850s was sulphur, and the anthracite deposits of eastern Pennsylvania were relatively free of this impurity.

The delay in the use of coke in the United States may now be placed in perspective. Ironmasters in the United States were adopting the new technology based on mineral fuel. Because of a difference in resources, however, they adopted it in a different form from that used originally in Britain. And because the innovations permitting use of the American resources, i.e. anthracite coal, were not introduced until the 1830s, they adopted it after a delay. The deposits of anthracite in the United States are very highly concentrated in a few counties of north-eastern Pennsylvania, and the high cost of transport to the far reaches of the large American economy prevented the use of this fuel from completely eclipsing pig iron production with charcoal. Consequently, pig iron made with charcoal still accounted for about a third of the total at the start of the Civil War.

By this time, the cost of transportation had declined enough to permit pig iron made with anthracite to compete with pig iron made with charcoal in places other than the eastern seaboard. But as Hunter noted, more changes were in progress at the close of the ante-bellum years than those in the cost of transport. In a closely related development, the demand for iron was beginning to shift. The plentiful use of castings typical of the 1840s had begun to wane; the demand for rails had begun to rise (Temin, 1964, chs. 1 and 2). The economy was also expanding geographically, and this led in turn to the discovery of new resources. We seek to know which of these developments made the use of coke in western blast furnaces more attractive than the importation of pig iron made with anthracite from eastern Pennsylvania.

Price comparisons again carry the key. The price of pig iron made with coke had to rise relative to the price of other types of pig iron if coke rather than one of the other fuels was to be used. If this price rise occurred because of a change in demand, as Hunter asserted, a price differential should have been maintained between pig iron made with anthracite and pig iron made with coke. The former class of iron was of higher quality than the latter before the Civil War, and there is no evidence that its quality deteriorated after that time. Only if the quality of pig iron made with coke improved until it was equal to that of pig iron made with anthracite could their prices become equal.

Yet this is precisely what happened. The extant price data for the Civil War years are obscure due to the paucity of comparable sources and the rapid changes in prices in these years. When prices do become available on a comparable and sustained basis, in the early 1870s, they show that pig iron made with coke and pig iron made with anthracite sold for the same price. In fact, the popular method of quoting prices was to refer to pig iron by grade, including iron made with anthracite and with coke in a single classification.[1] The quality of pig iron made with coke had improved to the point where this type of iron was interchangeable with pig iron made with anthracite. There had been a change in the supply curve of pig iron made with coke, in other words – a change that enabled producers of pig iron with this fuel to supply more value to consumers with t he same expenditure.

1. The first regular market report appeared in the *Engineering and Mining Journal*, 1874, p. 55. It assumed its complete form a month later, in ibid., 1874, pp. 134–5, when the two types of mineral fuel pig iron were classified together in the Pittsburgh report. An earlier report of sales appeared in *Iron Age*, 1873, p. 19. It showed the prices of the two types of mineral fuel pig iron at about the same level.

To what can we attribute this change in supply? The exploitation of high-quality coking coal is the obvious candidate. The modern observer of the ante-bellum iron industry cannot help but be struck by the absence of one of the most famous names of the American iron industry. I refer to Connellsville, the home of the metallurgical coke known by its name and famous for both its hardy physical structure and its relative chemical purity. This variety of coking coal was used in the expansion of coke pig iron production after the Civil War; its freedom from sulphur may be taken as the cause of the improvement in the quality of pig iron made with coke. The Connellsville coal region was initially discovered in the 1840s, but its exploitation is usually dated from 1859 – the date of the construction of the first blast furnace in Pittsburgh designed specifically to use this coal (Eavenson, 1939, pp. 165–76; Swank, 1892, pp. 476–7). This deposit was newly discovered in the boom of the 1840s and did not become known in the few years of the boom. The following decade did not witness a rise of pig iron production above its previous peak, and the inducement to use new resources was not strong. Only when the production of pig iron began to rise again at the end of the ante-bellum era was the new source of coke exploited.

While this provides an explanation for the improvement in the quality of pig iron made with coke, it is not clear that the discovery of new coal fields can be made a simple function of the expansion of the economy. This was certainly a potent factor, but it is also possible that ironmasters increased their search for new coal on the basis of increasing technical knowledge. Quite possibly, western ironmasters only came to realize that poor coal was the source of their difficulties in the course of the 1850s, their lack of exploration previously being determined as much by ignorance of what to look for as ignorance of where to look. We find an awareness of the difficulties of using sulphurous coal in these years, but we also find early writers criticized by later ones for their over-optimism on the usefulness of many coal deposits (Lesley, 1856, p. 29; Overman, 1854, p. 130). The process by which Connellsville coke was brought into use, therefore, retains some of its mystery even now.

The changing supply curve for pig iron made with coke thus explains the increased use of coke in this country. When a supply of bituminous coal became available that would produce an iron equal in quality to that made with anthracite, the relative price of coke pig iron rose and its production was encouraged. The growth of the transportation network, therefore, encouraged the universal use of mineral fuel in the United States iron industry through promoting geographical exploration and the exploitation of new resources, rather than by improving the competitive position of anthracite pig iron, as might have been anticipated in 1850. In addition, technical developments after the Civil War reduced the cost of using coke relative to the cost of using anthracite, and coke came to be the universal fuel for American blast furnaces (Temin, 1964, chs. 7, 9 and Appendix B).

But what about the change in demand noticed by Hunter? If the preceding argument is correct, the switch to coke would have taken place whether or not there was a change in the character of the demand for pig iron around 1860. Such a change, however, could have helped the transition, and we may ask whether or not it did.

The most important component of demand from this point of view is the demand for rails, as the quality of iron used was not a major concern in this area. Robert Fogel (1963) has estimated that the proportion of pig iron used for the production of rails in the ante-bellum era was less than 10 per cent. This low figure, coupled with the absence of a rising trend near 1860, suggests that the effects of changing demand may be neglected. Albert Fishlow has revised Fogel's estimates, and his data show that the proportion of pig iron used for rails was rising in the 1850s, reaching a level of about 20 per cent by the end of the decade (Fishlow, 1963, ch. 3).[2] This estimate gives greater scope to the influence of demand, but even it does not say anything about the actual effect.

The problem is that even if the production of rails did use an increasing proportion of the pig iron produced in the late 1850s,

2. Fishlow also estimates the non-rail demand for iron coming from the railroads. While the magnitude of the total railroad demand for iron is important for other purposes, it is not relevant to a discussion of technological change in the iron industry. Many of the products used by the railroads did not differ from products used elsewhere in the nature of their demand for iron, and the volume of their production may be neglected.

this could have been the effect of an increased supply of low-cost pig iron – made with coke – as easily as it could have been the cause of the increased production of pig iron with coke. The resolution of this problem requires the specification of the relevant demand and supply functions. Among the demand functions that need to be specified is the demand for rails, including the demand for rails to be used to replace worn-out rails, and the nature of this demand (together with the reciprocal supply of scrap coming from the worn-out rails themselves) is precisely the point on which Fogel and Fishlow differ. The problem does not appear to be easily soluble, and at this point we cannot assess the extent to which changes in the composition of demand helped the transition from charcoal to coke in the blast furnaces of the American iron industry. What the preceding argument has shown is that this help was not needed to effect the transition observed.

Can we conclude, then, that the railroad was not important for the American iron industry? The answer is no. This discussion has been concerned with changes in the production of pig iron and has discussed that part of the iron industry that made this material. Rails, however, were made from wrought iron, and the effects of rail production were to be found in the branch of the industry that made wrought iron (and later steel) from pig iron rather than in the branch that made pig iron from ore. The proportion of wrought iron produced used for rails was much higher than the proportion of pig iron so used, and the production of rails had a direct impact on the activities of rolling mills – in contrast to the indirect effect that has been claimed for blast furnaces. The production of rails led to the growth of integrated iron works and the use of the three-high mill in the ante-bellum era and to the exploitation of the Bessemer process in the years following the Civil War. The use of the Bessemer process was an important change for the iron industry: among its effects were several important changes in the production of pig iron after 1870 that may reasonably be attributed to the demand for rails.[3]

References

ASHTON, T. S. (1924), *Iron and Steel in the Industrial Revolution*, Manchester University Press.

3. 'Hard driving' may be included in this category. See Temin (1964), pt. 2.

EAVENSON, H. N. (1939), 'The early history of the Pittsburgh coal bed', *Western Pennsylvania Hist. Mag.*, September.

FISHLOW, A. (1963), 'The economic contribution of American railroads before the Civil War', Unpublished Doctoral Dissertation, Harvard University.

FOGEL, R. W. (1963), 'Railroads and American economic growth: essays in econometric history', Unpublished Doctoral Dissertation, Johns Hopkins University.

HABAKKUK, H. J. (1962), *American and British Technology in the Nineteenth Century*, Cambridge University Press.

HUNTER, L. C. (1928), 'A study of the iron industry of Pittsburgh before 1860', Unpublished Doctoral Dissertation, Harvard University.

HUNTER, L. C. (1929), 'The influence of the market upon technique in the iron industry in western Pennsylvania up to 1860', *J. econ. bus. Hist.*, February, pp. 241–81.

LESLEY, J. P. (1856), *A Manual of Coal and its Topography*, J. B. Lippincoth.

OVERMAN, F. (1854), *The Manufacture of Iron*, 3rd edn, H. C. Baird.

SWANK, J. M. (1892), *History of the Manufacture of Iron in All Ages*, 2nd edn, Franklin.

TEMIN, P. (1964), 'A history of the American iron and steel industry from 1830 to 1900', Unpublished Doctoral Dissertation, MIT.

14 E. Mansfield

Technical Change and the Rate of Imitation[1]

E. Mansfield, 'Technical change and the rate of imitation',
Econometrica, October 1961, pp. 741–66.

1 Introduction

Once an innovation is introduced by one firm, how soon do others
in the industry come to use it? What factors determine how rapidly
they follow? The importance of these questions has long been
recognized. For example, Mason, Clark, Dunlop and others
pointed out in 1941 the need for studies to investigate how
rapidly '. . . an innovation spreads from enterprise to enterprise'
(National Bureau of Economic Research, 1943). Eighteen years
later, the need is perhaps even more obvious than it was then.

This paper summarizes some theoretical and empirical findings
regarding the rate at which firms follow an innovator. A simple
model is presented to help explain differences among innovations
in the rate of imitation. The model is then tested against data
showing how rapidly firms in four quite different industries
came to use twelve innovations. The results – though by no
means free of difficulties – seem quite encouraging, and should
help to fill a significant gap in the literature concerning technical
change.

The plan of the paper is as follows. Section 2 lists the twelve

1. The work on which this report is based was supported initially by
research funds of the Graduate School of Industrial Administration and
then by a contract with the National Science Foundation. It is part of a
broader study of research and technological change that I am conducting
under this contract. The paper has benefited from discussions with a number
of my colleagues at Carnegie Institute of Technology, particularly A.
Meltzer, F. Modigliani, J. Muth and H. Wein (now at Michigan State
University). A version of it was presented at the December 1959 meeting of
the Econometric Society. My thanks also go to the many people in industry
who provided data and granted me interviews. Without their co-operation,
the work could not have been carried out.

innovations and shows how rapidly firms followed the innovator in each case. Sections 3 to 5 present and test a deterministic model constructed to explain the observed differences among these rates of imitation. A stochastic version is discussed in section 6. Some of the study's limitations are pointed out in section 7 and its conclusions are summarized in section 8.

2 Rates of imitation

This section describes how rapidly the use of twelve innovations spread from enterprise to enterprise in four industries – bituminous coal, iron and steel, brewing and railroads. The innovations are the shuttle car, trackless mobile loader and continuous mining machine (in bituminous coal); the by-product coke oven, continuous wide strip mill and continuous annealing line for tin plate (in iron and steel); the pallet-loading machine, tin container and high-speed bottle filler (in brewing); and the diesel locomotive, centralized traffic control and car retarders (in railroads).

These innovations were chosen because of their outstanding importance and because it seemed likely that adequate data could be obtained for them. Excluding the tin container, all were types of heavy equipment permitting a substantial reduction in costs. The most recent of these innovations occurred after the Second World War; the earliest was introduced before 1900. In practically all cases, the bulk of the development work was carried out by equipment manufacturers and patents did not impede the imitation process.[2]

Figure 1 shows the percentage of major firms that had introduced each of these innovations at various points in time. To

2. For descriptions of these innovations (and in some cases historical material), see Association of Iron and Steel Engineers (1941) for the strip mill, Camp and Francis (1940) for continuous annealing and the by-product coke oven, annual issues of *Coal Age* on mechanization and American Mining Congress for the innovations in coal, *Fortune* (1936) for the tin container, *American Brewer* (1954) for pallet loaders and other issues for high speed fillers, Jewkes, Sawers and Stillerman (1958) for the diesel locomotive, Mansfield and Wein (1958, pp. 292–3) for car retarders, and Union Switch and Signal Co. (1931) for centralized traffic control. Actually, continuous annealing was often no cheaper than previous techniques, but it was required by customer demands.

avoid misunderstanding, note three things regarding the data in Figure 1.

1. Because of difficulties in obtaining information concerning smaller firms and because in some cases they could not use the innovation in any event, only firms exceeding a certain size (given in the Appendix) are included.[3]

2. The percentage of firms having introduced an innovation, regardless of the scale on which they did so, is given. The possible objections to this are largely removed by the fact that these innovations had to be introduced on a fairly large scale. By using them at all, firms made a relatively heavy financial commitment.[4]

3. In a given industry, most of the firms included in the case of one innovation are also included for the others. Thus the data for each of the innovations are quite comparable in this regard.

Two conclusions regarding the rate of imitation emerge from Figure 1. First, the diffusion of a new technique is generally a rather slow process. Measuring from the date of the first success-

3. For the innovations in the steel industry, we imposed the particular size limits cited in the Appendix because, according to interviews, it seemed very unlikely that firms smaller than this would have been able to use them. For the innovations in the coal and brewing industries, there were no adequate published data concerning the dates when particular firms first introduced them, and we had to get the information directly from the firms. It seemed likely that 'non-response' would be a considerable problem among the smaller firms. This consideration, as well as the fact that the smallest firms often could not use them, led us to impose the rather arbitrary lower limits on size shown in the Appendix. For the railroad innovations, we took firms large enough to use car retarders and centralized traffic control and, to ensure comparability, the same size limits were used for the diesel locomotive.

4. The only alternative would be to take the date when a firm first used the innovation to produce some specified percentage of its output. In almost every case, such data were not published and it would have been extremely difficult, if not impossible, to obtain them from the firms. To install a strip mill, by-product coke ovens, continuous annealing or car retarders, a firm had to invest many millions of dollars. Even for shuttle cars, trackless mobile loaders, canning equipment and pallet loaders, the investment (although less than $100,000 usually) was by no means trivial in these industries.

ful commercial application,[5] it took twenty years or more for all the major firms to install centralized traffic control, car retarders, by-product coke ovens and continuous annealing. Only in the case of the pallet-loading machine, tin container and continuous mining machine did it take ten years or less for all the major firms to install them.

Second, the rate of imitation varies widely. Although it sometimes took decades for firms to install a new technique, in other cases they followed the innovator very quickly. For example, it took about fifteen years for half of the major pig-iron producers to use the by-product coke oven, but only about three years for half of the major coal producers to use the continuous mining machine. The number of years elapsing before half the firms had introduced an innovation varied from 0·9 to fifteen, the average being 7·8.

3 A deterministic model

Why were these firms so slow to install some innovations and so quick to install others? What factors seem to govern the rate of imitation? In this section, we construct a simple deterministic model to explain the results in Figure 1. In sections 4 and 5, we test this model and see the effects of introducing some additional variables into it.

The following notation is used. Let n_{ij} be the total number of firms on which the results in Figure 1 for the jth innovation in the ith industry are based[6] ($j = 1,2,3$; $i = 1,2,3,4$). Let $m_{ij}(t)$ be the number of these firms having introduced this innovation at time t, π_{ij} be the profitability of installing this innovation relative to that of alternative investments, and S_{ij} be the investment required to install this innovation as a per cent of the average total assets of these firms. More precise definitions of π_{ij} and S_{ij} are provided in section 4. Let $\lambda_{ij}(t)$ be the proportion of

5. Note that we measure how quickly other firms followed the one that first *successfully* applied the technique. Others may have tried roughly similar things before but failed. By a successful application, we mean one where the equipment was used commercially for years, not installed and quickly withdrawn.

6. That is, the total number of firms for which we have data. See Table 1 for the n_{ij}.

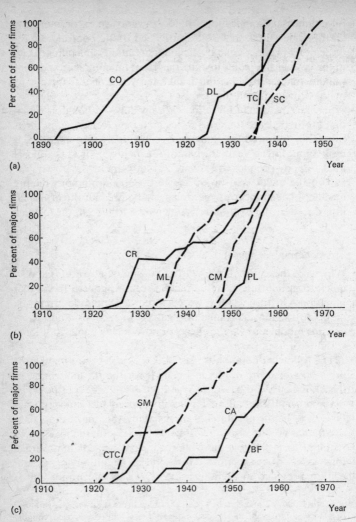

Figure 1 Growth in the percentage of major firms that introduced twelve innovations, bituminous coal, iron and steel, brewing and railroad industries, 1890–1958
(a) By-product coke oven (CO), diesel locomotive (DL), tin container (TC), and shuttle car (SC).

(b) Car retarder (CR), trackless mobile loader (ML), continuous mining machine (CM), and pallet-loading machine.
(c) Continuous wide-strip mill (SM), centralized traffic control (CTC), continuous annealing (CA), and high-speed bottle filler (BF).
Note: For all but the by-product coke oven and tin container, the percentages given are for every two years from the year of initial introduction. Zero is arbitrarily set at two years prior to the initial introduction in these figures (but not in the analysis). The length of the interval for the by-product coke oven is about six years and for the tin container it is six months. The innovations are grouped into the three sets shown above to make it easier to distinguish between the various growth curves.

'hold-outs' (firms not using this innovation) at time t that introduce it by time $t+1$, i.e.

$$\lambda_{ij}(t) = \frac{m_{ij}(t+1) - m_{ij}(t)}{n_{ij} - m_{ij}(t)}. \qquad 1$$

Our basic hypothesis can be stated quite simply. We assume that the proportion of 'hold-outs' at time t that introduce the innovation by time $t+1$ is a function of (1) the proportion of firms that already introduced it by time t, (2) the profitability of installing it, (3) the size of the investment required to install it, and (4) other unspecified variables. Allowing the function to vary among industries, we have

$$\lambda_{ij}(t) = f_i\left(\frac{m_{ij}(t)}{n_{ij}}, \pi_{ij}, S_{ij}, \dots\right). \qquad 2$$

In the following few paragraphs, we take up the presumed effects of variation in $m_{ij}(t)/n_{ij}$, π_{ij} and S_{ij} on $\lambda_{ij}(t)$ and the reasons for interindustry differences in the function.

First, one would expect that increases in the proportion of firms already using an innovation would increase $\lambda_{ij}(t)$. As more information and experience accumulate, it becomes less risky to begin using it.[7] Competitive pressures mount and 'bandwagon'

7. The profitability of installing each of these innovations was viewed at first with considerable uncertainty. For example, there was great uncertainty about maintenance costs for diesel locomotives, 'down-time' for continuous mining machines, the safety of centralized traffic control, and the useful life of by-product coke ovens. The perceived risks seldom disappeared after only a few firms had introduced them; in some cases according to interviews, it took many years. This helps to account for the fact (noted above) that the imitation process generally went on rather slowly.

effects occur. Where the profitability of using the innovation is very difficult to estimate, the mere fact that a large proportion of its competitors have introduced it may prompt a firm to consider it more favorably. Both interviews with executives in the four industries and the data in Figure 1 indicate that this is the case.[8]

Second, the profitability of installing the innovation would also be expected to have an important influence on $\lambda_{ij}(t)$. The more profitable this investment is relative to others that are available, the greater is the chance that a firm's estimate of the profitability will be high enough to compensate for whatever risks are involved and that it will seem worthwhile to install the new technique rather than to wait. As the difference between the profitability of this investment and that of others widens, firms tend to note and respond to the difference more quickly. Both the

---

This also suggests the potential importance of two variables not recognized explicitly in **2**: (a) the extent of the initial uncertainty concerning the profitability of an innovation (depending in part on the extent of prior field testing by manufacturers), and (b) the rate at which this uncertainty declined. For some types of innovations, a few installations and a relatively short period of use can cut the risks to little or nothing. For other types, installations under many sorts of conditions and a long period of use (to be sure of useful life, maintenance costs, etc.) are required. See section 7 for further discussion.

Note that all these innovations are new processes or (in the case of the tin container) a packaging innovation. This model would not be applicable to some innovations, like an entirely new product, where, as more firms produce it, it becomes less profitable for others to do so. There is no evidence of significant decreases of this sort in these cases. (E.g. as more firms introduced diesel locomotives, this did not make it less profitable for others to do so.) Moreover, the model presumes that π_{ij} is appreciably greater than unity (which is the case here). Finally, the data support the hypothesis in the text (see note 8).

8. Beginning with the date when $m_{ij}(t) = 1$, we computed $\lambda_{ij}(t)$ and $m_{ij}(t)/n_{ij}$, using the intervals in the footnote to Figure 1 as time units and stopping when $m_{ij}(t) = n_{ij}$. Then we calculated the correlation between $\lambda_{ij}(t)$ and $m_{ij}(t)/n_{ij}$. The correlation coefficients were 0·77 (continuous mining machine), 0·85 (by-product coke oven), 0·49 (tin container), 0·65 (centralized traffic control), 0·85 (continuous wide strip mill), 0·66 (diesel locomotive), 0·52 (car retarders), 0·55 (bottle fillers), 0·96 (pallet loaders), 0·81 (continuous annealing), 0·46 (trackless mobile loader) and 0·94 (shuttle car). Using a one-tailed test (which is appropriate here), all coefficients but those for the continuous mining machine, bottle filler, and trackless mobile loader are significant (0·05 level). All are positive.

interviews and the few other studies regarding the rate of imitation suggest that this is so (Griliches, 1957).[9]

Third, for equally profitable innovations, $\lambda_{ij}(t)$ should tend to be smaller for those requiring relatively large investments. One would expect this on the grounds that firms tend to be more cautious before committing themselves to such projects and that they often have more difficulty in financing them. According to the interviews, this factor is often important.

Finally for equally profitable innovations requiring the same investment, $\lambda_{ij}(t)$ is likely to vary among industries. It might be higher in one industry than in another because firms in the former industry have less aversion to risk, because markets are more keenly competitive, because the attitude of the labor force toward innovation is more favorable, or because the industry is healthier financially. Casual observation suggests that such inter-industry differences may have a significant effect on $\lambda_{ij}(t)$.

Returning to equation 2, we act as if the number of firms having introduced an innovation can vary continuously rather than assume only integer values, and we assume that $\lambda_{ij}(t)$ can be approximated adequately within the relevant range by a Taylor's expansion that drops third and higher order terms. Assuming that the coefficient of $(m_{ij}(t)/n_{ij})^2$ in this expansion is zero (and the data in Figure 1 generally support this),[10] we have

---

Of course, others have stated similar hypotheses before. E.g. Schumpeter (1939) noted that 'accumulating experience and vanishing obstacles' smooth the way for imitators, and Coleman, Katz and Menzel (1957), noted a 'snow-ball' effect. For what it is worth, almost all the executives we interviewed considered this effect to be present. The number of installations of the innovation or the number of firms using it might have been used rather than the proportion of firms, but the latter seems to work quite well.

9. Another recent study of the rate of imitation is found in Yance (1957). These papers focus on only one innovation (hybrid corn and the diesel locomotive). Of course, the interfirm variation in profitability, as well as the average, could influence $\lambda_{ij}(t)$.

10. To test this assumption for these innovations, we used $(m_{ij}(t)/n_{ij})^2$ as an additional independent variable in the regression described in note 8 and used the customary analysis of variance to determine whether this resulted in a significant increase in the explained variation. For all but continuous annealing, car retarders and the diesel locomotive, it does not (and in these cases, the increase is often barely significant). Hence, in most cases, there is no evidence that this coefficient is non-zero.

$$\lambda_{ij}(t) = a_{i1} + a_{i2}\frac{m_{ij}(t)}{n_{ij}} + a_{i3}\pi_{ij} + a_{i4}S_{ij} + a_{i5}\pi_{ij}\frac{m_{ij}(t)}{n_{ij}} +$$

$$+ a_{i6}S_{ij}\frac{m_{ij}(t)}{n_{ij}} + a_{i7}\pi_{ij}S_{ij} + a_{i8}\pi_{ij}^2 + a_{i9}S_{ij}^2 + ..., \quad \mathbf{3}$$

where additional terms contain the unspecified variables in **2**. Thus,

$$m_{ij}(t+1) - m_{ij}(t)$$
$$= \{n_{ij} - m_{ij}(t)\}\left[a_{i1} + a_{i2}\frac{m_{ij}(t)}{n_{ij}} + ... + a_{i9}S_{ij}^2 + ... \right]. \quad \mathbf{4}$$

Assuming that time is measured in fairly small units, we can use as an approximation the corresponding differential equation[11]

$$\frac{dm_{ij}(t)}{dt} = \{n_{ij} - m_{ij}(t)\}\left[Q_{ij} + \varphi_{ij}\frac{m_{ij}(t)}{n_{ij}} \right], \quad \mathbf{5}$$

the solution of which is

$$m_{ij}(t) = \frac{n_{ij}[\exp\{l_{ij} + (Q_{ij} + \varphi_{ij})t\} - (Q_{ij}/Q_{ij})]}{1 + \exp\{l_{ij} + (Q_{ij} + \varphi_{ij})t\}}, \quad \mathbf{6}$$

where l_{ij} is a constant of integration, Q_{ij} is the sum of all terms in **3** not containing $m_{ij}(t)/n_{ij}$, and

$$\varphi_{ij} = a_{i2} + a_{i5}\pi_{ij} + a_{i6}S_{ij} + ... \quad \mathbf{7}$$

Of course, φ_{ij} is the coefficient of $m_{ij}(t)/n_{ij}$ in **3**.

To get any further, we must impose additional constraints on the way $m_{ij}(t)$ can vary over time. One simple condition we can impose is that, as we go backward in time, the number of firms having introduced the innovation must tend to zero,[12] i.e.

$$\lim_{t \to -\infty} m_{ij}(t) = 0. \quad \mathbf{8}$$

11. This is like some approximations commonly used in capital theory. E.g. one often replaces equations like $x(t+1) - x(t) = rx(t)$ with $\dot{x}(t) = rx(t)$.

12. Of course, other conditions could be imposed in addition or instead. For example in section 6, we take as given the date when a particular number of firms had installed an innovation and we force $m_{ij}(t)$ to equal that number at that date. But **8** is all we need for present purposes. Note that it implies that Q_{ij} is zero. The data described in note 8 are consistent with this, but even if Q_{ij} were non-zero but small (and it certainly could not be large), **9** should be a reasonably good approximation. Note too that, if the model holds, $\varphi_{ij} > 0$.

Using this condition, it follows that

$$m_{ij}(t) = n_{ij}[1 + \exp\{-(l_{ij} + \varphi_{ij}t)\}]^{-1}. \qquad \textbf{9}$$

Thus, the growth over time in the number of firms having introduced an innovation should conform to a logistic function, an S-shaped growth curve frequently encountered in biology and the social sciences.[13]

If **9** is correct, it can be shown that the rate of imitation is governed by only one parameter – φ_{ij}.[14] Assuming that the sum of the unspecified terms in **7** is uncorrelated with π_{ij} and S_{ij} and that it can be treated as a random error term,

$$\varphi_{ij} = b_i + a_{i5}\pi_{ij} + a_{i6}S_{ij} + z_{ij}, \qquad \textbf{10}$$

where b_i equals a_{i2} plus the expected value of this sum and z_{ij} is a random variable with zero expected value. Hence, the expected value of φ_{ij} in a particular industry is a linear function of π_{ij} and S_{ij}.

To sum up, the model leads to the following two predictions. First, the number of firms having introduced an innovation, if plotted against time, should approximate a logistic function. Second, the rate of imitation in a particular industry should be higher for more profitable innovations and innovations requiring relatively small investments. More precisely, φ_{ij}, a measure of the rate of imitation, should be linearly related to π_{ij} and S_{ij}.

13. Note that things are simplified here by the fact that all firms we consider eventually introduced these innovations. (A few went out of business first, but not because of the appearance of the innovation. See the Appendix.) Had this not been the case, it would have been necessary either to provide a mechanism explaining the proportion that did not do so or to take it as given. Of course, by taking only the larger firms, we made sure that all could use these innovations. The smaller firms (that were not potential users) are omitted. For the high-speed bottle filler, we make the reasonable assumption that all the major firms will ultimately introduce it. Note that the argument here is different from that generally used in biology to arrive at the logistic function and that it explicitly includes variables affecting its shape.

14. It seems reasonable to take, as a measure of the rate of imitation, the time span between **1** the date when 20 per cent (say) of the firms had introduced an innovation, and **2** the date when 80 per cent (say) had done so. According to the model, this time span equals $2 \cdot 77 \ \varphi_{ij}^{-1}$ and is therefore independent of l_{ij}. If, rather than 20 and 80, we take P_1 and P_2, it can be shown that the time span equals
$$\varphi_{ij}^{-} \ \ln\{(1-P_1)P_2/P_1(1-P_2)\}.$$

Table 1 Parameters, Estimates and Root-Mean-Square Errors: Deterministic and Stochastic Models

Innovation	Parameters[a]						
	n_{ij}	π_{ij}	S_{ij}	d_{ij}	g_{ij}	t^{*}_{ij}	δ_{ij}
Diesel locomotive	25	1·59	0·015	35	1·00	1925	1
Centralized traffic control	24	1·48	0·024	0	1·50	1926	1
Car retarders	25	1·25	0·785	0	1·50	1924	0
Continuous wide strip mill	12	1·87	4·908	30	4·50	1924	0
By-product coke oven	12	1·47	2·083	10	4·00	1894	1
Continuous annealing	9	1·25	0·554	9	4·25	1936	1
Shuttle car	15	1·74	0·013	9	1·25	1937	0
Trackless mobile loader	15	1·65	0·019	6	2·50	1934	1
Continuous mining machine	17	2·00	0·301	8	1·00	1947	1
Tin container	22	5·07	0·267	0	6·50	1935	1
High-speed bottle filler	16	1·20	0·575	10	2·25	1951	1
Pallet-loading machine	19	1·67	0·115	0	2·25	1948	1

Innovation	Estimates and root mean square errors[b]					
	$\hat{\imath}_{ij}$	$\hat{\varphi}_{ij}$	r_{ij}	$\hat{\theta}_{ij}$	Error (det.)	Error (st.)
Diesel locomotive	— 6·64	0·20	0·89	0·30	2·13	5·63
Centralized traffic control	— 7·13	0·19	0·94	0·24	1·52	3·44
Car retarders	— 3·95	0·11	0·90	0·17	2·08	5·02
Continuous wide strip mill	—10·47	0·34	0·95	0·42	0·83	0·90
By-product coke oven	— 1·47	0·17	0·98	0·18	0·16	0·84
Continuous annealing	— 8·51	0·17	0·93	0·22	0·74	1·42
Shuttle car	—13·48	0·32	0·95	0·45	0·86	2·03
Trackless mobile loader	—13·03	0·32	0·97	0·39	0·71	1·66
Continuous mining machine	—24·96	0·49	0·98	0·59	0·81	2·22
Tin container	—84·35	2·40	0·96	2·64	1·34	3·00
High-speed bottle filler	—20·58	0·36	0·97	0·40	0·56	0·95
Pallet-loading machine	—29·07	0·55	0·97	0·63	1·10	1·58

(a) For definitions of these parameters, see sections 3 (n_{ij}), 4 (π_{ij} and S'_{ij}) and 5 (d_{ij}, g_{ij}, t^{*}_{ij} and δ_{ij}).
(b) For definitions of these measures, see sections 4 ($\hat{\imath}_{ij}$, $\hat{\varphi}_{ij}$, r_{ij} and error (det.)) and 6 ($\hat{\theta}_{ij}$ and error (st.)).
Source: See the Appendix and notes, 6, 15, 16, 17, 20, 22, 25, 29, 32, 34.

4 Tests of the model

We test this model in two steps: (a) by estimating φ_{ij} and l_{ij} and determining how well **9** fits the data, and (b) by seeing whether the expected value of φ_{ij} seems to be a linear function of π_{ij} and S_{ij}. The results of these tests suggest that the model can explain the results in Figure 1 quite well.

To carry out the first step, note that, if the model is correct, it follows from **9** that

$$\ln\left[\frac{m_{ij}(t)}{n_{ij}-m_{ij}(t)}\right] = l_{ij}+\varphi_{ij}t. \qquad\qquad \textbf{11}$$

Measuring time in years (from 1900), treating this as a regression equation, and using least squares, after properly weighting the observations (Berksen, 1953), we derive estimates of l_{ij} and φ_{ij} (Table 1).[15] To see how well **9** can represent the data, we insert these estimates into **9** and compare the calculated increase over time in the number of firms having introduced each innovation with the actual increase.

Judging merely by a visual comparison, the calculated growth curves generally provide reasonably good approximations to the actual ones. However, the 'fit' is not uniformly good. For the railroad innovations, there is evidence of serial correlation among the residuals, and the approximations are less satisfactory than for the other innovations. Table 1 contains two rough measures of 'goodness-of-fit': the root-mean-square deviation of actual from the computed number of firms having introduced the innovation[16] and the coefficient of correlation between

15. When $\ln \{m_{ij}(t)/n_{ij}-m_{ij}(t)\}$ was infinite, the observation was omitted. The observations were one year apart for the continuous wide strip mill, continuous mining machine and centralized traffic control, six years apart for the by-product coke oven, one month apart for the tin containers, and two years apart for the others. For the high-speed bottle filler, we used all the data available thus far, but the imitation process is not yet complete (Figure 1).

16. For each innovation, the difference between the computed and actual values of $m_{ij}(t)$ was obtained for $t = t_{ij}^{*}$, $t_{ij}^{*}+1$, ...; t_{ij}^{**}, where time is measured in the units described in note 15, t_{ij}^{*} is the first date when $m_{ij}(t) = 1$, and t_{ij}^{**} is the first date when $m_{ij}(t) = n_{ij}$. Then the root-mean-square of these differences was obtained. Note that this is a measure of absolute, not relative, error, and when comparing the results for different innovations, take account of differences in n_{ij}.

$\ln\{m_{ij}(t)n_{ij}-m_{ij}(t)\}$ and t.[17] They seem to bear out the general impression that **9** represents the data for most of the innovations quite well.[18]

To carry out the second step, we assume that a_{i5} and a_{i6} do not vary among industries.[19] Thus **10** becomes

$$\varphi_{ij} = b_i + a_5\pi_{ij} + a_6 S_{ij} + z_{ij}, \qquad 12$$

and, if we assume that errors in the estimates of φ_{ij} are uncorrelated with π_{ij} and S_{ij}, we have

$$\hat{\varphi}_{ij} = b_i + a_5\pi_{ij} + a_6 S_{ij} + z'_{ij}, \qquad 13$$

where $\hat{\varphi}_{ij}$ is the estimate of φ_{ij} derived above. Assuming that z'_{ij} is distributed normally with constant variance, standard tests used for linear hypotheses can be applied to determine whether a_5 and a_6 are non zero.

Before discussing the results of these tests, we must describe how π_{ij} and S_{ij} are measured. We obtained from as many of the firms as possible estimates of the pay-out period for the initial installation of the innovation and the pay-out period required from investments. They were derived primarily from correspondence, but published materials were used where possible.[20]

17. This coefficient is labeled r_{ij} in Table 1. Note that it is based on weighted observations (the weights being those suggested by Berkson, 1953).

18. Of course, the logistic function is not the only one that might represent the data fairly adequately. (E.g. the normal cumulative distribution function would probably do as well.) And, since the actual curve has to be roughly S-shaped, it is not surprising that it fits reasonably well. There seemed to be little point in attempting any formal 'goodness-of-fit' tests here. The number of firms considered in each case is quite small, and the tests would not be very powerful. All that we conclude from Table 1 is that **9** provides a reasonably adequate description of the data and hence that φ_{ij} is generally reliable as a measure of the rate of imitation.

19. Because of the small number of observations, we are forced to make this assumption. If data were available for more innovations, it would not be necessary. Of course, if interindustry differences in these coefficients are statistically independent of π_{ij} and S_{ij}, a_5 and a_6 can be regarded as averages and there is no real trouble.

20. A letter was sent to each firm asking (a) how long it took for the initial investment in the innovation to pay for itself, and (b) what the required pay-out period was during the relevant period. (For a definition of the pay-out period, see Swalm (1958). Both the realized and required pay-out periods are before taxes.) Replies were received from 50 per cent of the rail roads and

Then the average pay-out period required by the firms (during the relevant period) to justify investments divided by the average pay-out period for the innovation was used as a measure of π_{ij}.[21] To measure S_{ij}, we used the average initial investment in the innovation as a percentage of the average total assets of the firms (during the relevant period).[22] Table 1 contains the results.

Using these rather crude data, we obtained least squares

coal producers and 20 per cent of the steel companies and breweries. The estimates for the railroad innovations are probably most accurate since the firms referred us to fairly reliable published studies (see the Appendix). The data for the coal innovations are probably quite good too. The estimates for the steel innovations and the tin container are probably less accurate because they occurred so long ago and because fewer firms provided data. Despite the fact that the averages are probably more accurate than the individual figures and that the results were checked with other sources in interviews (see the Appendix), the estimates of the π_{ij} are fairly rough.

21. For relatively long-lived investments, the reciprocal of the pay-out period is a fairly adequate approximation to the rate of return (Gordon, 1955; Swalm 1958). Hence this measure of π_{ij} is approximately equal to the average rate of return derived (ex post) from the innovation divided by the average rate of return firms required (ex ante) to justify investments. Of course, it would be more appropriate for firms to recognize that introducing it next year, the year after, etc. are the alternatives to introducing it now and to make an analysis like Terborgh's (1949). But as he points out, most firms seem to make these decisions on the basis of the 'rate of return' or 'pay-out period'; and thus it seems reasonable to use such measures in a study (like this) where we try to explain behavior, not prescribe it.

Even so, this measure is only an approximation. (a) The rate of return from the investment in an innovation is measured ex post, not ex ante. (b) The average rate of return that could have been realized by the 'hold-outs' at a particular point in time probably differed from the average rate of return actually realized by all firms. Our estimates are based implicitly on the factor prices and age of old equipment (where replacement occurred) that prevailed when the innovation was installed, and these (and other) factors, as well as the innovation itself, do not remain unchanged. According to the interviews, the average return that could have been realized probably varied over time about an average that was highly correlated with our estimate, and hence the latter provides a fair indication of the level. But the parameters describing the temporal variation about this level are among the 'other' variables in (b). See section 7 and note 39.

22. For the steel, coal and brewing industries, we obtained the total assets of as many of these firms as possible (during the relevant period) from *Moody's*. For the railroads, the 1936 reproduction costs in Klein (1951) were used for the firms. Estimates of the approximate investment required to install each innovation were obtained primarily from the inter-

estimates of the parameters in **13** and tested whether they were zero. The resulting equation is

$$\hat{\varphi}_{ij} = \left\{ \begin{array}{c} -0.29 \\ -0.57 \\ -0.52 \\ -0.59 \end{array} \right\} + 0.530\,\pi_{ij} - 0.027\,S_{ij} \quad (r = 0.997), \qquad \textbf{14}$$
$$ (0.015) (0.014)$$

where the top figure in the brackets pertains to the brewing industry, the next to coal, the following to steel, and the bottom figure pertains to the railroads. The coefficients of π_{ij} and S_{ij} have the expected signs (indicating that increases in π_{ij} and decreases in S_{ij} increase the rate of imitation), and both differ significantly from zero. As expected, there are significant inter-industry differences, the rate of imitation (for given π_{ij} and S_{ij}) being particularly high in brewing. These differences seem to be broadly consistent with the hypothesis often advanced that the rate of imitation is higher in more competitive industries, but there are too few data to warrant any real conclusion on this score.[23]

The scatter diagram in Figure 2 shows that **14** represents the data surprisingly well. When corrected for the relatively few degrees of freedom, the correlation coefficient is 0.997. Of course, one point (the tin container) strongly affects the results. But if that point is omitted, the interindustry differences remain much the same, the coefficients of π_{ij} and S_{ij} keep the same signs (the

views (described in the Appendix), and each was divided by the average total assets of the relevant firms. Since the investment could vary considerably, the results are only approximate.

23. For a more extended statement of this hypothesis, see Robinson (1956). We encounter here the usual difficulty in measuring the 'extent of competition'. But most economists would undoubtedly agree that brewing and coal are more competitive than steel and railroads (and the average value of b_i is larger in the former industries). When six of my colleagues were asked to rank them by 'competitiveness', they put brewing first, coal second, iron and steel third, and railroads fourth. If these ranks are correlated with the estimates of b_i in **14**, the rank correlation coefficient is positive (0.80), but not statistically significant. (The 0.05 probability level is used throughout this paper.) There are too few industries to allow a reasonably powerful test of this hypothesis even if our procedures were refined somewhat, but for most rankings that seem sensible, the data are consistent with the hypothesis (the correlation is positive).

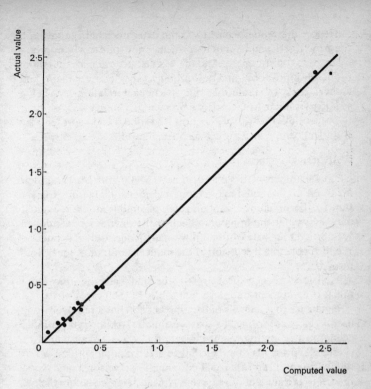

Figure 2 Plot of actual φ_{ij} against that computed from equation **14**, twelve innovations (source: Table 1 and equation **14**)
Note: The difference between an actual and computed value of φ_{ij} is equal to the vertical difference between the point and the 45° line.

latter becoming non-significant), and the correlation coefficient, corrected for degrees of freedom, is still 0·97.[24]

24. Five points should be noted. (a) The tin container is a somewhat different type of innovation from the others and the profitability data for it are probably least reliable (see note 20). Thus, besides its being an extreme point in Figure 2, there are other possible grounds for excluding it, and it is reassuring to note that its exclusion has so little effect on the results. (b) Because of the way it is constructed, the model can only be expected to work if π_{ij} and S_{ij} remain within certain bounds. If π_{ij} is close to one or S_{ij} is very large or both, it is likely to perform poorly. (c) One-tailed tests are used to determine the significance of the coefficients of π_{ij} and S_{ij}. (d) It is note-

Hence, the model seems to fit the data quite well. In general, the growth in the number of users of an innovation can be approximated by a logistic curve. And there is definite evidence that more profitable innovations and ones requiring smaller investments had higher rates of imitation, the particular relationship being strikingly similar to the one we predicted. Though it is no more than a simple first approximation, the model can represent the empirical results in Figure 1 surprisingly well.

5 Additional factors

Of course, other factors may also have been important, and their inclusion in the model may permit a significantly better explanation of the differences among rates of imitation. Moreover, it may show that the apparent effect of the previously mentioned variables on the rate of imitation is partly due to their influence. We turn now to a discussion of the influence of four other factors on $\lambda_{ij}(t)$.

First, one might expect $\lambda_{ij}(t)$ to be smaller if the innovation replaces equipment that is very durable. In such cases, there is a good chance that a firm's old equipment still has a relatively long useful life, according to past estimates. Although rational economic calculation might indicate that replacement would be profitable, firms may be reluctant to scrap equipment that is not fully written off and that will continue to serve for many years. If so, d_{ij} – the number of years that typically elapsed before the old equipment was replaced (before the innovation appeared) – may be one of the excluded variables in 2, and hence $\hat{\varphi}_{ij}$ may be a linear function of π_{ij}, S_{ij} and d_{ij}.[25] If d_{ij} is included in 14, we find that

worthy that $\hat{\varphi}_{ij}$, not a very obvious measure of the rate of imitation, should be linearly related to π_{ij} and S_{ij}, as the model predicts. Using the result in note 14, the number of years elapsing between when 20 per cent and when 80 per cent introduced the innovation can easily be used instead as the dependent variable in 14. The resulting relationship is non-linear. (e) This model may also help to explain the empirical results Griliches (1957) obtained in his excellent study of the regional acceptance of hybrid corn, but for which he provided no formal theoretical underpinning.

25. If we carry d_{ij} from 2 on and make the same assumptions we did with regard to π_{ij} and S_{ij}, it follows that φ_{ij} should be a linear function of π_{ij}, S_{ij} and d_{ij}. This, of course, is also true for the other variables discussed in this section. Estimates of d_{ij} were obtained primarily from the interviews.

$$\hat{\varphi}_{ij} = \begin{Bmatrix} -0\cdot28 \\ -0\cdot56 \\ -0\cdot51 \\ -0\cdot57 \end{Bmatrix} + 0\cdot528\,\pi_{ij} - 0\cdot020\,S_{ij} - 0\cdot0017\,d_{ij} \quad (r = 0\cdot997).\ \mathbf{15}$$
$$\phantom{\hat{\varphi}_{ij} =} (0\cdot015) \quad\ (0\cdot015) \quad\ (0\cdot0014)$$

Though there is some apparent tendency for the rate of imitation to be lower in cases where very durable equipment had to be replaced, it is not statistically significant. (For the values of d_{ij} and the other factors discussed below, see Table 1.)

Second, one might expect $\lambda_{ij}(t)$ to be higher if firms are expanding at a rapid rate. If they are convinced of its superiority, the innovation will be introduced in the new plants built to accommodate the growth in the market. If there is little or no expansion, its introduction must often wait until the firms decide to replace existing equipment. If g_{ij} – the annual rate of growth of industry sales during the period – is included in **14**,

$$\hat{\varphi}_{ij} = \begin{Bmatrix} -0\cdot32 \\ -0\cdot56 \\ -0\cdot53 \\ -0\cdot58 \end{Bmatrix} + 0\cdot484\,\pi_{ij} - 0\cdot025\,S_{ij} + 0\cdot042\,g_{ij} \quad (r = 0\cdot998).\ \mathbf{16}$$
$$\phantom{\hat{\varphi}_{ij} =} (0\cdot032) \quad\ (0\cdot013) \quad\ (0\cdot026)$$

Thus, there is some apparent tendency for the rate of imitation to be higher where output was expanding at a very rapid rate, but it is not statistically significant.[26]

If an innovation was largely a supplement or addition to old plant (like centralized traffic control and car retarders), if it served a different purpose than the old equipment (like canning equipment vs. bottling equipment), and if it displaced only labor (like pallet loaders), we let d_{ij} equal zero. In such cases, there appeared to be no equipment whose durability could influence the decision significantly. Note that the age distribution and number of units of old equipment are also important in determining how long it takes before the first one 'wears out'. It would also have been preferable to have included the profitability of replacing old equipment of various ages rather than just the average figure used. But the available data would not permit this. For further discussion, see Mansfield (1959).

26. Of course, the effect of g_{ij} may depend on whether old equipment must be replaced, how durable it is, the difference between the profitability of replacement and of installing the innovation in new plant, the extent of excess capacity at the beginning of the period, the size of a plant relative to the size of the market, etc. (For some relevant discussion, see Scitovsky, 1956.) The effect of this factor, like d_{ij}, reflects the possible unwillingness of firms to scrap existing equipment. It would have been preferable to have

Third, $\lambda_{ij}(t)$ may have increased over time. This hypothesis has been advanced by economists on numerous occasions (Jerome, 1934, p. xxv; Mack, 1941, p. 295). Presumably, the reasons for such a trend would be the evolution of better communication channels, more sophisticated techniques to evaluate machine replacement, and more favorable attitudes toward technological change. If t_{ij} – the year (less 1900) when the innovation was first introduced[27] – is included in **14**,

$$\hat{\varphi}_{ij} = \begin{Bmatrix} -0\cdot37 \\ -0\cdot64 \\ -0\cdot55 \\ -0\cdot63 \end{Bmatrix} + 0\cdot535\,\pi_{ij} - 0\cdot027\,S_{ij} + 0\cdot0014\,t_{ij}\ (r = 0\cdot997). \ \mathbf{17}$$
$$\phantom{\hat{\varphi}_{ij} = }\quad (0\cdot015) \quad\ (0\cdot016) \quad\ \ (0\cdot0014)$$

There is some apparent tendency for the rate of imitation to increase over time, but it is not statistically significant.

Finally, one might suppose that $\lambda_{ij}(t)$ would be influenced by the phase of the business cycle during which the innovation was first introduced. Let δ_{ij} equal one if the innovation is introduced in the expansion phase and zero if it is introduced in the contraction phase.[28] Including δ_{ij} in **14**, we have

$$\hat{\varphi}_{ij} = \begin{Bmatrix} -0\cdot26 \\ -0\cdot55 \\ -0\cdot48 \\ -0\cdot57 \end{Bmatrix} + 0\cdot530\,\pi_{ij} - 0\cdot033\,S_{ij} - 0\cdot022\,\delta_{ij}\ (r = 0\cdot997) \ \mathbf{18}$$
$$\phantom{\hat{\varphi}_{ij} = }\quad (0\cdot017) \quad\ (0\cdot020) \quad\ (0\cdot045)$$

and the effect of δ_{ij} turns out to be non-significant.

Thus, our results regarding these additional factors are largely inconclusive. Though each might be expected to have some effect, their inclusion in the analysis does not lead to a significantly

---

used figures on the profitability of using the innovation in new plant as well to replace equipment of various ages, but data were not available. Note too that g_{ij} may be affected by the appearance of the innovation. For a description of the data on g_{ij}, see the Appendix. If S_{ij} is omitted from **15–18**, the coefficients of d_{ij}, g_{ij}, etc. remain non-significant.

27. I.e. $t_{ij} = t_{ij}^* - 1900$. See note 31.

28. We used the National Bureau's reference dates (given in Moore, 1958), to determine whether t_{ij}^* was a year of contraction or recovery. The residuals in Figure 2 do not seem to be affected by whether or not an innovation was being accepted sometime during the depression of the 1930s.

better explanation of the observed differences in $\hat{\varphi}_{ij}$. Their apparent effects are in the expected direction, but data for more innovations will be required before one can be reasonably sure of the persistence and magnitude of their influence. There is no evidence that the effects of the previously considered variables on the rate of imitation are due to the operation of these factors. When these other factors are included, the coefficients of π_{ij} and S_{ij} and the interindustry differences remain relatively unchanged (though the coefficient of S_{ij} becomes non-significant).

6 A stochastic version of the model

In this section, we present and test a somewhat more sophisticated, stochastic version of the model. For the jth innovation in the ith industry, let $P_{ij}(t,k)$ be the probability that any one of the 'hold-outs' at time t will introduce it by time $t+k$, and assume (for small k) that

$$P_{ij}(t,k) = \theta_{ij}\frac{m_{ij}(t)}{n_{ij}}k, \qquad\qquad 19$$

$$\theta_{ij} = b_i' + a_5'\pi_{ij} + a_6'S_{ij} + z_{ij}'', \qquad\qquad 20$$

where the coefficients in 20 are analogous to those in 12. To see how closely this resembles the deterministic model, note that the *expected* increase in the number of users between time t and time $t+1$,

$$\{n_{ij} - m_{ij}(t)\}\theta_{ij}\frac{m_{ij}(t)}{n_{ij}},$$

is almost identical with the expression given before the *actual* increase in the number of users (see 4 and 5). The only (apparent) difference is that the terms corresponding to Q_{ij} are assumed to be zero from the start here, whereas in section 3 this followed from 8.[29]

Whereas our task in the deterministic model was to determine how the actual number of users grows over time, our problem here is to see how the expected number grows. To do so, we first obtain an expression for $P_{ij}^r(t)$ – the probability that at time t

29. To simplify things, we also assume from the beginning that the coefficients of π_{ij} and S_{ij} do not vary among industries (cf. 20). In the deterministic model we introduced this assumption later in the analysis.

there are exactly r firms in the ith industry that have not yet introduced the jth innovation. From **19**,

$$P_{ij}^r(t+k) = P_{ij}^r(t)\left[1 - r\frac{\theta_{ij}}{n_{ij}}(n_{ij}-r)k\right] +$$
$$+ P_{ij}^{r+1}(t)(r+1)\frac{\theta_{ij}}{n_{ij}}(n_{ij}-r-1)k + o(k),$$

where $o(k)$ represents terms that, if divided by k, tend to zero as k vanishes.[30] And if we subtract $P_{ij}^r(t)$ from both sides, divide by k, and let k tend to zero, the following differential-difference equation results:

$$\dot{P}_{ij}^r(t) \begin{cases} = -\dfrac{\theta_{ij}}{n_{ij}}r(n_{ij}-r)P_{ij}^r(t) + \dfrac{\theta_{ij}}{n_{ij}}(r+1)(n_{ij}-r-1)P_{ij}^{r+1}(t), \\ \qquad\qquad\qquad\qquad\qquad\qquad\qquad\quad \text{if } r < n_{ij}-1, \\ = -\dfrac{\theta_{ij}}{n_{ij}}r P_{ij}^r(t), \qquad\qquad\qquad\quad\ \text{if } r = n_{ij}-1, \quad \mathbf{21} \end{cases}$$

30. To derive this expression, note that

$$P_{ij}^r(t+k) = \sum_{x=r}^{n_{ij}} P_{ij}^x(t)\, D_{ij}^{x-r}(t),$$

where $D_{ij}^{x-r}(t)$ is the probability that $x-r$ firms begin using the innovation between time t and time $t+k$, given that $n_{ij}-x$ firms are using it at time t. From **19**,

$$D_{ij}^0(t) = [1-(\theta_{ij}/n_{ij})(n_{ij}-r)k]^r;$$

$$D_{ij}^1(t) = (r+1)\frac{\theta_{ij}}{n_{ij}}(n_{ij}-r-1)k\left[1-\frac{\theta_{ij}}{n_{ij}}(n_{ij}-r-1)k\right]^{r-1};$$

etc. Thus,

$$P_{ij}^r j(t+k) = P_{ij}^r(t)\left[1-\frac{\theta_{ij}}{n_{ij}}(n_{ij}-r)k\right]^r +$$
$$+ P_{ij}^{r+1}(t)(n+1)\frac{\theta_{ij}}{n_{ij}}(n_{ij}-r-1)\left[1-\frac{\theta_{ij}}{n_{ij}}(n_{ij}-r-1)k\right]^{r-1} + o(k)$$
$$= P_{ij}^r(t)\left[1-r\frac{\theta_{ij}}{n_{ij}}(n_{ij}-r)k\right] +$$
$$+ P_{ij}^{r+1}(t)\left[(r+1)\frac{\theta_{ij}}{n_{ij}}(n_{ij}-r-1)k\right] + o(k).$$

Note two things, (a) The assumption that the probability in **19** is the same for all 'hold-outs' is obviously only a convenient simplification. Without it, the analysis becomes extremely difficult. (b) We assume that the decisions of the 'hold-outs' between time t and time $t+k$ are independent. Hence, the probability that one 'hold-out' will introduce it between time t and time $t+k$ is

$$\{n_{ij}-m_{ij}(t)\}\theta_{ij}m_{ij}(t)k/n_{ij} + o(k),$$

and the probability that none will introduce it is

$$1 - \{n_{ij}-m_{ij}(t)\}\theta_{ij}m_{ij}(t)k/n_{ij} + o(k).$$

with the initial conditions that $P^r_{ij}(t^*_{ij})$ equals unity for $r = n_{ij}-1$ and zero otherwise, and where $\dot{P}^r_{ij}(t)$ is the time derivative of $P^r_{ij}(t)$ and t^*_{ij} is the date (taken as given)[31] when the innovation was first introduced. From **21**, it follows (Bailey, 1955) that $M_{ij}(t^*_{ij}+V)$ – the expected number of firms using the innovation at time $t^*_{ij}+V$ – equals

$$n_{ij} - \sum_{r=1} \frac{(n_{ij}-1)!}{(n_{ij}-r-1)!(r-1)!} \times$$
$$\times \left[(n_{ij}-2r)^2 - \frac{\theta_{ij}}{n_{ij}} V + 2 - (n_{ij}-2r) \sum_{u=r}^{n_{ij}-r-1} u^{-1} \right] \times$$
$$\times e^{-r(n_{ij}-r)(\theta_{ij}/n)V}, \quad \textbf{22}$$

where r runs up to $(\frac{1}{2}n_{ij}-1)$ for n_{ij} odd and up to $\frac{1}{2}n_{ij}$ for n_{ij} even.[32]

Thus, the stochastic version of the model leads to the following two propositions. First, the expected number of firms having introduced an innovation at any date subsequent to t^*_{ij} should be given by **22**. Second, θ_{ij} – the parameter that, for given n_{ij}, determines the expected rate of imitation – should be a linear function of π_{ij} and S_{ij}, the intercept of the function differing among industries.

To see how well the first proposition seems to hold, we estimate θ_{ij}, insert it, n_{ij}, and t^*_{ij} into **22**, and compare the growth over time in the computed number of users with the actual growth curve. Table 1 contains a rough measure of 'goodness-of-fit': the root-mean-square deviation of the actual from the computed number of users.[33] Although the agreement between the actual and

31. For all but the tin container, we use 31 December of the year during which the innovation was first introduced by one of these firms (or if we know the month when it was first introduced, we use that 31 December which was closest in time to it). For the tin container, we use 1 February 1935. See Table 1 for the year in which each t^*_{ij} falls.

32. When n_{ij} is even, some adjustment has to be made to this expression. See Bailey (1955) for the adjustment and for a much more extensive discussion of the argument.

33. To estimate θ_{ij}, we used the statistic suggested by Bailey (1955, p. 41) to estimate θ_{ij}/n_{ij} and multiplied the result by n_{ij}. This estimate of θ_{ij} is biased, but we can show that the bias is approximately equal to $\theta_{ij}/(n_{ij}-1)$, which is usually quite small. These estimates, together with the tables in Mansfield and Hensley (1960), were used to calculate the growth over time

computed growth curves is sometimes quite good, Table 1 shows that **22** is never so accurate as **9** in representing the data. Perhaps this is due in part to differences in the way parameters are estimated as well as differences in the model.[34]

To test the second proposition, we assume that errors in the estimates of θ_{ij} are uncorrelated with π_{ij} and S_{ij} and hence that

$$\hat{\theta}_{ij} = b_i' + a_5' \pi_{ij} + a_6' S_{ij} + z_{ij}''', \qquad \textbf{23}$$

where $\hat{\theta}_{ij}$ is our estimate of θ_{ij} and z_{ij}''' is a random error term. Then, proceeding as we did in sections 4 and 5, we get much the same sort of results as those obtained there. (1) We estimate the parameters in **23**, find that these estimates are almost identical with those given in **14** for the analogous parameters,[35] and conclude that the resulting equation represents the data very well ($r = 0.997$). (2) We introduce the additional variables discussed in section 5 into **23** and find that their coefficients have the expected signs, but that (except for the coefficient of t_{ij}) all are non-significant.

Thus, the model, in either its deterministic or stochastic form, seems to represent the empirical results in Figure 1 quite well. Although **22** does not fit the data as well as **9** in the deterministic version, the purpose and interpretation of these equations are quite different and hence this may not be very surprising.[36] With regard to the explanation of differences in the rate of imitation, **23**, like **13** in the deterministic model, provides an excellent

in the expected number of users. The root-mean-square errors are computed in the way described in note 16.

34. Two parameters (l_{ij} and φ_{ij}) are fitted to the data in the deterministic model whereas only θ_{ij} is fitted here. However, this is only part of the explanation (see note 36). Note too that the differences between the computed and actual curves are by no means random for a number of these innovations. The computed curves sometimes lie below the actual ones.

35. The only appreciable difference is that the coefficient of π_{ij} is 0.580 rather than 0.530.

36. In the deterministic model, we fit **9** to the data in such a way as to minimize differences between the actual and computed curves. In this model, we use the data to estimate the curve of *averages* that would result if the process were repeated indefinitely. Whereas the point in the former case was to minimize differences between the computed and actual curves, this was not the point here and hence it is not surprising that the deterministic model seems superior in this respect.

fit. The stochastic version of the model seems somewhat more reasonable to me, but it is difficult to tell at this point whether it is a significant improvement over the simpler, deterministic version.

Before concluding, one final point should be noted regarding the dispersion about the expected value in 22. This dispersion is often relatively large and hence a prediction of the rate of imitation based on 22 would be fairly crude, even if the model were correct and θ_{ij} were known in advance. To illustrate this, consider an innovation where $n_{ij} = 12$, $t^*_{ij} = 1924$ and $\theta_{ij} = 0.42$. Table 2 shows the expected number of firms having introduced it at various points in time and the relatively large standard deviation of the distribution about this expectation. Of course, despite this variation, the model could set useful upper and lower bounds on how long the entire process would take.[37]

7 Limitations

In evaluating these results, the limitations of the data, methods and scope of the investigation must be taken into account. For example, although its scope far exceeds that of the few studies previously conducted, it nonetheless is limited to only four industries and to only a few innovations in each one. Before concluding, it seems worthwhile to point out some of these limitations in more detail.

First, there is the matter of scope. Although the four industries included here vary with respect to market structure, size of firm, type of customer, etc., they can hardly be viewed as a cross-section of American industry. For example, none is a relatively young industry with a rapidly changing technology. In addition, the innovations included here are all important, they generally required fairly large investments, and the imitation process was not impeded by patents. To check and extend these results, similar studies should be carried out for other industries and other types of innovations. Moreover, international comparisons might be attempted.

Second, the data are not always as precise as one would like. For example, much of the data regarding the profitability of installing an innovation had to be obtained from questionnaires

37. For further discussion of the distribution about these expected values, see Mansfield and Hensley (1960) and the literature cited there.

Table 2 Expected Value and Standard Deviation of the Number of Firms Having Introduced an Innovation, Assuming that $n_{ij} = 12$, $t_{ij}^* = 1924$ and $\theta_{ij} = 0.42$, 1924–39

Date (31 December)	Expected number	Standard deviation
1924	1·0	—
1925	1·5	0·9
1926	2·1	1·4
1927	2·7	1·6
1928	3·6	2·3
1929	4·5	2·7
1930	5·5	2·9
1931	6·4	3·1
1932	7·2	3·2
1933	8·0	3·1
1934	8·8	2·9
1935	9·4	2·6
1936	10·0	2·4
1937	10·5	2·1
1938	10·8	1·8
1939	11·1	1·5

Source: Tables in Mansfield and Hensley (1960). Small errors due to interpolation are present in both the expected values and standard deviations.

and interviews with firms. Although these estimates were checked against others obtained from suppliers of the equipment, trade journals, etc., they are probably fairly rough. Similarly, the estimates of the pay-out period required for investments and the durability of old equipment are only rough. Because of these errors of measurement, the regression coefficients in **14** to **18** are probably biased somewhat toward zero.[38]

38. It can easily be shown that measurement errors, if random, have this effect. Another limitation of the analysis is that the model can only be expected to hold for relatively profitable innovations. Certainly, it will not hold in cases where $\pi_{ij} < 1$ and it may do poorly if π_{ij} is not appreciably greater than unity. But this limitation is not very serious because if π_{ij} does not exceed unity, the innovation almost certainly will not become generally accepted, and if π_{ij} is not much greater than unity, the innovation is probably not very important.

Third, the data measure the rate of imitation among large firms only. As noted above, firms not exceeding a certain size (specified in the Appendix) were excluded. Moreover, the rate at which firms imitated an innovator, not the rate at which a new technique displaced an old one or the rate at which investment in a new technique mounted, is considered here. Although these are related topics also being studied, they were not taken up in this paper.

Fourth, various factors other than those considered here undoubtedly exerted some influence on the rate of imitation. Variation over time in the profitability of introducing an innovation (due to improvements, the business cycle, etc.) is one potentially important factor that is omitted.[39] Others are the sales and promotional efforts made by the producers of the new equipment and the extent of the risks firms believed they assumed at first by introducing an innovation.[40] I could find no satisfactory way to measure them, but perhaps further research will obviate

39. One cannot tell with any accuracy how, and to what extent, the profitability varied in each of these cases. But, to illustrate the factors at work, note a few broad changes that occurred in the case of three of the innovations. (a) The appearance of improved diesel locomotives in the thirties and forties increased the profitability of introducing them and hastened their acceptance. In addition, first costs declined, the relative prices of coal and oil changed, and the composition of the 'hold-outs' varied. (b) The demand for toluene during the First World War undoubtedly accelerated the imitation process in the case of the by-product coke oven. (c) The Depression reduced the profitability of introducing centralized traffic control, whereas the wartime traffic boom increased its profitability. For further discussion, see note 21. Of course, the relative profitability may have varied less than the absolute profitability.

40. In each of these cases, there was considerable promotional effort by the producers of the new equipment and considerable help given to the firms that used it. E.g. Semet-Solvay and Koppers helped to finance by-product coke plants and provided personnel, General Motors helped to train diesel operators, kept maintenance men at hand, etc. For some discussion of the perceived risks, see note 7.

Another factor that could be very important for some innovations is the presence of patents. Still another is the development of a marked improvement in the innovation. E.g. an innovation may be applicable at first to only a few firms and a significant improvement is required before others use it. In a case like this, the model is clearly not applicable (Mansfield, 1959). In each of these cases, improvements occurred, but they were of a more gradual and less significant nature.

this difficulty and disclose the effects of these and other such factors on the rate of imitation. I suspect that the results underestimate their influence and that data for more innovations would show that the unexplained variation (due to their effects) is greater than is indicated in Figure 2.

8 Conclusion

In their discussions of technological change, economists have often cited the need for more research regarding the rate of imitation. Because an innovation will not have its full economic impact until the imitation process is well under way, the rate of imitation is of considerable importance. It is unfortunate that so little attention has been devoted to it and that consequently we know so little about the mechanism governing how rapidly firms come to use a new technique.

My purpose in this paper has been to present and test a simple model designed to help explain differences among process innovations in the rate of imitation. This model is built largely around one hypothesis – that the probability that a firm will introduce a new technique is an increasing function of the proportion of firms already using it and the profitability of doing so, but a decreasing function of the size of the investment required.[41] When confronted with data for twelve innovations, this model seems to stand up surprisingly well. As expected, the rate of imitation tended to be faster for innovations that were more profitable and that required relatively small investments. An equation of the form predicted by the model can explain practically all of the variation among the rates of imitation.

As expected, there were also interindustry differences in the rate of imitation. We have too few industries really to test the hypothesis often advanced that the rate of imitation is faster in more competitive industries, but the differences seem to be generally in that direction. When several other factors are included in the model, the empirical results are largely inconclusive. There was some apparent tendency for the rate of imitation to be higher when the innovation did not replace very durable

41. This is the stochastic version of the model. In the deterministic version, we use the proportion of 'hold-outs' that introduce it rather than this probability.

equipment, when the firms' output was growing rapidly and when the innovation was introduced in the more recent past. But it was almost always statistically non-significant.

In view of the relatively small number of innovations and the uneven quality of the data, these results are quite tentative. But even so, they should be useful to economists concerned with the dynamics of firm behavior and the process of economic growth. To understand how economic growth is generated, we must know more about the way innovations occur and how they become generally accepted. Further attempts should be made to check and extend the results and to pursue other approaches to the particular area considered here. We need to know much more about this and other aspects of technological change.

Appendix
Basic data

Iron and steel industry. For the continuous wide strip mill, all firms having more than 140,000 tons of sheet capacity in 1926 were included; for the by-product coke oven, all firms with over 200,000 tons of pig iron capacity in 1901 were included; and for continuous annealing of tin plate, the nine major producers of tin plate in 1935 were included. In the case of the strip mill and coke oven, a few of these firms merged or went out of business before installing them, and there was no choice but to exclude them.

The date when each firm first installed a continuous wide strip mill was taken from the Association of Iron and Steel Engineers (1941). Similar data for the coke oven were obtained from various editions of the *Directory of Iron and Steel Works* of the American Iron and Steel Institute and issues of the *Iron Trade Review* and *Iron Age*. The date when each firm installed continuous annealing lines was obtained from correspondence with the firms.

The size of the investment required and the durability of replaced equipment came from interviews. (The estimates of the pay-out periods were also checked there. cf. note 20.) The interviews (each about two hours long) were with major officials of three steel firms, the president and research manager of a firm that builds strip mills and continuous annealing lines, officials of a firm that builds coke ovens, and representatives of a relevant engineering association and of a trade journal. The data on growth of output were annual industry growth rates and were for sheets during 1926–37 (strip-mill), pig iron during 1900–25 (coke oven), and tin plate during 1939–56 (continuous annealing). They were taken from the *Bituminous Coal Annual* of the Bituminous Coal Institute, Association of Iron and Steel Engineers (1941) and *Annual Statistical Reports* of the American Iron and Steel Institute.

Railroad industry. For centralized traffic control, all Class 1 line-haul roads with over five billion freight ton-miles in 1925 were included. For the diesel locomotives and car retarders, essentially the same firms were included. (The Norfolk and Western, a rather special case, was replaced by the New Haven and Lehigh Valley in the case of the diesel locomotive. Some important switching roads were substituted in the case of car retarders, an innovation in switching techniques.) An entire system is treated here as one firm.

The date when a firm first installed centralized traffic control was usually derived from a questionnaire filled out by the firm. For those that did not reply, estimates by K. Healy (1954) were used. The date when each firm first installed diesel locomotives was determined from various editions of the Interstate Commerce Commission's *Statistics of Railways.* The date when each firm first installed car retarders was taken from various issues of *Railway Age.*

For centralized traffic control and car retarders, some of the pay-out periods were estimates published in the *Signal Section, Proceedings of Association of American Railroads,* and the rest were obtained from questionnaires filled out by the firms. All estimates of the pay-out period for the diesel locomotive and the pay-out period required for investment were obtained from questionnaires. Information regarding the size of the investment required and the durability of old equipment was obtained primarily from interviews with eight officials of six railroads (ranging from president to chief engineer) and three officials of a signal manufacturing firm and a locomotive manufacturing firm. The data on growth of output were annual growth rates for total freight ton-miles during 1925–41 (diesel locomotive) and 1925–54 (centralized traffic control and car retarders). They were taken from the *Statistics of Railways*.

Bituminous coal industry. Practically all firms producing over four million tons of coal in 1956 (according to McGraw-Hill's *Keystone Coal Buyers Manual*) were included. A few firms that did strip mining predominantly were excluded in the case of the continuous mining machine, and a few had to be excluded in the case of the shuttle car and trackless mobile loader because they would not provide the necessary data. The date when each firm first introduced these types of equipment was usually obtained from questionnaires filled out by the firms, but in the case of the continuous mining machine, data for two firms that did not reply were derived from the *Keystone Coal Buyers Manual*.

Data regarding the size of the investment required and the durability of old equipment were obtained from interviews with two vice-presidents of coal firms, several executives of firms manufacturing the equipment, employees of the Bureau of Mines and representatives of an independent coal research organization. The data on growth of output were annual growth rates for bituminous coal production during 1934–51 (trackless mobile loader), 1937–51 (shuttle car) and 1947–56 (continuous mining machine). They were taken primarily from the *Bituminous Coal Annual*.

Brewing industry. We tried to include all breweries with more than one million dollars in assets in 1934 (according to the *Thomas Register*), but several would not provide the necessary data and they could not be obtained elsewhere. The date when a firm first installed each type of equipment was usually taken from a questionnaire that it filled out, but in a few cases it was provided by manufacturers of the equipment or articles in the *Brewers' Journal*. The size of the necessary investment and the durability of equipment that was replaced were determined from interviews with a number of officials in two breweries and sales executives of two can companies. The data on growth of output were annual growth rates for beer production during 1935–7 (tin containers) and 1950–58 (pallet-loading machine and high-speed bottle filler). They were taken from the *Brewers' Almanac* and

Business Week (20 June 1959). The 1950–58 figures refer only to half of these larger firms.

References

AMERICAN MINING CONGRESS (annual), *Coal Mine Modernization Yearbook*.

ASSOCIATION OF IRON AND STEEL ENGINEERS (1941), *The Modern Strip Mill*, publisher unknown.

BAILEY, N. (1955), 'Some problems in the statistical analysis of epidemic data', *J. Roy. Stat. Soc.* B, vol. 17, pp. 35–68.

BERKSEN, J. (1953), 'A statistically precise and relatively simple method of estimating the bio-assay with quantal responses, based on the logistic function', *J. Amer. Stat. Assn.*, vol. 48, pp. 565–99.

CAMP, J., and FRANCIS, C. (1940), *The Making, Shaping and Treating of Steel*, US Steel Co.

COLEMAN, J., KATZ, F., and MENZEL, H. (1957), 'The diffusion of an innovation among physicians', *Sociometry*, vol. 20, pp. 253–70.

COMMITTEE ON PRICE DETERMINATION (1943), *Cost Behavior and Price Policy*, National Bureau of Economic Research.

GORDON, M. (1955), 'The payoff period and the rate of profit', *J. Bus.*, vol. 28, pp. 253–60.

GRILICHES, Z. (1957), 'Hybrid corn: An exploration in the economics of technological change', *Econometrica*, vol. 25, pp. 501–22.

HEALY, K. (1954), 'Regularization of capital investment in railroads', *Regularization of Business Investment*, National Bureau of Economic Research.

JEROME, H. (1934), *Mechanization in Industry*, National Bureau of Economic Research.

JEWKES, J., SAWERS, D., and STILLERMAN, R. (1958), *The Sources of Invention*, St Martin's Press.

KLEIN, L. (1951), 'Studies in investment behavior', *Conference on Business Cycles*, National Bureau of Economic Research.

MACK, R. (1941), *The Flow of Business Funds and Consumer Purchasing Power*, Columbia University Press.

MANSFIELD, E. (1959), 'Acceptance of technical change: The speed of response of individual firms', Carnegie Institute of Technology.

MANSFIELD, E., and HENSLEY, C. (1960), 'The logistic process: Tables of the stochastic epidemic curve and applications', *J. Roy. Stat. Soc.*, Series B, vol. 22, pp. 332–47.

MANSFIELD, E., and WEIN, H. (1958), 'A model for the location of a railroad classification yard', *Manag. Sci.*, vol. 4, pp. 292–313.

MOORE, G. (1958), 'Measuring recessions', *J. Amer. Stat. Assoc.*, vol. 53, pp. 259–316.

ROBINSON, J. (1956), *The Accumulation of Capital*, Irwin.

SCHUMPETER, J. (1939), *Business Cycles*, McGraw-Hill.

SCITOVSKY, T. (1956). 'Economies of scale and European integration', *Amer. econ. Rev.*, vol. 46, pp. 79–91.

SWALM, R. (1958), 'On calculating the rate of return of an investment', *J. indust. Eng.*, vol. 9, pp. 99–103.

TERBORGH, G. (1949), *Dynamic Equipment Policy*, McGraw-Hill.

UNION SWITCH AND SIGNAL COMPANY (1931), *Centralized Traffic Control*.

YANCE, J. (1957). 'Technological change as a learning process: The dieselization of the railroads', unpublished.

Part Four
Long-Term Consequences of Technological Change

The statement that technological change has been an important factor in generating the long-term economic development of the wealthy nations of the world would command widespread, perhaps universal, assent. But what numerical value can we assign to 'important'? Ideally, one would like to have a list of all factors which have contributed to economic growth and to be able to assign, for particular time-periods in individual countries, a precise numerical value to represent the contribution of each. We are, unfortunately, a long way from the possession of such 'finely-tuned' information, and the limitations inherent in the data and the methodological and conceptual difficulties of all measurement procedures render such a goal perhaps forever unattainable. Nevertheless, our understanding of both technological change and economic growth would be enormously improved if we could repose some confidence in estimates even of the orders of magnitude of individual factors.

The attempt to quantify the contribution of technological change to economic growth began in a serious way with the publication of the first two papers reprinted in this section: first the one by Abramovitz in 1956, followed by Solow's paper in 1957. Both of these papers explore the quantitative importance of technological change to the long-term economic growth of the American economy. The two papers differ in many ways: with respect to time-periods, coverage and basic methodology. Yet the authors concur in the conclusion that only a very small portion of the long-term growth in American *per capita* output can be accounted for by an increasing quantity of capital and labour inputs. If we distinguish between the growth in *per capita* output due to (a) employing *more* inputs, and (b) producing a larger output *per unit of input*, the Abramovitz and Solow papers suggest forcefully that the latter has been far more significant than the former. Although there was some tendency to label the rise in productivity as 'technological change', such a label was not justified, as Abramovitz was careful to point out. The extremely large residual with which both authors were left after attempting to measure the growth in output *per capita* which was attributable to rising inputs *per capita*, encompassed a wide range of possible causes of improved efficiency other than technological change. In fact, the methodologies were such that the residual captured *all* causes of rising output *per capita other* than rising inputs *per capita*. The unexplained growth in resource productivity, as Abramovitz put it, is a 'measure of our ignorance', which turned out to be surprisingly large.

These startling results provoked a wide response on the part of economists wakened, as it were, from their 'dogmatic slumber'. Indeed, the response was so great that it would be fair to describe the attempt to explain the residual as the leading 'growth industry' of the economics profession during the 1960s. Denison's article summarizes his heroic effort at quantification of the residual. He examines the growth performance of American national income between 1929 and 1957 and, with the aid of some strong assumptions, attempts to quantify the contribution of each of a number of variables to this growth. In addition to his estimates of the contribution of changes in quantities of capital and labour inputs, Denison attempts to adjust for quality changes in labour inputs, particularly those attributable to education and the effects of shorter hours on quality. Moreover, he decomposes the residual into several component parts, the most important of which turn out to be advances in knowledge and economies of scale. Denison also performs similar calculations for the period 1909–29.

Finally, Griliches suggests a different way of attacking the residual, an approach of general applicability which he uses to examine the sources of productivity growth in American agriculture for the period 1940–60. Griliches argues that the very notion of a production function '. . . is not very useful if it is not a *stable* function, if there are very large unexplained shifts in it'. His approach is to allow the production function to remain constant and to explain the measured changes in output in terms of changes in quantities of inputs, identifiable changes in *qualities* of inputs (both labour and capital), and economies of scale.

15 M. Abramovitz

Resource and Output Trends in the United States
since 1870[1]

M. Abramovitz, 'Resource and output trends in the United States
since 1870', *American Economic Review, Papers and Proceedings*,
May 1956, pp. 5–23.

Introduction

This paper is a very brief treatment of three questions relating to
the history of our economic growth since the Civil War: (a) How
large has been the net increase of aggregate output *per capita*, and
to what extent has this increase been obtained as a result of
greater labor or capital input on the one hand and of a rise in
productivity on the other? (b) Is there evidence of retardation,
or conceivably acceleration, in the growth of *per capita* output?
(c) Have there been fluctuations in the rate of growth of output,
apart from the short-term fluctuations of business cycles, and, if
so, what is the significance of these swings?

The answers to these three questions, to the extent that they
can be given, represent, of course, only a tiny fraction of the
historical experience relevant to the problems of growth. Even
so, anyone acquainted with their complexity will realize that no
one of them, much less all three, can be treated satisfactorily in a
short space. I shall have to pronounce upon them somewhat
arbitrarily. My ability to deal with them at all is a reflection of
one of the more important, though one of the less obvious, of the
many aspects of our growing wealth; namely, the accumulation
of historical statistics in this country during the last generation.

For the most part, the figures which I present or which underlie

1. I should like to thank Professor Simon Kuznets and Mr J. W. Kendrick
who made available to me certain unpublished estimates of national product,
productivity, capital stock and hours of work. Their contributions are
further described in the notes to Table 1. I am grateful to Richard A.
Easterlin, Solomon Fabricant, J. W. Kendrick and G. H. Moore for their
critical review of the manuscript and to Mrs Charlotte Boschan for assisting
in its preparation.

my qualitative statements are taken directly from tables of estimates of national product, labor force, productivity and the like compiled by others. In a few cases I have ventured to compute ratios or extend the tables forward or backward by combining estimates. But no original estimates depending on the compilation or reworking of primary data are included.

The period since 1870 has an important unifying characteristic in that throughout these eighty years the economy has been growing in response to the complex of cumulative forces which we generally call industrialization. It is quite clear, however, that 1870 was not the beginning of the process of industrialization in this country. The proportion of gainful workers in agriculture fell from 71 per cent in 1820 to 64 per cent in 1850. It fell another 10 percentage points by 1870. Steam transport by water and rail was already common when the period begins. The proportion of the gainfully employed engaged in manufacturing and construction rose from 12 to 21 per cent between 1820 and 1870. Real *per capita* output rose significantly during the 1850s. It was set back by the Civil War, but aggregate output well-nigh doubled from 1850 to 1870.[2] The data before 1870 – and still more before 1850 – are highly dubious, but it seems clear that the period since 1870 does not include the entire era of industrialization and rapid income rise in this country. We are, in an important sense, dealing with a period arbitrarily delimited by the availability of fairly reliable comprehensive figures.

It may be of some use if I try to state at the very beginning the three main conclusions of my paper. First, between the decade 1869–78 and the decade 1944–53, net national product *per capita* in constant prices approximately quadrupled, while population more than tripled. The source of the great increase in net product per head was not mainly an increase in labor input per head, not even an increase in capital per head, as these resource elements are conventionally conceived and measured. Its source must be sought principally in the complex of little understood forces which caused productivity, that is, output per unit of utilized resources, to rise.

Second, it is not clear that there has been any significant trend

2. These are W. I. King's figures (1915, Table 23), as deflated by Simon Kuznets (1952, p. 240).

in the rates of growth of total output and of output per head. It is true that national product estimates, on their face, suggest some decline in the rates of growth – somewhat more clearly for total output; somewhat less clearly for output *per capita*. It is doubtful, however, whether the data can be accepted with confidence for this purpose and still more doubtful whether the apparent retardation in growth, such as it is, represents the effect of persistent forces. In so far as one can observe a decline in the rate of growth, its source is not in the productivity of resources, which has continued to grow at a steady, perhaps an accelerating, pace. Its source has been a decline in the rate of growth of labor input per head and of capital input per head.

Third, the rate of growth of output has not been even. In addition to ordinary business cycles, the rate of growth has risen and fallen since 1870 in long waves of approximately twenty years' duration. Preliminary study suggests that these waves represent, in the main, surges in productivity or resource supply rather than in the proportion of our resources employed. An adequate understanding both of the history of our growth and of our prospects during the next generation depends on our ability to determine whether these surges and relapses are to some significant degree truly recurrent or wholly fortuitous.

The average rate of growth, 1869–1953

My first problem has to do with the overall expansion of our economy since 1870. My principal criterion of growth is net national product *per capita* in 1929 prices, and since I use Kuznets's data, I follow him in measuring the increase by comparing average product and related data for labor, capital and so on, for the decade 1869–78 with that for the decade 1944–53.[3] Comparisons based on such decade averages eliminate most but, of course, not all the effects of business cycles, which might otherwise serve to distort somewhat our impressions of the long-term rate of growth. They do not protect our measures from the effects

3. Professor Kuznets has very kindly permitted me to use his newly revised estimates extended to 1953. These are, as yet, unpublished, but very similar figures are published in 'Long-Term Changes' (1952). The broad concepts on which the data are based and the methods of estimate are described in that volume, pp. 29–34. The latter have been altered in certain details in ways which Professor Kuznets will describe in a later publication.

of fluctuations longer in duration than business cycles, the so-called 'secular swings', which I shall discuss later. It would be better to calculate rates of growth from properly derived trend values. But in measures for a period as long as eighty years, when growth was so rapid, the distortion resulting from secular swings will not prevent us from seeing the broad outlines of the picture, and I judged it unnecessary to calculate statistical trend lines for this purpose.

1. Net national product in the decade 1944–53 stood about thirteen times as high as it had in 1869–78 (Table 1). This increase implies an average rate of growth of 3·5 per cent per annum. Population, however, more than tripled in the same period. Net product *per capita*, therefore, approximately quadrupled, implying an average rate of growth of 1·9 per cent per annum.

These calculated rates of increase are only rough approximations of the figures we are really after. Long-term estimates of national products are inevitably marred by statistical weaknesses, biases and uncertainties of conception (Kuznets, 1952, pp. 33–47). We must accept the fact that even the most comprehensive and consistent measures of our rate of expansion must be treated with a great deal of reserve.

2. The quadrupling – more or less – of net national product *per capita* resulted in part from an increase in the input of resources *per capita* and in part from a rise in the productivity, that is, the output per unit, of representative units of resources. However, the shares of these two elements, in so far as they can be separated, were very different. The input of resources per head of the population appears to have increased relatively little while the productivity of resources increased a great deal. How does this arise?

The input of resources is usually conceived to consist of labor services, including salaried management, and property or capital services, to which is attached the contributions of entrepreneurship made in connection with the investment of capital in industry. If we measure labor services in man-hours, as is usually done, we find that labor input *per capita* declined slightly between the seventies and the present. This resulted from the counteraction of two trends. The labor force ratio, that is, the ratio of labor force to population, grew about 25 per cent as a result of

Table 1 Measures of US Economic Growth,
1869–78 to 1944–53

		Relatives for 1944–53 (1869–78 = 100)
1.	Net national product	1325
2.	Population	334
3.	Net national product *per capita*	397
4.	Labor force	423 (393)
5.	Ratio: labor force to population	127 (118)
6.	Employment	427 (396)
7.	Ratio: employment to population	128 (119)
8.	Standard hours	73
9.	Man-hours	312 (290)
10.	Man-hours *per capita*	94 (87)
11.	Capital	993
12.	Capital *per capita*	297
13.	Index of total input of resources	381 (361)
14.	Index of input *per capita*	114 (108)
15.	Net national product per employed worker	310 (334)
16.	Net national product per man-hour	426 (458)
17.	Net national product per capital unit	134
18.	Index of net national product per unit of total input	348 (367)

Figures in parentheses exclude armed forces.

All the figures in this table, unless otherwise noted, were drawn from series of averages for overlapping decades running 1869–78, 1874–83, etc.

The units of the data from which the relatives were calculated are shown in the notes to each line.

Line:

1. Newly revised estimates by Simon Kuznets (billions of dollars in 1929 prices) (1961, pt B).

2. Kuznets (1961, pt E). Decade averages computed from annual data underlying five-year moving averages to be published.

3. Line 1 ÷ line 2 (1929 dollars per person).

4. See line 2.

5. Line 4 ÷ line 2 (per cent).

6. Line 4 less estimated unemployment (millions) as follows: *1869–78 to 1884–93:* from J. Schmookler (1952, Table 3, col. 2).

1889–98 to 1939–48: by applying unemployment percentage from Kuznets (1952, Table 10, col. 1) to his estimates of the civilian labor force and adding armed forces. From 1889–1918, the labor force figures were first divided into agricultural and non-agricultural segments. The unemployment percentages, which for those years represent only non-agricultural unemployment, were applied to the latter only.

1944–53: By applying ratio of civilian employment to civilian labor force as estimated by Bureau of the Census (1955, p. 56) to Kuznets's estimate of civilian labor force and adding armed forces.

7. Line 6÷line 2 (per cent).

8. *1869–78 to 1939–48:* from Kuznets (1952, Table 7, col. 1). *1944–53:* extrapolated on the basis of estimates kindly supplied to the author by J. W. Kendrick. (Hours per week.)

9. Line 6×line 8 (millions of man-hours per week).

10. Line 9÷line 2 (weekly hours *per capita*).

11. *1874–83 to 1939–48:* Kuznets (1952, Table 11, col. 3). Single figures are provided once each decade, 1879 to 1939, for years running 1879 1889, etc. In addition there are figures for 1934 and 1944. The given data are assumed to represent averages for decades whose central points they approximate (1879 for 1874–83, etc.). Overlapping decades interpolated where necessary by straight line arithmetic interpolations from both preceding and succeeding observations. The two results were then averaged.

1869–78: Extrapolated from 1874–83 by movement of estimates by Schmookler (1952, Table 5, col. 3).

1944–53: Extrapolated from 1939–48 on basis of estimates kindly supplied by J. W. Kendrick (billions of dollars in 1929 prices).

12. Line 11÷line 2 (dollars per person).

13. Weighted index of relatives (1919–28 = 100), combining man-hours×3 and capital×1. Weights represent the relative values of service incomes and property incomes respectively as estimated by J. W. Kendrick for 1929 and supplied to author. Kendrick's relative weights were, more precisely, 72:28.

14. Weighted index of relatives (1919−28 = 100), combining man-hours per capita and capital *per capita* with weights as in line 13.

15. Line 1÷line 6 (dollars per employed in 1929 prices).

16. Line 1÷line 9 (dollars per man-hour).

17. Line 1÷line 11 (cents per dollar of capital).

18. Index of NNP÷index of total input of resources (1919−28 = 100).

changes in the age composition of the population, because of the shift of people from farms to cities, and because the great increase in the participation of women in work offset the withdrawal of young people to school and of elderly men to earlier retirement. On the other hand, the reduction in working hours more than counterbalanced the increase in the labor force ratio.[4]

4. See C. D. Long (1958, ch. 11). While there may have been some difference in the percentage of unemployment between the 1870s and the 1950s, the great decline in working time per member of the labor force was due to a reduction in hours of work. The change in working hours recorded in our table is based on a series appearing in Kuznets's 'Long-Term

The physical volume of capital, of course, increased much more rapidly than population. An estimate of total capital, which takes account of land, structures, producers' durable equipment, inventories and net foreign claims, increased to nearly ten times its size seventy-five years ago. Capital per head of the population approximately tripled.[5]

What has been the increase in the input of all resources *per capita*? Suppose we combine our indexes of labor input *per capita* and of capital supply *per capita* with weights proportionate to the base period incomes going to labor and property, respectively. If we may equate productivity with earnings, we obtain a combined index of resources which has a particular meaning. It tells us how net national product *per capita* would have grown

Changes' extended an extra decade on the basis of Kendrick's figures. But other estimates make the long-term decline somewhat less or more. For comparison, the following alternatives are of interest:

	Base period	Given year or period	Index of average hours in given year (base = 100)
1. Kuznets, standard hours	1869–78	1944–53	73
2. Dewhurst and Fichlander, actual hours	1870–80	1950	62
3. Barger, actual hours in commodity production	1869–79	1949	83
4. Barger, actual hours in distribution	1869–79	1949	66
5. Kuznets, standard hours	1894–1903	1944–53	79
6. Kendrick, actual hours	1899	1953	83

Sources:

Line 1. Kuznets (1952, Table 7). Figures extended from 1939–48 to 1944–53 on the basis of estimates kindly supplied by J. W. Kendrick.

Line 2. Dewhurst and Associates (1955), Appendix 20–24.

Line 3. Barger (1955, Table 5).

Line 4. Same as line 3.

Line 5. Same as line 1.

Line 6. Supplied by J. W. Kendrick.

5. Estimates of capital wealth are extremely rough and must be treated with great reserve. While there is no doubt that capital increased much faster than population, we may well doubt whether the relative increase was just that suggested by the figures. Our figures are based on the table presented by Kuznets for the years 1879–1944 (1952, Table 11). See notes to

had the productivity of resources remained constant at base period levels while only the supplies of resources per head increased. Such an index, based on the twenties, rises only some 14 per cent between the seventies and the last decade. To account for the quadrupling of net national product *per capita*, the productivity of a representative unit of all resources must have increased some 250 per cent. This seems to imply that almost the entire increase in net product *per capita* is associated with the rise in productivity. This result may arise in some part from our choice of a base period. We chose a fairly recent base period, 1919–28, close to the valuation base of the national product estimates, 1929. Since the relative importance of service and property incomes remains fairly stable over the entire period (Kuznets, 1952, pp. 135–7), and since capital increased far more rapidly than labor, the price of a unit of capital service must have fallen over time compared with that of a unit of labor. The choice of a fairly recent year as a base for our relatives in effect means weighting each unit of capital by a relatively low price.

Table 1. The figures may be compared with R. W. Goldsmith's estimates (1951, p. 18).

| | Relatives for 1944 | | |
| | Goldsmith (1900 = 100) | Kuznets (1899 = 100) | Ratio |
	1	2	2 ÷ 1
Land	133	208	1·86
Reproducible wealth*	271	344	1·27
Total	216	284	1·31

* Structures, producers' durable equipment, inventories and net foreign claims.

Neither Goldsmith's figures nor Kuznets's are free of serious difficulties due to weaknesses in the statistical source of capital data and to problems of valuation and deflation (Kuznets, 1952, pp. 79–80 and Goldsmith, 1951, and following comments by Kuznets). It is possible that the true increase of capital lies outside the range suggested by both sets of figures. Our figures make no allowance for changes in the service hours of capital comparable with that for labor. There is no statistical basis for such an adjustment. The decline in labor hours is not a reliable indication since capital is often operated on multiple shifts or even continuously. It is not clear whether such practices have grown or declined.

Experiment, however, indicates that choice of base is of minor importance for the question at hand. If we shift the base of the index of resources to 1869–78, the increase of total input between 1869–78 and 1944–53 becomes 44 per cent. If we compare this with the rise of net national product *per capita* in 1929 prices, the indicated rise in productivity is still much greater, 175 per cent. This calculation, however, overstates the importance of the shift in base. If we shift the base for our resource index to 1869–78, we should also value national product in the prices of that decade. This would, in all likelihood, make the trend of national product steeper and so indicate a greater increase in productivity than the 175 per cent mentioned above (Kuznets, 1952, pp. 44–7).

3. This result is surprising in the lopsided importance in which it appears to give to productivity increase, and it should be, in a sense, sobering, if not discouraging, to students of economic growth. Since we know little about the causes of productivity increase, the indicated importance of this element may be taken to be some sort of measure of our ignorance about the causes of economic growth in the United States and some sort of indication of where we need to concentrate our attention. Since it will do little good to provide a catalogue of the possible causes of the rise in efficiency, I shall merely add two notes which have to do with a proper understanding of calculations which resolve the growth of output into the growth of resources and productivity, respectively. They will, I hope, also take some of the edge off my conclusion and serve to put the importance of factor input in somewhat better perspective.

First, although input of resources *per capita* has not increased much, this does not mean that the increase of resources has not contributed significantly to the rise in output per head. Total input of labor and capital has increased a great deal; population more than tripled. The nearly constant number of man-hours *per capita*, therefore, meant a tripling of total man-hours. The tripling of capital per head meant a more than ninefold increase in total capital. The quadrupling of net national product *per capita* meant a twelvefold rise of total national product. But 'the division of labor is limited by the extent of the market'. If there is anything to the notion that when raw materials are plentiful resources and output will be connected according to a law of increasing returns to scale, then the great expansion of total resources must have contributed substantially to the increase in productivity.

Second, our calculations of resource inputs are based on usual definitions of labor supply and capital. These conventional methods of measuring resource inputs are faulty and, in the case of this country during the last seventy-five years, probably understate the increase in factor input. We therefore tend to overstate the rise in productivity.

On the side of labor, it is clear that the reduction in the importance of teenagers and old men in the labor force has concentrated employment in the age groups whose output per man is relatively high. It also seems likely that with the urbanization and commercialization of work there has been an increase in the intensity of labor. These changes may perhaps be offset by the augmented importance of women in the labor force. It seems possible, however, that a properly weighted index of man-hour input would have increased significantly over the period even if we leave out of account such matters as improvements in skill and managerial capacity which reflect training and other capital investment (Kuznets, 1952, p. 77).

On the side of capital, there is a chronic underestimate of investment and accumulated stock because, for purposes of measurement, we identify capital formation with the net increase of land, structures, durable equipment, commodity stocks and foreign claims. But underlying this conventional definition of investment is a more fundamental concept which is broader; namely, any use of resources which helps increase our output in future periods. And if we attempt to broaden the operational definition, then a number of additional categories of expenditures would have to be included, principally those for health, education and training,[6] and research. These are fairly obvious because one is conscious both of an income motivation and an income effect. But there are other classes of expenditures where motives are

6. A properly constructed index of labor input which gave due weight to the higher productivity of more highly educated or trained workers and to differences in vigor would be an alternative way to try to take these inputs into account.

mixed or disguised but which have at least the incidental effect of increasing productivity; namely, expenditures for food, clothing and some recreation. The fact is that, in a thoroughly commercialized economy, disposing of a large surplus above its requirements for minimum consumption, very few expenditures are wholly without the aim and effect of increasing income. If this is so, effective capital formation, broadly conceived, must be sought in certain types of consumption and governmental expenditures as well as in conventional net investment.

The point of these two comments is simply that the relation between the contributions of resource expansion and of productivity growth is more complicated than our conventional measures can reveal. Two morals may be drawn. First, the long-term expansion of the labor supply must be restudied so as to provide a measure of the value of its changing composition as well as its changing size. And the expansion of the capital stock must be restudied to take account of a broader conception of accumulated resources. It may well be that we shall find it inconvenient to merge these additional categories of accumulation with conventional capital. But whatever our terminology, we have to pay close attention to all the ways our society uses its resources to increase its future product.

When all due allowance for the concealed increase in resource expansion has been made, however, there will remain a huge area to be explained as an increase in productivity. Our capital stock of knowledge concerning the organization and technique of production has grown at a phenomenal pace. A portion of this increase – presumably an increasing proportion – is due to an investment of resources in research, education, and the like. This part we may possibly be able to attribute accurately to the input of these resources in so far as we learn to trace the connection between such investment in knowledge and its marginal social contribution, as distinct from those small parts of its value which can be privately appropriated. Beyond this, however, lies the gradual growth of applied knowledge which is, no doubt, the result of human activity, but not of that kind of activity involving costly choice which we think of as economic input. To identify the causes which explain not only the rate at which our opportunities to raise efficiency increase but also the pace at which we take advantage of those opportunities will, no doubt, remain the central problem in both the history and theory of our economic growth. The chief excuse for attempts to separate the measurable contributions of resources from those of productivity is to pose this problem as clearly as possible.

The trend of the rate of growth

From these measures of the net expansion of output and resources since the Civil War, I turn next to the often asked question: has our rate of growth been slowing up. The retardation of growth in Great Britain and in other leading industrial countries and our own experience in the thirties have made the possibility of retardation a source of widespread anxiety.

Unfortunately, the information now available does not permit us to make a secure answer. The sources of error and bias in national product estimates – already noticed in connection with the measures of expansion – apply with aggravated force when we try to compare rates of growth at different times. We can often guess the direction in which national product estimates are biased, but in most cases we do not now know whether a particular bias affected the figures more strongly in one decade than another. It is clear, for example, that our inability to take consistent account of household production makes the rate of growth of national product too high during a period in which household production was giving way to commercial production. It is probable also that the rate of transfer from home to business changed over time. But did the transfer proceed more rapidly in the last quarter of the nineteenth century than in the second quarter of the twentieth, and by how much? This is the question relevant to changes in the rate of growth. We cannot answer it with any confidence. It is certain, therefore, that any statements about a long tendency in the rate of growth of national product must be treated with the greatest reserve unless the drift is so large and so persistent that no likely combination of biases and errors could account for it. In my judgement, the drift of the figures is not so clear. It is, nevertheless, worth while to review them, partly to check the bases for much current interpretation and speculation and partly because it is interesting to try to allocate the apparent changes in output growth to inputs and productivity.

Taking the figures as they stand, they give some indication of a slowing down in the rate of growth over the course of the eighty-odd years since 1870. To see this, one has to take account not only of the ordinary business cycles, which generally run their course well within a decade, but also of the longer fluctuations which appear in the rate of growth of output. I shall have something more to say about these fluctuations in the next section. A smoothing of the data to eliminate both types of fluctuations suggests that total net product rose more rapidly during the last quarter of the nineteenth century than it did during the second quarter of the twentieth century. The apparent decline in the rate of growth of product *per capita* is less pronounced (see Figs. 1 and 2).[7]

Whatever the showing of the figures, however, it is not at all clear that they are accurate enough for the purpose or, if accurate, that they represent the work of persistent forces in the economy. The very high rate of growth in the last quarter of the nineteenth century reflects an exceptionally high rate of increase during the late seventies and early eighties. If we neglect this apparently remarkable decade and take into account the possibilities of error and bias, the rates of growth afford no significant indication of retardation until we reach the depression of the thirties.[8] The

7. Kuznets's original estimates of net national product, which appear in the form of decade averages of annual data for overlapping decades, may be taken to eliminate most of the effects of ordinary business cycles. The same may be said of the rates of change between the overlapping decade averages (essentially rates of change per quinquennium). If we then take five-item moving averages of these rates of change (end items weighted one-half), we average experience for a twenty-year period, which is probably long enough to eliminate most of the effects of the longer fluctuations in the rates of growth. Both the quinquennial rates of change and the moving averages are shown in the chart.

8. Compare Arthur F. Burns's conclusions for the period 1870–1930 based upon his study of physical output indexes. While he is highly skeptical about any conclusion which might be reached on the basis of the data available to him, he ventured to write: '. . . if there has been any decline in the rate of growth in the total physical production of this country, its extent has probably been slight, and it is even mildly probable that the rate of growth may have increased somewhat' (1934, p. 279). Since the retardation in the growth of the physical volume of production was almost certainly less than that in population, Burns felt it was still less probable that the growth of *per capita* output had been drifting downward.

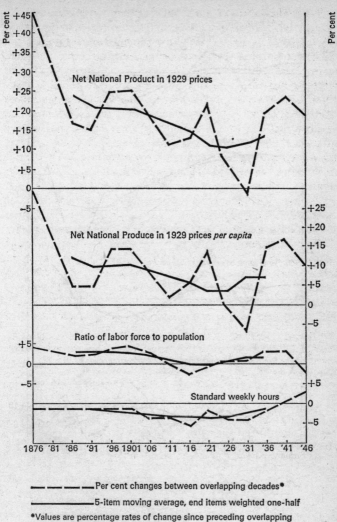

Figure 1 Trends in growth rates, 1869–1953

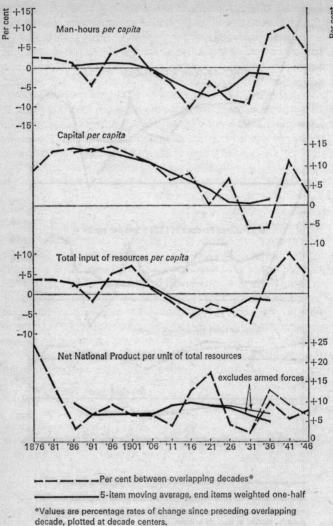

*Values are percentage rates of change since preceding overlapping decade, plotted at decade centers.

early figures of rapid growth are the least secure portions of the estimates. If valid, they may reflect a temporary surge of output.[9]

On the other hand, the low rate of growth in the second quarter of the present century is entirely a reflection of the Great Depression. The rates of expansion since 1934 are as high as in any earlier period other than the (possibly exceptional) period in the late seventies and early eighties. They would look still higher on the basis of the Commerce figures than they do on the basis of the Kuznets estimates.

Whether there has been a significant degree of persistent retardation in the growth of national product *per capita* would, therefore, seem to turn on the answers to two questions presently unanswerable. Do the various biases and weaknesses in the estimates make for an appearance of acceleration or retardation? Did the surge of the early years and the deep depression of the latter years represent fortuitous or persistent forces?

Whatever the answers to these important questions of history, it is possible to reach some conclusion with regard to the sources of the apparent retardation. Whatever tendency there may have been for growth of net product *per capita* to decline is traceable very largely, if not entirely, to a decline in the rate of growth of resources used per head of the population. Until the last two decades, which were years of accelerated growth both of input and output *per capita*, all the elements of resource input had grown less rapidly or declined more rapidly in later decades than in earlier. The ratio of labor force to population, which increased fairly steadily from 1870 until around 1910, thereafter fell, or grew very little, until the decade of the forties. With the exception of these recent years, hours of work fell at a more rapid rate during the 1900s than during the late 1800s. As a result, man-hours per head rose at a declining rate until the turn of the century and then fell at an increasing pace until the mid-thirties.

9. There is, indeed, some evidence that rates of growth were lower in the immediately preceding decades. After a discussion of W. I. King's older estimates for the period 1850–80, Professor Kuznets comments: '. . . the only safe comparison one can draw is that *per capita* real income did show some increase from 1850 to 1880, perhaps as much as 50 per cent or more, perhaps as little as 20 per cent or less'. This contrasts with Professor Kuznets's own estimate that *per capita* real income rose some 50 per cent in the single decade interval 1869–78 to 1879–88 (1952, p. 240).

One may add that the diversion of labor force to military purposes increased over time. So the decline in the rate of growth of civilian man-hours per head was even more pronounced than in that of total man-hours. In the thirties, of course, great unemployment was an aggravating element. The growth of capital per head, as conventionally measured, slowed down drastically. It rose at a constantly slower rate until the end of the twenties, and then declined during the depression. In spite of rapid growth during the last fifteen years, capital per head in the late forties was only a little more plentiful than in the twenties. Until relatively recent years, therefore, every major element of resources made for retardation in the growth of net product *per capita*. The combined index of resources *per capita* rose at a declining rate until the early 1900s and then fell at an increasing rate until the middle thirties.

It was these changes in the growth of resources per head which account for most, if not all, the retardation in the growth of net product *per capita* recorded in the estimates. Productivity per man-hour, on the other hand, has been rising at a fairly constant rate since the eighties, and this trend has dominated the movement of the productivity of all resources. The productivity of capital, taken alone, seemed to be falling until about the First World War. It has been rising since, a fact which has helped to maintain the rate of rise in the productivity of all resources. The essential constancy in the rate of rise of productivity is perhaps the most significant single fact which emerges from a review of our economic record since the Civil War.[10] Whether this reflects

10. It is a 'fact' heavily qualified by all the errors and biases in the national product figures and in the estimates of labor input and capital. Moreover, it measures both labor input and capital in a fashion which neglects some increase in labor input due to change in the age composition of the labor force and probably in the intensity of work. It also neglects the fact that a substantial volume of resources has been devoted to the improvement of intangible capital: technology, labor skills, health and organization. The rate of accumulation of such intangible capital may be increasing. It is a 'fact' which is somewhat bolstered by the showing of other overall measures of productivity. These measures, to which I refer below, are not based on data which are wholly independent of those on which I rely, but they involve some degree of independence and they are each calculated on a somewhat different plan: (a) John W. Kendrick's estimate of 'national output per unit of labor and capital combined', 1899–1953, shows no retardation in growth. Its rate of growth since 1919 is somewhat greater than it

an essentially unweakened capacity to increase the efficiency of our resources in the future is perhaps the most significant single question which requires an answer.

Fluctuations in the rate of growth

The trend of the rate of increase of national product, whether constant or slowly declining, is a generalization concerning our growth which abstracts from its fluctuations and pretends to describe only its persistent or underlying movement. But, of course, the growth of output in reality is anything but steady. It rarely runs in the same direction for many months and almost never for even two months at the same rate.

We have learned to think of these alterations of the rate of growth as in part accidental and in part systematic. Aside from seasonal fluctuations, the systematic movement principally identified in the past has been the short-term business cycle either in its minor or major variant. If, however, to reveal the secular trend in output we calculate moving averages for periods long enough to eliminate business cycles (nine-year moving averages, for example), the resulting curve of output for the period since 1870 still reveals striking fluctuations – not in the level of output but in its rate of growth. The curve mounts relatively steeply for a time and then exhibits retardation in a pattern which has repeated itself roughly every twenty years. The

---

was in the two earlier decades (National Bureau of Economic Research, 1955, p. 45). (b) The Twentieth Century Fund estimate of 'real private national income per private manhour', 1850–1952, has a trend which suggests a mild degree of acceleration (Dewhurst and Associates, 1955, pp. 39–42). (c) Jacob Schmookler's estimate of gross national product per combined unit of labor and capital, 1869–1938, shows no tendency to retardation in growth after the first decade (1952, Table 9). (d) Harold Barger's estimates of productivity per man-hour in commodity production (agriculture, mining and manufacturing) and distribution, 1869–1949, show either a steady rate of growth or else acceleration, whether taken individually or in combination. Since Barger's estimates are based on indexes of the physical volume of production in the four industrial branches, his figures are more nearly independent of our own than are the other alternatives. Barger's figures take no account of productivity in the service industries other than distribution. It is possible that a productivity index for the remainder of the service trades, if one could be devised, would change the picture (1955, pp. 37–41).

same observations may be made if one calculates rates of increase in decade averages of output for overlapping decades (see chart) (Kuznets, 1952, pp. 48–57). The possibility, therefore, arises that there is a significant cycle in the secular trend of output – meaning by this, movements which persist over a period longer than a business cycle – with an approximate duration of twenty years.[11]

In relatively recent times, the hypothesis of a twenty-year growth cycle starts with Kuznets's early work on secular trends in which he suggested the existence of fluctuations of this duration in the rate of growth of production of many individual commodities, in the rate of rise of many prices, and in several other types of time series (1930, ch. 4). The hypothesis was then taken up by Arthur F. Burns (1934, ch. 5), in which he showed not only that twenty-year growth cycles were characteristic of the output of many commodities but also that the cycle was general in the sense that the growth cycles of different commodities tended to concur in time and that they also appeared in indexes of aggregate industrial production. Burns also found his secular swings in non-agricultural prices, in shares traded, in business failures, and in patents issued. Finally Kuznets in later work has shown that the same swings appear in his long-term estimates of gross and net national product (1952, pp. 48–57), in labor productivity, in population and immigration (with a lag), and in residential construction (with a longer lag).[12] Unpublished work by Kuznets and Dorothy S. Thomas carries the subject further, particularly as regards population change, internal migration, construction, and certain financial series. Still others suggest the presence of a similar cycle in foreign countries.[13] Both Kuznets and Burns con-

11. Although my discussion is restricted to the twenty-year cycle, I do not mean to suggest that the secular trend of output may not be subject to other significant types of fluctuations. If it is, however, their period is too long to be distinguished clearly from the underlying trend in a review covering some seventy to eighty years.

12. Simon Kuznets and Ernest Rubin (1954, pp. 30–34). The findings of this paper are, to some extent, similar to those of Brinley Thomas.

13. See Walter Hoffman (1955; a translation of the German original published 1940), pt C. Brinley Thomas, 1954, especially chs. 7 and 8, argues that there were twenty-year cycles in the United States (and, to some extent, Canada and Australia) connected by immigration and capital

sidered their work only exploratory and neither was persuaded that the evidence so far accumulated established the existence of significant recurrence of movement; that is, of true cycles.

Kuznets finds three complete swings in the rate of growth in the period since 1870 and one incomplete swing – a rise beginning 1932 and (tentatively) reaching its peak in 1945.[14] The variation in the rate of growth between the expansion and contraction phases of the growth cycles is large compared with the average rate of growth itself. For example, in the period 1873–1926, that is, before the huge fluctuations associated with the Great Depression and the Second World War, the overall average rate of rise of GNP per worker was about 20 per cent per decade. But the average rate of growth in upswing periods was about five times as rapid as in the downswing periods. The average difference between the rate of growth in the upswing periods and that in the downswing periods was as large as the average rate of

---

movements to inverted cycles in Great Britain, Sweden and perhaps Germany. B. Weber and S. J. Handfield-Jones (1954) attempt to connect the long waves in Kuznets's figures for national product with successive waves of innovation in the application of steam power to industry and transport (1870–82), in the further extension of steam and steel and in the development of new resources (1894–1907), and in electricity, industrial chemicals and the internal combustion engine (1919–29).

14. The suggested chronology runs as follows:

Trough	Peak
1873	1884
1892	1903
1912	1926
1932	1945

The dates were determined by observing a graph of a nine-year moving average of GNP per worker in 1929 prices and locating the points at which the slope became significantly steeper and flatter. The first and last dates are set only tentatively until the data can be extended far enough backwards and forwards to confirm the position of the inflection point. This chronology was presented in an unpublished memorandum, 'Swings in the rate of secular growth', prepared for the Capital Requirements Study of the National Bureau (March, 1952). A similar chronology based on the movement of rates of change of net national product in 1929 prices between overlapping decades appears in Kuznets (1952, p. 55). An earlier chronology, based on the consensus of many commodity production series, but containing an extra cycle in the decade 1910–20, was presented by A. F. Burns (1934, p. 196). Since but few examples have as yet been traced in the American data, neither the average duration of the alleged cycle nor its variability can be considered established.

growth itself.[15] If we add the last long swing, which covers the Great Depression and the upswing of the forties, the size of the average fluctuation becomes very much greater than the average rate of growth.

The significance of these long swings is not yet established. At least two possibilities are present which would rob the observed fluctuations of most of their meaning. It may be that what we observe are only accidental variations in the severity or duration of ordinary business cycles, which assume the appearance of long swings when their effects are stretched out and smoothed by moving averages or some similar device. And even if it is true that the swings reflect forces which operate over periods longer than business cycles, it may still be true that these forces are predominantly irregular and haphazard.

These negative possibilities cannot now be dismissed. Indeed the influence upon the swings so far experienced in this country of substantial irregular forces was patent and undeniable. Thus it seems reasonable to attribute some significant responsibility for the swing beginning around 1873 to the recovery from the Civil War, for the swing beginning around 1912 and continuing through the twenties to the First World War, and for the swing beginning in 1932 and continuing into the forties to the Second World War. It would be impossible to try to review the considerable body of relevant evidence in the short space available to me. For purposes of this discussion, I can simply record my conviction that there is sufficient evidence to make the long-swing hypothesis worthy of closer investigation.[16]

15. These are geometric means weighted by the duration of phases. The data are from Kuznets's memorandum (1930).

16. Merely to indicate that this position has some tangible basis, one may cite the following:

1. In support of the proposition that the long swings are more than merely an illusory reflection of business cycles: (a) The persistence of long swings in figures arranged to show average levels in identified business cycles (Kuznets, 1952a). (b) The persistence of long swings in figures for business cycle peaks alone, which thus partially eliminate the effects of long and deep depressions (*ibid*). (c) The existence of long swings in British data which, at least for 1870–1914, appear to fluctuate inversely to the swings in this country, whereas the normal business cycle relation is positive (B. Thomas, 1954, ch. 7). (d) The fact that the period required for the exploita-

If supported by further study, the long-swing hypothesis promises to make a serious contribution to our understanding of economic change. I shall cite three reasons:

First, if it be true that the long swings reflect, in significant degree, the operation of systematic responses to either regular or irregular stimuli, then study of our past growth will best be organized in periods corresponding to the long swings. And a proper understanding of these waves of growth will presuppose an ability to separate the unique from the recurrent forces at work in each period.

Second, the long swings appear to represent fluctuations in productivity growth and in the increase of manpower and capital to a greater degree than business cycles whose most prominent characteristic is that they are fluctuations in the intensity with which resources are employed. (Before the Great Depression, quinquennial changes in the level of employment were not well

---

tion of major innovations or new territory is certainly longer than the five or six years associated with even major business expanisons. This does not account for the twelve or thirteen year long-swing expansions or for twenty-year cycles, but it argues for the presence of unsteady expansive stimuli which carry over from one business cycle to another.

2. In support of the view that the long swings exhibit at least some regular features, in addition to the impact of many irregular circumstances, confident assertion is prevented by lack of study and by the fact that US production data in fair quantity now reach back only to 1860 and, therefore, reveal only three and one-half long swings. Subject to these limitations, there are clear hints of regularities which suggest the presence of an internal structure with some stability. I refer only to certain prominent observations in published sources: (a) Burns's finding that during periods of long-swing expansion, the rates of growth of production of different commodities become increasingly different and that this dispersion of the rates of growth declines in long-swing contractions (1934, pp. 242–7). (b) Burns's finding that each period of long-swing expansion is followed by a business cycle depression of great severity, a finding which he tentatively connects with the increasing dispersion in the rate of expansion of individual industries during the upswing (1934, pp. 247–53). (c) Kuznets's and Rubin's finding (1954) confirming B. Thomas's finding (1954, chs. 7 and 8) concerning the lagged response of immigration to the rate of growth of output, and Kuznets's finding that the rate of increase of population showed a lagged response to economic growth (1952, p. 55). (d) The common finding (Kuznets and Rubin, 1954) that there is a lagged response of construction to population growth.

correlated with the long swings in the rate of growth of output, nor were the magnitude of the changes in employment percentages comparable in size with those in output (Kuznets, 1952, Tables 3 and 10). These facts also bear on the question of the independence of the longer swings from business cycles. It is not yet clear, however, that the unemployment figures are sufficiently accurate for the purpose, and the conclusion needs to be checked by further study.) Unless it turns out that fluctuations in the growth of productivity or of resource supply are themselves chiefly governed by business cycle movements, we must anticipate fluctuations in the rate of growth of output even if we succeed in maintaining employment at high levels. Since past fluctuations in the rate of growth were wide relative to its long-term average, projections of output looking forward a decade or two – such as are often made – would need to take into account the current phase of the long swing. This presupposes a capacity to define the recurrent features of long swings – something we cannot do today.

Finally, our past experience with long swings shows that every upswing in the rate of growth has terminated in a depression of great severity. This may, as Burns tentatively suggested (see footnote 16 above), be connected with a tendency for growth to become increasingly unbalanced as the upswing proceeds, presumably leading to a decline of investment in the overexpanded industries. Or a mere slowing down of the rate of growth of output for any reason may lead to a reduction of investment, as one variant of the Harrod–Domar theory suggests. In either case, there is reason to expect that whenever our rate of progress begins to slow down markedly, forces will also be present making for serious depression. Such depressions will not necessarily be experienced in view of the role government may play in counteracting them. But certainly the wisdom and energy of the government will be put to a severe test. The experience with long swings suggests that our liability to severe depression may be a normal part of a swing in the rate of growth, which may itself be due, in part, to recurrent causes. If these could be identified and better understood, our ability to prepare for, and to meet, the emergency of depression would undoubtedly be enhanced.

References

BARGER, H. (1955), *Distribution's Place in the American Economy since 1869*, National Bureau of Economic Research.

BUREAU OF THE CENSUS (1955), *Survey of Current Business*, Washington, DC.

BURNS, A. F. (1934), *Production Trends in the United States since 1870*, National Bureau of Economic Research.

DEWHURST and ASSOCIATES (1955), *America's Needs and Resources: A New Survey*, Twentieth Century Fund.

GOLDSMITH, R. W. (1951), 'Derivation of a perpetual inventory of national wealth since 1896', *Studies in Income and Wealth*, vol. 14, National Bureau of Economic Research.

HOFFMAN, W. (1955), *British Industry, 1700–1950*, translated from the German of 1940, Blackwell.

KING, W. I. (1915), *The Wealth and Income of the People of the United States*, Macmillan.

KUZNETS, S. (1930), *Secular Movements in Production and Prices*, Houghton Mifflin.

KUZNETS, S. (1952), 'Long-term changes', *Income and Wealth of the United States*, Bowes & Bowes.

KUZNETS, S. (1952a), *Swings in the Rate of Secular Growth*, unpublished memorandum.

KUZNETS, S. (1961), *Capital in the American Economy*, Princeton University Press.

KUZNETS, S., and RUBIN, E. (1954), *Immigration and the Foreign Born*, National Bureau of Economic Research.

LONG, C. D. (1958), *The Labor Force under Changing Employment and Income*, National Bureau of Economic Research.

NATIONAL BUREAU OF ECONOMIC RESEARCH (1955), *35th Annual Report*, Washington, DC.

SCHMOOKLER, J. (1952), 'The changing efficiency of the American economy, 1869–1938', *Rev. Econ. Stat.*, vol. 34, pp. 214–31.

THOMAS, B. (1954), *Migration and Economic Growth*, Cambridge University Press.

WEBER, B., and HANDFIELD-JONES, S. J. (1954), 'Variations in the rate of economic growth in the USA, 1869–1938', *Oxford Econ. Papers*, vol. 6, pp. 101–31.

16 R. Solow

Technical Change and the Aggregate Production Function[1]

R. Solow, 'Technical change and the aggregate production function',
Review of Economics and Statistics, August 1957, pp. 312-20.

In this day of rationally designed econometric studies and super input–output tables, it takes something more than the usual 'willing suspension of disbelief' to talk seriously of the aggregate production function. But the aggregate production function is only a little less legitimate a concept than, say, the aggregate consumption function, and for some kinds of long-run macro-models it is almost as indispensable as the latter is for the short run. As long as we insist on practicing macroeconomics we shall need aggregate relationships.

Even so, there would hardly be any justification for returning to this old-fashioned topic if I had no novelty to suggest. The new wrinkle I want to describe is an elementary way of segregating variations in output per head due to technical change from those due to changes in the availability of capital per head. Naturally, every additional bit of information has its price. In this case the price consists of one new required time series, the share of labor or property in total income, and one new assumption, that factors are paid their marginal products. Since the former is probably more respectable than the other data I shall use, and since the latter is an assumption often made, the price may not be unreasonably high.

Before going on, let me be explicit that I would not try to justify what follows by calling on fancy theorems on aggregation and index numbers.[2] Either this kind of aggregate economics

1. I owe a debt of gratitude to Dr Louis Lefeber for statistical and other assistance and to Professors Fellner, Leontief and Schultz for stimulating suggestions.
2. Mrs Robinson (1954) in particular has explored many of the profound difficulties that stand in the way of giving any precise meaning to the quantity of capital, and I have thrown up still further obstacles (*Review of*

appeals or it doesn't. Personally I belong to both schools. If it does, I think one can draw some crude but useful conclusions from the results.

Theoretical basis

I will first explain what I have in mind mathematically and then give a diagrammatic exposition. In this case the mathematics seems simpler. If Q represents output and K and L represent capital and labor inputs in 'physical' units, then the aggregate production function can be written as

$$Q = F(K, L; t). \qquad 1$$

The variable t for time appears in F to allow for technical change. It will be seen that I am using the phrase 'technical change' as a shorthand expression for *any kind of shift* in the production function. Thus slowdowns, speedups, improvements in the education of the labor force, and all sorts of things will appear as 'technical change'.

It is convenient to begin with the special case of *neutral* technical change. Shifts in the production function are defined as neutral if they leave marginal rates of substitution untouched but simply increase or decrease the output attainable from given inputs. In that case the production function takes the special form

$$Q = A(t)f(K, L), \qquad 1a$$

and the multiplicative factor $A(t)$ measures the cumulated effect of shifts over time. Differentiate 1a totally with respect to time and divide by Q and one obtains

$$\frac{\dot{Q}}{Q} = \frac{\dot{A}}{A} + A \frac{\partial f}{\partial K} \frac{\dot{K}}{Q} + A \frac{\partial f}{\partial L} \frac{\dot{L}}{Q}$$

where dots indicate time derivatives. Now define

$$w_k = (\partial Q / \partial K) K / Q$$

Economic Studies, vol. 23, no. 2). Were the data available, it would be better to apply the analysis to some precisely defined production function with many precisely defined inputs. One can at least hope that the aggregate analysis gives some notion of the way a detailed analysis would lead.

and

$$w_L = (\partial Q / \partial L) L / Q$$

the relative shares of capital and labor, and substitute in the above equation (note that $\partial Q / \partial K = A \, \partial f / \partial K$, etc.) and there results:

$$\frac{\dot{Q}}{Q} = \frac{\dot{A}}{A} + w_K \frac{\dot{K}}{K} + w_L \frac{\dot{L}}{L}. \qquad 2$$

From time series of $\dot{Q}/Q$, w_k, $\dot{K}/K$, w_L and $\dot{L}/L$ or their discrete year-to-year analogues, we could estimate $\dot{A}/A$ and thence $A(t)$ itself. Actually an amusing thing happens here. Nothing has been said so far about returns to scale. But if all factor inputs are classified either as K or L, then the available figures always show w_K and w_L adding up to one. Since we have assumed that factors are paid their marginal products, this amounts to assuming the hypotheses of Euler's theorem. The calculus being what it is, we might just as well assume the conclusion, namely that F is homogeneous of degree one. This has the advantage of making everything come out neatly in terms of intensive magnitudes. Let $Q/L = q$, $K/L = k$, $w_L = 1 - w_K$; note that $\dot{q}/q = \dot{Q}/Q - \dot{L}/L$ etc., and 2 becomes

$$\frac{\dot{q}}{q} = \frac{\dot{A}}{A} + w_K \frac{\dot{k}}{k}. \qquad 2a$$

Now all we need to disentangle the technical change index $A(t)$ are series for output per man-hour, capital per man-hour, and the share of capital.

So far I have been assuming that technical change is neutral. But if we go back to 1 and carry out the same reasoning we arrive at something very like 2a, namely

$$\frac{\dot{q}}{q} = \frac{1}{F} \frac{\partial F}{\partial t} + w_k \frac{\dot{k}}{k}. \qquad 2b$$

It can be shown, by integrating a partial differential equation, that if $\dot{F}/F$ is independent of K and L (actually under constant returns to scale only K/L matters) then 1 has the special form 1a and shifts in the production function are neutral. If in addition $\dot{F}/F$ is constant in time, say equal to a, then $A(t) = e^{at}$ or in discrete approximation $A(t) = (1+a)^t$.

The case of neutral shifts and constant returns to scale is now easily handled graphically. The production function is completely represented by a graph of q against k (analogously to the fact that if we know the unit-output isoquant, we know the whole map). The trouble is that this function is shifting in time, so that if we observe points in the (q, k) plane, their movements are compounded out of movements along the curve and shifts of the curve. In Figure 1, for instance, every ordinate on the curve

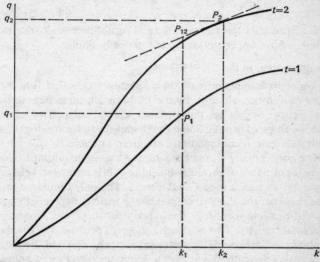

Figure 1

for $t = 1$ has been multiplied by the same factor to give a neutral upward shift of the production function for period 2. The problem is to estimate this shift from knowledge of points P_1 and P_2. Obviously it would be quite misleading to fit a curve through raw observed points like P_1, P_2 and others. But if the shift factor for each point of time can be estimated, the observed points can be corrected for technical change, and a production function can then be found.[3]

3. Professors Wassily Leontief and William Fellner independently pointed out to me that this 'first-order' approximation could in principle be improved. After estimating a production function corrected for technical change (see below), one could go back and use it to provide a second approximation to the shift series, and on into further iterations.

The natural thing to do, for small changes, is to approximate the period 2 curve by its tangent at P_2 (or the period 1 curve by its tangent at P_1). This yields an approximately corrected point P_{12}, and an estimate for $\Delta A/A$, namely $\overline{P_{12}P_1}/q_1$. But

$$k_1 P_{12} = q_2 - \partial q/\partial k \Delta k$$

and hence $\overline{P_{12}P_1} = q_2 - q_1 - \partial q/\partial k \Delta k = \Delta q - \partial q/\partial k \Delta k$
and $\Delta A/A = P_{12}P_1/q_1 = \Delta q/q - \partial q/\partial k(k/q)\Delta k/k =$
$$\Delta q/q - w_k \Delta k/k$$

which is exactly the content of **2a**. The not-necessarily-neutral case is a bit more complicated, but basically similar.

An application to the US: 1909–49

In order to isolate shifts of the aggregate production function from movements along it, by use of **2a** or **2b**, three time-series are needed: output per unit of labor, capital per unit of labor, and the share of capital. Some rough-and-ready figures, together with the obvious computations, are given in Table 1.

The conceptually cleanest measure of aggregate output would be real net national product. But long NNP series are hard to come by, so I have used GNP instead. The only difference this makes is that the share of capital has to include depreciation. It proved possible to restrict the experiment to private non-farm economic activity. This is an advantage (a) because it skirts the problem of measuring government output, and (b) because eliminating agriculture is at least a step in the direction of homogeneity. Thus my q is a time-series of real private non-farm GNP per man-hour, Kendrick's valuable work.

The capital time-series is the one that will really drive a purist mad. For present purposes, 'capital' includes land, mineral deposits, etc. Naturally I have used Goldsmith's estimates (with government, agricultural and consumer durables eliminated). Ideally what one would like to measure is the annual flow of capital services. Instead one must be content with a less Utopian estimate of the stock of capital goods in existence. All sorts of conceptual problems arise on this account. As a single example, if the capital stock consisted of a million identical machines and if each one as it wore out was replaced by a more durable machine of the same annual capacity, the stock of capital as measured would surely increase. But the maximal flow of capital services would be constant. There is nothing to be done about this, but something must be done about the fact of idle capacity. What belongs in a production function is capital in use, not capital in place. Lacking any reliable year-by-year measure of the utilization

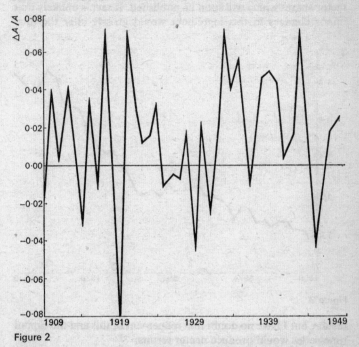

Figure 2

of capital I have simply reduced the Goldsmith figures by the fraction of the labor force unemployed in each year, thus assuming that labor and capital always suffer unemployment to the same percentage. This is undoubtedly wrong, but probably gets closer to the truth than making no correction at all.[4]

The share-of-capital series is another hodge-podge, pieced

4. Another factor for which I have not corrected is the changing length of the work-week. As the work-week shortens, the intensity of use of existing capital decreases, and the stock figures overestimate the input of capital services.

together from various sources and *ad hoc* assumptions (such as Gale Johnson's guess that about 35 per cent of non-farm entrepreneurial income is a return to property). Only after these computations were complete did I learn that Edward Budd of Yale University has completed a careful long-term study of factor shares which will soon be published. It seems unlikely that minor changes in this ingredient would grossly alter the final

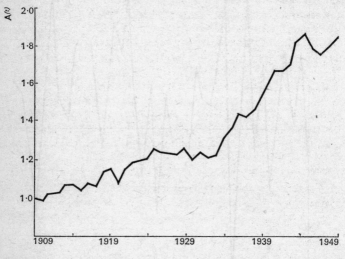

Figure 3

results, but I have no doubt that refinement of this and the capital time-series would produce neater results.

In any case, in **2a** or **2b** one can replace the time-derivatives by year-to-year changes and calculate $\Delta q/q - w_K \Delta k/k$. The result is an estimate of $\Delta F/F$ or $\Delta A/A$, depending on whether these relative shifts appear to be neutral or not. Such a calculation is made in Table 1 and shown in Figure 2. Thence, by arbitrarily setting $A(1909) = 1$ and using the fact that $A(t+1) = A(t)(1 + \Delta A(t)/A(t))$ one can successively reconstruct the $A(t)$ time series, which is shown in Figure 3.

I was tempted to end this section with the remark that the $A(t)$ series, which is meant to be a rough profile of technical change, at least looks reasonable. But on second thought I decided that I had very little prior notion of what would be 'reasonable' in this context. One notes with satisfaction that the trend is strongly upward; had it turned out otherwise I would not now be writing this paper. There are sharp dips after each of the World Wars; these, like the sharp rises that preceded them, can easily be rationalized. It is more suggestive that the curve shows a distinct levelling-off in the last half of the 1920s. A sustained rise begins again in 1930. There is an unpleasant saw-tooth character to the first few years of the $\Delta A/A$ curve, which I imagine to be a statistical artifact.

The outlines of technical change

The reader will note that I have already drifted into the habit of calling the curve of Figure 2 $\Delta A/A$ instead of the more general $\Delta F/F$. In fact, a scatter of $\Delta F/F$ against K/L (not shown) indicates no trace of a relationship. So I may state it as a formal conclusion that over the period 1909–49, shifts in the aggregate production function netted out to be approximately neutral. Perhaps I should recall that I have defined neutrality to mean that the shifts were pure scale changes, leaving marginal rates of substitution unchanged at given capital–labor ratios.

Not only is $\Delta A/A$ uncorrelated with K/L, but one might almost conclude from the graph that $\Delta A/A$ is essentially constant in time, exhibiting more or less random fluctuations about a fixed mean. Almost, but not quite, for there does seem to be a break at about 1930. There is some evidence that the average rate of progress in the years 1909–29 was smaller than that from 1930–49. The first twenty-one relative shifts average about 9/10 of 1 per cent per year, while the last nineteen average $2\frac{1}{4}$ per cent per year. Even if the year 1929, which shows a strong downward shift, is moved from the first group to the second, there is still a contrast between an average rate of 1·2 per cent in the first half and 1·9 per cent in the second. Such *post hoc* splitting-up of a period is always dangerous. Perhaps I should leave it that there is some evidence that technical change (broadly interpreted) may have accelerated after 1929.

The overall result for the whole forty years is an average upward shift of about 1·5 per cent per year. This may be compared

Table 1 Data for Calculation of $A(t)$

Year	% labor force employed (1)	Capital stock ($ mill.) (2)	Col. 1 × Col. 2 (3)	Share of property in income (4)	Priv. non-farm GNP per man-hour (5)	Employed capital per man-hour (6)	$\Delta A/A$ (7)	$A(t)$ (8)
1909	91·1	146,142	133,135	0·335	$0.623	$2.06		1·000
1910	92·8	150,038	139,235	0·330	0.616	2.10	—0·017	0·983
1911	90·6	156,335	141,640	0·335	0.647	2.17	0·039	1·021
1912	93·0	159,971	148,773	0·330	0.652	2.21	0·002	1·023
1913	91·8	164,504	151,015	0·334	0.680	2.23	0·040	1·064
1914	83·6	171,513	143,385	0·325	0.682	2.20	0·007	1·071
1915	84·5	175,371	148,188	0·344	0.669	2.26	—0·028	1·041
1916	93·7	178,351	167,115	0·358	0.700	2.34	0·034	1·076
1917	94·0	182,263	171,327	0·370	0.679	2.21	—0·010	1·065
1918	94·5	186,679	176,412	0·342	0.729	2.22	0·072	1·142
1919	93·1	189,977	176,869	0·354	0.767	2.47	—0·013	1·157
1920	92·8	194,802	180,776	0·319	0.721	2.58	—0·076	1·069
1921	76·9	201,491	154,947	0·369	0.770	2.55	0·072	1·146
1922	81·7	204,324	166,933	0·339	0.788	2.49	0·032	1·183
1923	92·1	209,964	193,377	0·337	0.809	2.61	0·011	1·196
1924	88·0	222,113	195,460	0·330	0.836	2.74	0·016	1·215
1925	91·1	231,772	211,198	0·336	0.872	2.81	0·032	1·254
1926	92·5	244,611	226,266	0·327	0.869	2.87	—0·010	1·241
1927	90·0	259,142	233,228	0·323	0.871	2.93	—0·005	1·235
1928	90·0	271,089	243,980	0·338	0.874	3.02	—0·007	1·226
1929	92·5	279,691	258,714	0·332	0.895	3.06	0·020	1·251
1930	88·1	289,291	254,865	0·347	0.880	3.30	—0·043	1·197
1931	78·2	289,056	226,042	0·325	0.904	3.33	0·024	1·226
1932	67·9	282,731	191,974	0·397	0.879	3.28	0·023	1·198
1933	66·5	270,676	180,000	0·362	0.869	3.10	0·011	1·211

Year	1	2	3	4	5	6	7	8
1934	70·9	262,370	186,020	0·355	0·921	3·00	0·039	1·298
1935	73·0	257,875	188,201	0·351	0·943	2·87	0·059	1·349
1936	77·3	254,875	197,018	0·357	0·982	2·72	−0·010	1·429
1937	81·0	257,076	208,232	0·340	0·971	2·71	0·021	1·415
1938	74·7	259,789	194,062	0·331	1·000	2·78	0·048	1·445
1939	77·2	257,314	198,646	0·347	1·034	2·66	0·050	1·514
1940	80·6	258,108	207,987	0·357	1·082	2·63	0·044	1·590
1941	86·8	262,940	228,232	0·377	1·122	2·58	0·003	1·660
1942	93·6	270,063	252,779	0·356	1·136	2·64	0·016	1·665
1943	97·4	269,761	262,747	0·342	1·180	2·62	0·071	1·692
1944	98·4	265,483	261,235	0·332	1·265	2·63	0·021	1·812
1945	96·5	261,472	252,320	0·314	1·296	2·66	−0·044	1·850
1946	94·8	258,051	244,632	0·312	1·215	2·50	−0·017	1·769
1947	95·4	268,845	256,478	0·327	1·194	2·50	0·016	1·739
1948	95·7	276,476	264,588	0·332	1·221	2·55	0·024	1·767
1949	93·0	289,360	269,105	0·326	1·275	2·70	—	1·809

Notes and sources

Column 1: Percentage of labor force employed. 1909–26, from Douglas, *Real Wages in the United States* (Boston and New York, 1930), p. 460. 1929–49, calculated from *The Economic Almanac, 1953–4* (New York, 1953), pp. 426–8.

Column 2: Capital stock. From Goldsmith, *A Study of Saving in the United States*, vol. 3 (Princeton, 1956), pp. 20–21, sum of columns 5, 6, 7, 9, 12, 17, 22, 23, 24.

Column 3: 1×2.

Column 4: Share of property in income. Compiled from *The Economic Almanac*, pp. 504–5; and Burkhead (1953), pp. 192–219. Depreciation estimates from Goldsmith, p. 427.

Column 5: Private non-farm GNP per man-hour, 1939 dollars. Kendrick's data, reproduced in *The Economic Almanac*, p. 490.

Column 6: Employed capital per man-hour. Column 3 divided by Kendrick's man-hour series, ibid.

Column 7: $\Delta A/A = \Delta 5/5 - 4 \times \Delta 6/6$.

Column 8: From 7.

[The last seven observations in column 8, for the years 1943–9, are incorrect due to an arithmetic slip in the calculations. This was pointed out in a note by Hogan (1958, pp. 407–11).]

with a figure of about 0·75 per cent per year obtained by Stefan Valavanis-Vail (1955) by a different and rather less general method, for the period 1869–1948. Another possible comparison is with the output-per-unit-of-input computations of Jacob Schmookler (1952) which show an increase of some 36 per cent in output per unit of input between the decades 1904–13 and 1929–38. Our $A(t)$ rises 36·5 per cent between 1909 and 1934. But these are not really comparable estimates, since Schmookler's figures include agriculture.

As a last general conclusion, after which I will leave the interested reader to his own impressions, over the forty-year period output per man-hour approximately doubled. At the same time, according to Figure 2, the cumulative upward shift in the production function was about 80 per cent. It is possible to argue that about one-eighth of the total increase is traceable to increased capital per man-hour, and the remaining seven-eighths to technical change. The reasoning is this: real GNP per man-hour increased from \$0·623 to \$1·275. Divide the latter figure by 1·809, which is the 1949 value for $A(t)$, and therefore the full shift factor for the forty years. The result is a 'corrected' GNP per man-hour, net of technical change, of \$0·705. Thus about eight cents of the sixty-five cent increase can be imputed to increased capital intensity, and the remainder to increased productivity.[5]

Of course this is not meant to suggest that the observed rate of technical progress would have persisted even if the rate of investment had been much smaller or had fallen to zero. Obviously much, perhaps nearly all, innovation must be embodied in new plant and equipment to be realized at all. One could imagine this process taking place without net capital formation as old-fashioned capital goods are replaced by the latest models so that the capital–labor ratio need not change systematically. But this raises problems of definition and measurement even more formidable than the ones already blithely ignored. This whole area of interest has been stressed by Fellner.

For comparison, Solomon Fabricant (1954) has estimated that

5. For the first half of the period, 1909–29, a similar computation attributes about one-third of the observed increase in GNP per man-hour to increased capital intensity.

over the period 1871–1951 about 90 per cent of the increase in output *per capita* is attributable to technical progress. Presumably this figure is based on the standard sort of output-per-unit-of-input calculation.

It might seem at first glance that calculations of output per unit of resource input provide a relatively assumption-free way of measuring productivity changes. Actually I think the implicit load of assumptions is quite heavy, and if anything the method proposed above is considerably more general.

Not only does the usual choice of weights for computing an aggregate resource-input involve something analogous to my assumption of competitive factor markets, but in addition the criterion output divided by a weighted sum of inputs would seem tacitly to *assume* (a) that technical change is neutral, and (b) that the aggregate production function is *strictly* linear. This explains why numerical results are so closely parallel for the two methods. We have already verified the neutrality, and as will be seen subsequently, a strictly linear production function gives an excellent fit, though clearly inferior to some alternatives.[6]

The aggregate production function

Returning now to the aggregate production function, we have earned the right to write it in the form **1a**. By use of the (practically unavoidable) assumption of constant returns to scale, this can be further simplified to the form

$$q = A(t)f(k,1), \qquad\qquad 3$$

which formed the basis of Figure 1. It was there noted that a simple plot of q against k would give a distorted picture because of the shift factor $A(t)$. Each point would lie on a different member of the family of production curves. But we have now

6. For an excellent discussion of some of the problems, see Abramovitz (1956). Some of the questions there raised could in principle be answered by the method used here. For example, the contribution of improved quality of the labor force could be handled by introducing various levels of skilled labor as separate inputs. I owe to Professor T. W. Schultz a heightened awareness that a lot of what appears as shifts in the production function must represent improvement in the quality of the labor input, and therefore a result of real capital formation of an important kind. Nor ought it be forgotten that even straight technical progress has a cost side.

provided ourselves with an estimate of the successive values of the shift factor. (Note that this estimate is quite *independent* of any hypothesis about the exact shape of the production function.) It follows from **3** that by plotting $q(t)/A(t)$ against $k(t)$ we reduce all the observed points to a *single* member of the family of curves in Figure 1, and we can then proceed to discuss the shape of

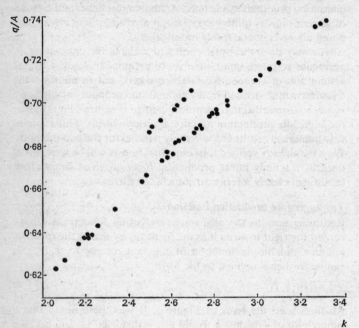

Figure 4

$f(k,1)$ and reconstruct the aggregate production function. A scatter of q/A against k is shown in Figure 4.

Considering the amount of *a priori* doctoring which the raw figures have undergone, the fit is remarkably tight. Except, that is, for the layer of points which are obviously too high. These maverick observations relate to the seven last years of the period, 1943–9. From the way they lie almost exactly parallel to the main scatter, one is tempted to conclude that in 1943 the aggregate production function simply shifted. But the whole earlier procedure was designed to purify those points from shifts in the function, so that way out would seem to be closed. I suspect the explanation may lie in some systematic incomparability of the capital-in-use series. In particular during the war there was almost certainly a more intensive use of capital services through two- and three-shift operation than the stock figures would show, even with the crude correction that has been applied. It is easily seen that such an underestimate of capital inputs leads to an overestimate of productivity increase. Thus in effect each of the affected points should really lie higher and toward the right. But further analysis shows that, for the orders of magnitude involved, the net result would be to pull the observations closer to the rest of the scatter.

At best this might account for 1943–5. There remains the postwar period. Although it is possible that multi-shift operation remained fairly widespread even after the war, it is unlikely that this could be nearly enough to explain the whole discrepancy.[7] One might guess that accelerated amortization could have resulted in an underestimate of the capital stock after 1945. Certainly other research workers, notably Kuznets and Terborgh, have produced capital stock estimates which rather exceed Goldsmith's at the end of the period. But for the present, I leave this a mystery.

In a first version of this paper, I resolutely let the recalcitrant observations stand as they were in a regression analysis of Figure 4, mainly because such casual amputation is a practice I deplore in others. But after some experimentation it seemed that to leave them in only led to noticeable distortion of the results. So, with some misgivings, in the regressions that follow I have omitted the observations for 1943–9. It would be better if they could be otherwise explained away.

Figure 4 gives an inescapable impression of curvature, of persistent but not violent diminishing returns. As for the possibility of approaching capital-saturation, there is no trace on this gross product level, but even setting aside all other difficulties, such a scatter confers no particular license to guess about what happens at higher K/L ratios than those observed.

7. It is cheering to note that Professor Fellner's (1956, p. 92) new book voices a suspicion that the post-war has seen a substantial increase over pre-war in the prevalence of multi-shift operation.

As for fitting a curve to the scatter, a Cobb–Douglas function comes immediately to mind, but then so do several other parametric forms, with little to choose among them.[8] I can't help feeling that little or nothing hangs on the choice of functional form, but I have experimented with several. In general I limited myself to two-parameter families of curves, linear in the parameters (for computational convenience), and at least capable of exhibiting diminishing returns (except for the straight line, which on this account proved inferior to all others).

The particular possibilities tried were the following:

$$q = a + \beta k \qquad \qquad \textbf{4a}$$
$$q = a + \beta \log k \qquad \qquad \textbf{4b}$$
$$q = a - \beta/k \qquad \qquad \textbf{4c}$$
$$\log q = a + \beta \log k \qquad \qquad \textbf{4d}$$
$$\log q = a - \beta/k. \qquad \qquad \textbf{4e}$$

Of these, **4d** is the Cobb–Douglas case; **4c** and **4e** have upper asymptotes; the semi-logarithmic **4b** and the hyperbolic **4c** must

Table 2

Curve	a	β	r
4a	0·438	0·091	0·9982
b	0·448	0·239	0·9996
c	0·917	0·618	0·9964
d	−0·729	0·353	0·9996
e	−0·038	0·913	0·9980

cross the horizontal axis at a positive value of k and continue ever more steeply but irrelevantly downward (which means only that some positive k must be achieved before any output is forthcoming, but this is far outside the range of observation); **4e** begins at the origin with a phase of increasing returns and ends with a phase of diminishing returns – the point of inflection occurs at $k = \beta/2$ and needless to say all our observed points come well to the right of this.

8. A discussion of the same problem in a different context is to be found in Prais and Houthakker (1955, pp. 82–8). See also Prais (1952–3, pp. 87–104).

The results of fitting these five curves to the scatter of Figure 4 are shown in Table 2.

The correlation coefficients are uniformly so high that one hesitates to say any more than that all five functions, even the linear one, are about equally good at representing the general shape of the observed points. From the correlations alone, for what they are worth, it appears that the Cobb–Douglas function **4d** and the semi-logarithmic **4b** are a bit better than the others.[9]

Since all of the fitted curves are of the form $g(y) = a + \beta h(x)$, one can view them all as linear regressions and an interesting test of goodness of fit proposed by Prais and Houthakker (1955, p. 51) is available. If the residuals from each regression are arranged in order of increasing values of the independent variable, then one would like this sequence to be disposed 'randomly' about the regression line. A strong 'serial' correlation in the residuals, or a few long runs of positive residuals, alternating with long runs of negative residuals, would be evidence of just that kind of smooth departure from linearity that one would like to catch. A test can be constructed using published tables of critical values for runs of two kinds of elements.

This has been done for the linear, semi-logarithmic and Cobb–Douglas functions. The results strongly confirm the visual impression of diminishing returns in Figure 4, by showing the linear

9. It would be foolhardy for an outsider (or maybe even an insider) to hazard a guess about the statistical properties of the basic time series. A few general statements can be made, however. (a) The natural way to introduce an error term into the aggregate production function is multiplicatively: $Q = (1+u)F(K,L;t)$. In the neutral case it is apparent that the error factor will be absorbed into the estimated $A(t)$. Then approximately the error in $\Delta A/A$ will be $\Delta u/1+u$. If u has zero mean, the variance of the estimated $\Delta A/A$ will be approximately $2(1-\rho)$ var u, where ρ is the first autocorrelation of the u series. (b) Suppose that marginal productivity distribution does not hold exactly, so that $K/Q\,\partial Q/\partial K = w_k + v$, where now v is a random deviation and w_k is the share of property income. Then the error in the estimated $\Delta A/A$ will be $v\,\Delta k/k$, with variance $(\Delta k/k)^2$ var v. Since K/L changes slowly, the multiplying factor will be very small. The effect is to bias the estimate of $\Delta A/A$ in such a way as to lead to an overestimate when property receives less than its marginal product (and k is increasing). (c) Errors in estimating $A(t)$ enter in a relatively harmless way so far as the regression analysis is concerned. Errors of observation in k will be more serious and are likely to be large. The effect will of course be to bias the estimates of β downward.

function to be a systematically poor fit. As between **4b** and **4d** there is little to choose.[10]

A note on saturation

It has already been mentioned that the aggregate production function shows no signs of levelling off into a stage of capital-saturation. The two curves in Table 2 which have upper asymptotes (c and e) happen to locate that asymptote at about the same place. The limiting values of q are, respectively, 0·92 and 0·97. Of course these are both true asymptotes, approached but not reached for any finite value of k. It could not be otherwise: no analytic function can suddenly level off and become constant unless it has always been constant. But on the other hand, there is no reason to expect nature to be infinitely differentiable. Thus any conclusions extending beyond the range actually observed in Figure 4 are necessarily treacherous. But, tongue in cheek, if we take 0·95 as a guess at the saturation level of q, and use the *linear* function **4a** (which will get there first) as a lower-limit guess at the saturation level for k, it turns out to be about 5·7, more than twice its present value.

But all this is in terms of *gross output*, whereas for analytic purposes we are interested in the *net* productivity of capital. The difference between the two is depreciation, a subject about which I do not feel able to make guesses. If there were more certainty about the meaning of existing estimates of depreciation, especially over long periods of time, it would have been better to conduct the whole analysis in terms of net product.

However, one can say this. Zero net marginal productivity of capital sets in when gross marginal product falls to the 'marginal rate of depreciation', i.e. when adding some capital adds only enough product to make good the depreciation on the increment of capital itself. Now in recent years NNP has run a bit over 90 per cent of GNP, so capital consumption is a bit under 10 per cent of gross output. From Table 1 it can be read that capital per unit of output is, say, between 2 and 3. Thus annual depreciation is between 3 and 5 per cent of the capital stock.

10. The test statistic is R, the total number of runs, with small values significant. For **4a**, $R = 4$; for **4b**, $R = 13$. The 1 per cent critical value in both cases is about 9.

Capital-saturation would occur whenever the gross marginal product of capital falls to 0·03–0·05. Using **4b**, this would happen at K/L ratios of around 5 or higher, still well above anything ever observed.[11]

Summary

This paper has suggested a simple way of segregating shifts of the aggregate production function from movements along it. The method rests on the assumption that factors are paid their marginal products, but it could easily be extended to monopolistic factor markets. Among the conclusions which emerge from a crude application to American data, 1909–49, are the following:

1. Technical change during that period was neutral on average.

2. The upward shift in the production function was, apart from fluctuations, at a rate of about 1 per cent per year for the first half of the period and 2 per cent per year for the last half.

3. Gross output per man-hour doubled over the interval, with $87\frac{1}{2}$ per cent of the increase attributable to technical change and the remaining $12\frac{1}{2}$ per cent to increased use of capital.

4. The aggregate production function, corrected for technical change, gives a distinct impression of diminishing returns, but the curvature is not violent.

References

ABRAMOVITZ, M. (1956), 'Resource and output trends in the US since 1870', *Amer. econ. Rev., Pap. Proc.*, vol. 46.

BURKHEAD, J. (1953), 'Changes in the functional distribution of income', *J. Amer. Stat. Assoc.*, vol. 48.

FABRICANT, S. (1954), 'Economic progress and economic change', *34th Annual Report of the National Bureau of Economic Research*, New York.

FELLNER, W. (1956), *Trends and Cycles in Economic Activity*, Holt, Rinehart & Winston.

11. And this is under relatively pessimistic assumptions as to how technical change itself affects the rate of capital consumption. A warning is in order here: I have left Kendrick's GNP data in 1939 prices and Goldsmith's capital stock figures in 1929 prices. Before anyone uses the βs of Table 2 to reckon a yield on capital or any similar number, it is necessary to convert Q and K to a comparable price basis, by an easy calculation.

HOGAN, W. P. (1958). 'Technical progress and production functions', *Rev. Econ. Stat.*, November.

PRAIS, S. J. (1952–3), 'Non-linear estimates of the Engel curves', *Rev. Econ. Stud.*, vol. 20.

PRAIS, S. J., and HOUTHAKKER, H. S. (1955), *The Analysis of Family Budgets*, Cambridge University Press.

ROBINSON, J. (1954), 'The production function and the theory of capital', *Rev. Econ. Stud.*, vol. 21.

SCHMOOKLER, J. (1952), 'The changing efficiency of the American economy, 1869–1938', *Rev. Econ. Stat.*, August.

VALAVANIS-VAIL, S. (1955), 'An econometric model of growth, USA 1869–1953', *Amer. econ. Rev., Pap. Proc.*, vol. 45.

17 E. Denison

United States Economic Growth

E. Denison, 'United States economic growth', *Journal of Business*,
April 1962, pp. 109–21.

The title of this article would once have introduced a glowing
description of enormous past advances in living standards and of
the further achievements confidently anticipated for the future.
Today, it is more often a title for expressions of alarm. This alarm
assumes a dual character.

There is widespread apprehension that markets will be unable
to absorb the enormous quantity of goods and services that future
growth of our productive potential is expected to provide, and
that economic growth will bring a persistent rise in technological
unemployment. Widespread in the thirties, this concern continues
and in the past few years has again become intense.

A quite different fear, one seldom expressed before, has
emerged during the past decade: the fear that the growth of our
productive capacity itself will be inadequate. The public is
warned that the Soviet Union, Western Europe and Japan are
growing more rapidly than the United States, and that our world
position must deteriorate if these trends continue. It is cautioned
that urgent public and private needs cannot be met unless United
States growth accelerates. It is told that, far from responding to
this challenge, in the past four years this country has allowed the
rate of growth of its national product to fall 'even' below its
historic 3 per cent rate.

To consider these two fears, it is absolutely essential to distinguish between the growth of the nation's 'potential' production, its ability or capacity to produce marketable goods and
services, and changes in the ratio of actual production to 'potential' production. The growth of potential production depends on
changes in the quantity and quality of available labor and capital,
the advance of knowledge and similar factors, while the ratio of

actual to potential production is governed mainly by the relationship between aggregate demand and potential production.

The fact that the growth of actual national product since 1956 or 1957 has fallen short of our average past record stems from the partial failure of United States business cycle or economic stabilization policy, not in policies affecting the growth of our productive potential. Since 1956 or 1957 our ability to produce has increased at least as fast as it did, on the average, in the past. Moreover, it is now clear that this failure is not related in significant degree to any change in the rate of technological progress or any structural change in the economy. It is a failure in meeting the old problem of equating changes in aggregate demand with changes in productive potential and unit costs. The recent study by Edward D. Kalachek and James W. Knowles demonstrates this about as definitively as any empirical proposition in economics is likely to be established.[1] I am convinced that, given appropriate fiscal and monetary policy, the maintenance of high employment need not be made more difficult in this country by the rate at which productive potential advances; rather, if there is any connection, rapid growth of output per man eases the problem by allowing a more rapid advance of wage rates without inflation.

We should move vigorously to reduce unemployment and use our productive capacity fully, and to continue to do so in the future. The loss of income and other costs imposed upon those unemployed or working short hours, upon proprietors and others dependent on profits, is ample reason to do so. The main tools open to the federal government are fiscal and monetary policy, and I believe they are adequate.

Greater success here would also stimulate the growth of productive potential. The more or less automatic effect upon the *long-term* growth rate of bringing actual production closer to the potential would be much less than is often supposed – I would put it on the order of 0·2 percentage points in the growth rate – but this is not a small amount. More important, if we should wish to take positive measures to accelerate growth, maintenance of

1. Joint Economic Committee, United States Congress, *Higher Unemployment Rates, 1957–60: Structural Transformation or Inadequate Demand*, Government Printing Office, 1961.

high employment is an essential prerequisite to their adoption. It will not seem sensible to make sacrifices to accelerate the expansion of our productive potential by working or saving more or by eliminating laws and practices that protect groups in our society at the cost of total output if we do not use the potential we have.

I shall not discuss employment policy further but will concentrate in this article on the long-term growth of our productive potential.

Sources of past growth

I come now to the first of four key questions that I shall discuss.

What are the sources of our past growth? From 1929 to 1957 the real national income or product increased at an average annual rate of 2·93 per cent.[2] I have tried to break this rate down among its sources. The results, shown in Table 1, are rough estimates and their derivation required some strong assumptions, but I think they provide correct perspective.

The table distinguishes broadly between the contribution of increases in factor inputs and increases in output per unit of input. To derive the former, I start (in the left-hand columns of Table 1) with estimates of the share breakdown of the national income. I estimate that in the 1929–57 period, the earnings of labor (including the labor of proprietors) represented 73 per cent of the national income; earnings of land, 4·5 per cent; and earnings of reproducible capital, 22·5 per cent. The last amount is divided in the table among types of capital.

The center section of the table provides estimates of the rate at which the various factor inputs increased, computed from indexes of their amount. The most familiar number here is that shown for employment on line 7. The series used, the Office of Business Economics estimates of persons engaged in production, increased at an average annual rate of 1·31 per cent from 1929 to 1957. Over the same period, as shown on line 9, annual hours of work declined at an average rate of 0·73 per cent a year. All the

2. This article is concerned with growth of real national income (or real net national product at factor cost). Its growth rate from 1909 to 1929, and 1929 to 1957, is taken to have been the same as that of real gross national product. Calculations and estimates used in this article are fully described in Denison (1962).

Table 1 Sources of Growth of Real National Income

Line	Source of growth	Share of national income (per cent distribution)		Growth rates (per cent per year)		Contribution to growth rate of real national income (percentage points)	
		1909–29	1929–57	1909–29	1929–57	1909–29	1929–57
1	Real national income	100·0	100·0ᵃ	2·82ᵇ	2·93	2·82ᵇ	2·93
2	Increase in total inputs, adjusted	—	—	2·24	1·99	2·26	2·00
3	Adjustment	—	—	-0·09	-0·11	—	—
4	Increase in total inputs, unadjusted	—	—	2·33	2·10	—	—
5	Labor, adjusted for quality change	68·9	73·0	2·30	2·16	1·53	1·57
6	Employment and hours	—	—	1·62	1·08	1·11	0·80
7	Employment	—	—	1·58	1·31	1·11	1·00
8	Effect of shorter hours on quality of a man-year's work	—	—	0·03	-0·23	0·00	-0·20
9	Annual hours	—	—	-0·34	-0·73	-0·23	-0·53
10	Effect of shorter hours on quality of a man-hour's work	—	—	0·38	0·50	0·23	0·33
11	Education	—	—	0·56	0·93	0·35	0·67
12	Increased experience and better utilization of women workers	—	—	0·10	0·15	0·06	0·11
13	Changes in age-sex composition of labor force	—	—	0·01	-0·01	0·01	-0·01
14	Land	7·7	4·5	0·00	0·00	0·00	0·00
15	Capital	23·4	22·5	3·16	1·88	0·73	0·43
16	Nonfarm residential structures	3·7	3·1	3·49	1·46	0·13	0·05
17	Other structures and equipment	14·6	15·0	2·93	1·85	0·41	0·28
18	Inventories	4·8	3·9	3·31	1·90	0·16	0·08
19	United States owned assets abroad	0·6	0·7	4·20	1·97	0·02	0·02

Line	Item						
20	Foreign assets in United States (an offset)	0·3	0·2	-1·85	1·37	0·01	0·00
21	Increase in output per unit of input			0·56	0·92	0·56	0·93
22	Restrictions against optimum use of resources					c	-0·07
23	Reduced waste of labor in agriculture					c	0·02
24	Industry shift from agriculture					c	0·05
25	Advance of knowledge					c	0·58
26	Change in lag in application of knowledge					c	0·01
27	Economies of scale – independent growth of local markets					c	0·07
28	Economies of scale – growth of national market					0·28	0·27

a For 1930–40 and 1942–6, interpolated distributions rather than the actual distributions for these dates were used. Estimates are 1929–58 averages.

b This rate, like that for 1929–57, derives from Department of Commerce estimates. Estimates by John W. Kendrick, based on adjustment to Department of Commerce concepts of estimates by Simon Kuznets, yield a growth rate of 3·17, which would result in a figure for output per unit of input (line 21) of 0·91.

c Not estimated.

other entries under labor represent an attempt to adjust for changes in its quality.

The question I first raise is what effect reduction in length of work day and work week had on effective labor input. There is agreement that, as the number of hours worked per man decreases from a very high level, output increases up to a point where total output per man is at a maximum. As hours worked decrease further, output per man declines but output per man-hour increases until a point of maximum output per man-hour is reached. Finally, even output per man-hour turns down. Curves illustrating this relationship, with the crucial points indicated by As, Bs and Cs, are familiar.

I have put numerical values on these points, and the particular values I use are the first of what I view as three key assumptions in the whole study. I assume that in 1929, when normal hours averaged 2529 a year, or 48·6 a week based on 52 weeks, they were at the point of maximum output per man, where a slight change in normal hours does not affect output per man. I assume that at the 1957 level of 2069 hours a year, or 39·8 a week, a slight change in hours is offset to the extent of 40 per cent by an opposite change in output per man-hour, so that a 1 per cent reduction in number of hours per man reduces output per man by 0·6 per cent. Corresponding changes at intermediate levels are established by proportional interpolation. This turns out to imply that the effect of the 18 per cent reduction in average hours over the entire 1929–57 period was to reduce output per man by 6 per cent, in comparison with what it would have been had hours not changed.[3] The relationship was extended upward for application in the pre-1929 period. If extended downward, it implies that the point of maximum output per man-hour would be reached at 33·9 hours a week based on 52 weeks.

My assumption yields the result, shown in line 10, that the quality of a man-hour's work increased at an average rate of 0·5 per cent a year as a direct consequence of the shortening of hours, and that the combined result of changes in employment

3. This result, that two-thirds of the effect of shorter hours was offset by greater efficiency of labor, is less sensitive than might be supposed to the level of hours at which points A, B and C are placed. The range of reasonable estimates appears to me to lie between 50 and 100 per cent.

and hours was an increase in labor input at an average rate of 1·08 per cent, as shown in line 6. The assumed effect of shorter hours on labor input is intended to include the fact that they may result in less intensive use of capital and land, so that when I come to those factors I make no adjustment for the effect of shorter hours.

The average quality of labor has been raised by a huge increase in its education. I estimate that the average male worker over twenty-five years of age in 1957 had spent four-fifths again as many days in school as had the average male worker in 1929, and two and a half times as many days as his 1910 counterpart.

To estimate the effect of improved education on the average quality of the labor force, I constructed for various dates distributions of male workers by the number of years of school they had completed. I also estimated the number of days of school attendance represented by a year's schooling, later assuming that the effect on the quality of labor of an increase in the number of days of school attended per year was the same as for a similar percentage increase in years of schooling. I then turned to census data for 1949 for typical income differentials by years of education among workers of the same age. At this point, I introduced the second key assumption of the study – namely, that three-fifths of the income differentials that appear when men of similar age are classified by years of education results from the *effect* of more education on the ability to contribute to production, while the remaining two-fifths reflects the tendency for individuals of greater natural ability and energy to continue their education and of other variables that are associated with, but not the result of, the amount of education. This three-fifths assumption, together with the observed income differentials by amount of education, and the estimates of years and days of school completed enabled me to compute an index of the effect of increased education on the ability of the average worker to contribute to production. My conclusion, as shown in line 11 of Table 1, is that from 1929 to 1957 the increase of education raised the average quality of the labor force at an average annual rate of 0·93 per cent a year. What this rate implies is that an increase of 80 per cent in the average amount of schooling raised the average quality of labor by 30 per cent.

The average annual earnings of women workers have increased

relative to those of adult males. I believe this mainly reflects the fact that so many women now make a career of work whereas they formerly were employed only for a short period before marriage or birth of a child. As a result women are, on the average, much more experienced and valuable employees. Reduced discrimination against employment of women in jobs that utilize their full potential may also have contributed. Increased experience and better utilization of women workers raised the average quality of the entire work force by 0·15 per cent a year, as shown in line 12.

Since, on the average, men earn more and presumably contribute more to production than women, and women more than children, changes in labor-force composition affect the average quality of the labor force. In the 1929–57 period changes were almost offsetting; they were slightly unfavorable, as shown by the −0·01 on line 13.

When indexes for the separate elements affecting the quantity and quality of labor are multiplied together, they yield a series for the quantity of labor input that is rather fully adjusted for quality change. It will not be fully apparent from this brief discussion, but what emerges is a series in which a year's work by an adult male in 1929 is taken as a standard of uniform quality. As shown in line 5, labor input so measured increased at an average annual rate of 2·16 per cent from 1929 to 1957.

The quantity of *land* available for use did not change during the period considered and, therefore, appears with a zero growth rate in line 14.

Capital input, which is restricted to privately owned capital, is measured in five parts. Input of structures and equipment other than residences is shown in line 17 to have increased at an average annual rate of 1·85 per cent, much less than the national product. Three things need to be said about this series. Firstly, capital input of structures and equipment is measured by the value of the gross stock in constant prices. Secondly, like all available series for capital, it does not reflect improvements in the quality of structures and equipment that are made possible by the advance of knowledge. Thirdly, because of an upward bias in price indexes used to deflate construction, the growth of this type of input probably is understated.

Line 16 and lines 18–20 measure the growth rates of the deflated value of the gross stock of residences, of the deflated value of inventories, of the deflated value of United States owned investments abroad and of foreign-owned investments in the United States.

The five separate series for the capital components are combined by use of income share weights to obtain a series for total capital input. Similarly, the indexes for labor, land and capital are combined by income share weights to obtain an index of the increase in total factor inputs. As shown in line 4, total factor input increased at an average annual rate of 2·10 per cent from 1929 to 1957. The income share weights used were changed each five years; the weights for these subperiods are not shown here.

The justification for use of income share weights to combine the various indexes, and also the use to which I shall next put them, derives from the marginal productivity explanation of the distribution of income. The reasoning is well known and, in this respect if no other, my procedure follows a well-trod path, so I shall not pause to explain it.

The right side of the table gives the number of percentage points in the total growth rate that I estimate was contributed by each source of growth. Except for three refinements that I will shortly describe, the upper portion of the right side of Table 1, referring to the increase in inputs, could be derived from the left and center sections by simple multiplication. For example, from 1929 to 1957 education improved the average quality of labor at an average annual rate of 0·93 per cent. Labor represented an average of 73 per cent of total inputs in this period. Consequently, 73 per cent of 0·93, or 0·68 percentage points, is the rate at which total inputs would have increased had there been no change in other elements affecting total input. It is also the rate at which the national product would have increased if, in addition, there had been no change in output per unit of input. The number given on the right-hand side of line 11, 0·67 percentage points, differs from the results of this simple multiplication for three reasons.

The first is that I did not actually use constant income weights based on the period as a whole but periodically changed the weights to give effect to changes in the relative marginal productivity of the factors. The second is the necessity of allowing for a peculiarity in the deflation of real product. The Department of Commerce real-product series assumes or implies that output per man or output per man-hour has not changed in general government, much of the household and non-profit institution sector, and most of the construction industry. In allowing a contribution for the increased quality of labor in these areas, and for increased capital output per man in the case of construction, I have counted increases in input for which no corresponding output increase is allowed. The amount involved is put at 0·11 percentage points in the growth rate in line 3. It has been allocated among the relevant sources and eliminated from their estimated contribution to growth. Third, although use throughout the calculations of growth rates rather than indexes or percentage changes almost eliminates the problem of statistical interaction, it does not quite do so. What remains has been allocated among the relevant sources.

The increase in inputs accounted for 2·00 percentage points, or 68 per cent, of the growth of total output from 1929 to 1957. I turn now to the allocation among sources of the remaining 0·93 percentage points that results from the growth of output per unit of input, or productivity.

Most of the possible sources considered do not appear in the table at all, either because they appeared not to have changed over the period or because their effect on the growth rate was calculated at less than 0·01. In a few cases it was impossible to decide whether changes were favorable or unfavorable to growth. Also, some changes that might affect a 'truer' measure of national income or product do not affect the national product as it is actually measured.

Various private and governmental restrictions prevent the optimum allocation and most efficient use of resources. Most did not change appreciably over the period, but some cost us more output in 1957 than in 1929. I estimate that they subtracted 0·07 points from the growth rate over that period, as shown in line 22. Line 23 refers to the fact, or what I believe to be the fact, that labor nominally employed but ineffectively utilized in agriculture was a smaller fraction of all labor, though a larger fraction of farm labor, in 1957 than in 1929. Line 24 arises because, even after eliminating excessive resources from the computation, resources in agriculture earned less than resources of equal quality in the rest of the economy in the base year of the real product estimates, 1954, because they were in oversupply. The shift out of agriculture thus contributed to a statistical rise in the national product estimated at 0·05 points in the growth rate.

Skip now to the last line, economies of scale associated with the growth of the national market. In the absence of any satisfactory procedure to arrive at a figure statistically, my effort here was to set down a number representing a sort of norm of expert opinion. I assume – and this is the third major assumption of the study – that in the 1929–57 period, economies of scale added 10 per cent to the increment to output that would otherwise be provided by all other sources. Consequently, I allocate one-eleventh of the total growth rate in the 1929–57 period, or 0·27 percentage points, to this source. I use a slightly higher fraction in the earlier period since economies of scale presumably decline as the size of the economy increases.

Much, probably most, of this contribution, of course, is the result of the expansion of local and regional markets that automatically accompanies the growth of the national economy. In addition, however, local markets grew independently as a result of increasing concentration of population and especially of the adaptation of the trade and service industries to the general ownership of automobiles. The contribution of this independent development is represented in line 27.

Finally, we come to the contribution to growth of the advance of knowledge and the speed with which it is incorporated into production. I believe that, as indicated in line 26, the change in the lag of the average practice behind the best known was of negligible importance.

The estimate in line 25 for the contribution of the advance of knowledge is obtained as a residual, and has the usual weakness of a residual. It is intended to measure the contribution to the growth rate of the advance of knowledge of all types relevant to production, including both managerial and technological knowledge. Many will find the contribution of 0·58 percentage points, or 20 per cent, of the total 1929–57 growth rate that I attribute to the advance of knowledge surprisingly small.

If expectation of a larger figure is based on previous studies, there is no basis for surprise. The term 'technological progress' has often been applied to all the sources of growth except changes in man-hours, land and capital. That definition would embrace everything in Table 1 except lines 7, 9, 14 and 15. These accounted for only three-tenths of the growth rate, so my estimates would leave seven-tenths of total growth for attribution to 'technological progress' if that broad definition were used. If the calculation were confined to the private economy, the fraction would be still larger. The main object of these calculations has been to divide up the contribution to growth of what has been vaguely termed technological progress.

Nevertheless, the figure for the contribution of knowledge is rather small even if expectations are based on *a priori* observation. The explanation is that much of what usually is thought of as the fruits of technological progress is simply not caught in the growth rate of the national product as measured because of the character of the price indexes used in deflation. There are two main points here.

Firstly, the introduction of new or better final products does not, in general, increase the measured national product. For example, when we say on the basis of the official estimates that total real consumption increased by 112 per cent from 1929 to 1957, we are comparing actual consumer purchases in 1929 with the sum of (a) products purchased in 1957 that were identical with those bought in 1929 and (b) the sum of products *not* available in 1929 valued in terms of the products that the resources used in their production *could* have provided in 1957 if used to produce the products that did exist in 1929. The estimates do not reflect either the improvements made in a great range of products without a corresponding change in their production costs, nor the vastly greater range of choice open to today's consumer.

Secondly, the replacement of retail outlets or of establishments providing services to consumers by more efficient types of retail outlets and service establishments does not, in general, raise the measured national product because the price indexes measure price changes in identical establishments.

In general, the advance of knowledge can contribute to the measured growth rate only by reducing production costs for already existing final products, or through improvements in business organization at levels other than those serving the final purchaser.

Once these characteristics of the output measure are understood, it is not, I think, surprising that the contribution of the advance of knowledge to the measured growth rate is not larger.

Space limitations prohibit extended discussion of the results of this part of the study, but the table speaks for itself. In summary, from 1929 to 1957 five sources contributed an amount equal to 101 per cent of the growth rate, out of a total of 109 per cent contributed by all sources making a positive contribution. These were: increased employment (34 per cent); increased education (23 per cent); increased capital input (15 per cent); the advance of knowledge (20 per cent); and economies of scale associated with the growth of the national market (9 per cent). The reduction of working hours accounted for −7 per cent of the total 'contribution' of −9 per cent to the growth rate provided by sources adverse to growth, and increased restrictions against the optimum use of resources for the remainder.

The breakdown in the 1909–29 period was, of course, different. Increases in capital and in employment contributed more than in the period after 1929, and improvements in the quality of the labor force, much less.

But whatever period we examine, it is clear that economic growth, occurring within the general institutional setting of a democratic, largely free-enterprise society, has stemmed and will stem mainly from an increased labor force, more education, more capital and the advance of knowledge, with economies of scale exercising an important, but essentially passive, reinforcing influence. Since 1929 the shortening of working hours has exercised an increasingly restrictive influence on the growth of output.

Future rate of growth

The second question is: *What rate of growth can we reasonably anticipate in the future?* Opinions on this vary, as they must, because the future is inherently uncertain. My own projection, derived by summing estimates of the contribution anticipated from each of the sources affecting past growth, is that if we are reasonably successful in maintaining high employment, avoid a major war, and otherwise maintain existing policies and conditions in fields affecting growth importantly, we can look forward to a growth rate in our productive potential of 3·33 per cent over the period from 1960 to 1980. The rate is about 3·5 per cent if we start the calculation from the recession-reduced actual national product in 1960 and assume 1980 will be a prosperous year. The 3·33 per cent rate for productive potential is almost one-seventh above the actual rate from 1929 to 1957. It implies an average annual percentage increase in potential output per person employed of 1·62 per cent per year, just slightly above that experienced from 1929 to 1957, and of 2·17 in output per man-hour.

The third question is: *Will the projected growth rate be high enough?* This obviously cannot be answered without establishing criteria, which I shall not attempt, but three things can be said.

Firstly, the growth rate projected would yield a large improvement in living standards unless the proportion of output required for defense increases enormously. If the fraction of national product devoted to consumption does not change, *per capita* consumption in 1980 would be above 1960 by 38–46 per cent on the basis of the two middle population projections of the Census Bureau.

Secondly, if it should be considered necessary or desirable to increase expenditures for defense and other essential public purposes, they could be doubled in twenty years without changing the proportion of national product they absorb, and by changing the proportion such expenditures could be enormously increased while still allowing a sizable advance in living standards.

Thirdly, a growth rate of 3·33 or 3·5 per cent is not likely to win any statistical growth race with the Soviet Union or any other industrial country that is presently substantially behind us in productivity and that has established institutional conditions equally favorable to growth. Mainly this is because the possibilities open to us of quick gains by imitation are so much more limited. Whether a statistical growth race is important to impress world opinion is a matter of dispute. But for us to accept a challenge for a growth-rate race with Russia or Japan would be as sensible as for Roger Bannister, the day he ran the four-minute mile, to have wagered a promising high-school sophomore on which of them could reduce his best time by the larger percentage.

I come now to the last and most interesting of the four questions that I shall consider. *Should we try to change the future growth rate, and if so what courses are open to us?*

The growth in potential that we get under present arrangements is largely the result of individual decisions based on a balancing of present and future income against present and future real costs.

Decisions made collectively through governments and other organizations also affect economic growth. In the past, however, most collective decisions that influence growth have not hinged on their effect on growth but on other considerations, even though exceptions to this generalization are not hard to find.

The issue raised in much of the current growth discussion is whether, as a society, we should deliberately adopt measures to achieve a rate of economic growth higher than that which will emerge as a result of individual decisions and of collective decisions not explicitly based on growth effects.

This is a real and legitimate issue. A large national product in the future is desirable, but measures to raise the growth rate significantly involve costs. Certainly, a democratic society is *entitled* to make a collective decision to use the instruments of government and other institutions to promote rapid growth, and there are many steps that might be taken. But to say that a democratic society *can* decide to accelerate growth is not the same as to say that it would be wise for it to do so. If such a decision is to represent a rational choice, it must be based on a comparison of the benefits with the costs that are imposed.

What choices are open to us if we wish to raise the growth rate over the next twenty years above what it would otherwise be? I shall indicate what I conclude would be necessary to raise the growth rate by 0·1 percentage point, as from 3·3 to 3·4 per cent. Such a change would yield a national product in 1980 higher than otherwise by about 2 per cent or $20 billion. Put the other way around, to change the growth rate over the next twenty years by 0·1 of a percentage point requires some action that will make the 1980 national product 2 per cent larger than it would be in the absence of that action.

This requires that we increase either the quantity or quality of the total input of labor, land and capital into the productive system or else increase its productivity.

To raise the national product 2 per cent by increasing inputs would require slightly less than a 2 per cent increase in total inputs because of the existence of economies of scale. I estimate total input in 1980 would have to be increased about 1·83 per cent. One way to do this would be to increase all kinds of input by 1·83 per cent. The other would be to increase only one kind of input by a larger percentage. I estimate that in 1954–8 labor comprised 77 per cent of total input, capital 20 per cent and land 3 per cent. It follows that we could raise *total* input by 1·83 per cent in 1980 if we could raise labor input alone by 2·4 per cent over what it would otherwise be, or capital input alone by 9·3 per cent, or land alone by 61·0 per cent.

Suppose we wish to add 0·1 to the growth rate by increasing the quantity or the average quality of *labor* input in 1980 by 2·4 per cent, over and above what it would otherwise be. This could be done if we wished, and could find ways, to achieve any of the following changes:

1. Prevent half the deaths that will otherwise occur from 1960 to 1980 among individuals less than sixty-five years of age.

2. Cut in half time lost from work because of sickness and accidents.

3. Draw into the labor force one-tenth of all able-bodied persons over twenty years of age who will not otherwise be working in 1980.

4. Double the rate of net immigration over the next twenty years.

5. Operate with a work week one hour longer than otherwise.

6. Eliminate two-thirds of the loss of work resulting from seasonal fluctuations in nonfarm production.

7. Reduce cyclical unemployment below what it would be otherwise by 2 per cent of the labor force – an impossibility unless the total unemployment rate would otherwise be above 4 per cent.

8. Add one and a half years to the average time that would otherwise be spent in school by everyone completing school between

now and 1980, or make an equivalent improvement in the quality of education.

To raise the growth rate one-tenth of a percentage point by increasing the *capital stock* more rapidly would require devoting an additional 1 per cent of the national income to net saving and investment throughout the next twenty years. This would be an increase of about one-sixth in the nation's net saving rate.

To increase *land* input offers no significant possibilities.

The alternative to increasing the quantity or quality of inputs is to increase productivity by accelerating the advance of knowledge or the efficiency with which the economy works. One important source of increase in productivity, the economies of scale that occur when the economy grows for other reasons, cannot be affected directly. I have taken it into account in estimating the yield from increasing inputs, and will also do so in examining other ways of increasing productivity. Let us consider the others.

My projection assumes the advance of knowledge will contribute 0·8 to the 1960–80 growth rate, more than in 1929–57. We could thus add 0·1 to the growth rate if we could raise by one-eighth the rate at which knowledge relevant to production advances. But many discoveries and inventions originate abroad, and many are not the result of deliberate research. On possible assumptions, we would have to increase by one-half the annual increment to knowledge that originates in the United States and is subject to being affected by deliberate action.

We could also add 0·1 to the growth rate over the next twenty years if we could reduce the lag of average production practices behind the best known by two and two-thirds years, in addition to any reduction that would otherwise take place. This would be a huge reduction in the world's most advanced country.

There are a number of smaller possibilities which we could combine to add 0·1 to the growth rate.

Thus, we might eliminate all the misallocation and wasteful use of resources that results from barriers to international trade (which I estimate costs us 1·5 per cent of the national income) *and* misallocation resulting from private monopoly in markets for products.

Or we might eliminate state resale price maintenance laws *and* racial discrimination in hiring. (I estimate these cost us 1 per cent and 0·8 per cent of the national income, respectively.)

Or we might shift to other uses of resources going into the production of unwanted or little-wanted farm products *and* also eliminate unemployment and underemployment resulting from long-term declines in individual industries and areas by re-employing workers immediately upon their becoming surplus.

Or we might eliminate all formal obstacles imposed by labor unions against use of the most efficient production practices *and* also consolidate local school districts and firms in regulated industries, particularly the railroads, wherever this would reduce unit costs.[4]

There are, of course, other possibilities but they appear small.

It is not at all clear that we know *how* to do some of these things; and even where we do, they involve costs. Some, such as those leading to more investment in private or public capital, or to a faster rate of advance in knowledge through more research, require that the nation consume less than it otherwise could. Others, such as diversion of resources to provide more education or better medical care, which are classified as consumption in the national product, require that the nation consume less in other forms. Still others, such as longer hours of work or enlargement of the labor force, require that more work be done. Except for increasing immigration, all of the changes that would permanently raise the growth rate by any considerable amount impose costs of one of these types.

Costs of this kind are not imposed by changes that would make the economy operate more efficiently with given resources and given knowledge. Also, the means by which such changes could be brought about are frequently obvious, often simply requiring the repeal of existing laws that prevent the best allocation and use of resources. From a broad standpoint, such changes consequently are particularly attractive, even though their possible stimulus to long-term growth is temporary and rather small. Even these, however, require some real or imagined sacrifice on the part of some members of society. Were it not so, these changes would already have been made.

4. Some of the smaller of the changes suggested would not change the growth rate as actually measured.

Decisions on whether or not to try to affect the growth rate by any of the means I have suggested cannot sensibly be made without full consideration of their costs.

Almost any policy to affect growth also has other consequences. Some of these consequences, such as improvement of health, or a better educated citizenry, will be widely accepted as desirable. Others, especially any appreciable sacrifice of individual freedoms, will be as widely regarded as undesirable. Still others, including notably changes in the distribution of income, will be regarded as desirable by some individuals and undesirable by others. Among all the policies that might be adopted that would affect growth, there are few indeed where the effect on growth is, or should be, the primary consideration in their appraisal.

A serious effort to stimulate growth significantly would not, in my opinion, concentrate on one or two approaches but would be broadly based. This view is reinforced in the case of steps to increase factor inputs by the phenomenon of diminishing returns. Large increases in either labor or capital input, but especially the latter, without increases in the other, would yield a proportional increase in the growth rate smaller than is implied by calculations used to arrive at the results I have just presented.

If there is one point to be stressed above all, it is that faster growth is not a free good, not something that can be achieved by wishing or by speeches. To change the growth rate requires that something be done differently, and this entails costs in every significant case. Whether the gain is worth the cost can be judged only by careful consideration of each particular proposal.

Reference

DENISON, E. F. (1962), *The Sources of Economic Growth in the United States and the Alternatives Before Us*, Committee for Economic Development.

18 Z. Griliches

The Sources of Measured Productivity Growth:
United States Agriculture, 1940–60[1]

Z. Griliches, 'The sources of measured productivity growth: United
States agriculture, 1940–60', *Journal of Political Economy*, August
1963, pp. 331–46.

1 Introduction

Public interest in economic growth as an objective policy has
stimulated a substantial number of studies into the causes of
economic growth and possible ways of affecting it. All of these
studies use the concept of an 'aggregate production function'
either explicitly or implicitly. This function can be represented in
a general way by

$$Y_t = g(X_t, u_t, T_t),$$

where Y_t is an index of physical output (or value added) in an
industry, X_t is a set of 'measurable' inputs, usually labor and
capital indexes, u_t is a random, or short-term, cyclical variable
such as weather in agriculture or unemployment in manufacturing, T_t is the 'level of technology', a postulated unobserved
'latent' variable usually to be inferred from the data in a residual
fashion, and $g(\ )$ is the function describing the correction between these variables. It is also often assumed that this function
is homogeneous of degree 1 in X (constant returns to scale), and
that the industry operates in perfectly competitive product and
factor markets and is in equilibrium (or at least was in equilibrium in the weight-base period). This last set of assumptions
allows one to approximate the relevant coefficients of the production function by relative factor shares, evading thereby a
direct estimation of it.

Several recent studies of the historical record of United States
growth have used such a framework together with conventional
measures of labor and capital and found that very little of the

1. This paper is part of a larger econometric study of sources of productivity growth. I am indebted to both the National Science Foundation and
the Ford Foundation for their generous support of this work.

growth in output can be accounted for by changes in these inputs, most of the growth being explained by the residual factor T – 'technical change'.[2] These findings have in turn encouraged further attempts to refine and improve the measurement of what appears to be the most important source of economic growth.[3]

This formulation of the problem and the direction in which the resulting research has evolved are, in my opinion, not very helpful to the understanding of growth. The whole concept of a production 'function' is not very useful if it is not a *stable* function, if there are very large unexplained shifts in it. Moreover, it does not further our understanding of growth to label the unexplained residual changes in output as 'technical change'. Nor does it help much to measure these changes accurately if we do not know what they are.

The purpose of this paper is to suggest an alternative, potentially more fruitful, approach to the problem and to illustrate it by reference to the growth of productivity in agriculture in the United States. According to this approach, changes in output are attributable to changes in the quantities and *qualities* of inputs, and to economies of scale, rather than to 'technical change', the production function itself remaining constant (at least over substantial stretches of time). Conventionally derived residual measures of productivity growth are viewed not as measures of technical change but rather as the result of errors in the measurement procedure. These 'errors' are due to several sources: (a) the list of variables affecting output may be misspecified, excluding some relevant factors from the calculations; (b) changes in included variables may be mismeasured, particularly if changes in their quality are disregarded; and (c) wrong weights may be used in estimating the contribution of changes in individual inputs to the growth in output. 'Correcting' such errors will, I believe, lead to a substantial reduction in what has been conventionally measured as growth in 'total factor productivity', reducing thereby the proportion of growth that had to be previously attributed to this essentially 'unexplained' category. Such an

2. Among others see Solow (1957), pp. 312–20, and Fabricant (1959).

3. All these studies also assume 'neutrality', that is, that changes in T do not affect the form of the relationship between Y and X. Some such assumption is necessary because of the latency of T_t.

approach does not, of course, remove technical change from the explanation of growth; it aspires, rather, to transform what is currently a catch-all residual variable into movements along a more general production function and into identifiable changes in the qualities of inputs.

This approach can be illustrated by bringing together the results of a number of studies of production relations and input quality change in United States agriculture.[4] These studies indicate that the main sources of conventionally measured productivity increases in United States agriculture during the 1940–60 period appear to have been:

1. Improvements in the quality of labor as a consequence of a rise in educational levels.

2. Improvements in the quality of machinery services that had been disguised by biases in the standard price indexes used to deflate capital equipment expenditures.

3. Under-estimation of the contribution of capital and over-estimation of the contribution of labor to output growth by the conventional factor-share based weights.

4. Economies of scale.[5]

A separate estimate of each of these sources of growth is derived from various pieces of data.[6] The main point of this study, however, is not to define technical change away but rather to explain it.[7]

4. See in particular Griliches (1960, 1963).

5. The imputation of part of the observed growth in output to the last two sources arises out of the denial of the conventional assumption of equilibrium.

6. In this, my approach is very similar to that taken by Denison (1962), except that I concentrate on providing more detailed and, I hope, better documented estimates for a smaller number of factors. My approach is also related to Solow's (1962) discussion of 'embodiment' of technical change in new capital, except that my concept of input quality change allows for embodiment in other inputs besides capital, and the estimation procedure does not require that the rate of embodiment or improvement in quality be constant over the period in question.

7. One could, of course, first compute the conventional measures of technical change and then proceed to explain them on the basis of this same set of factors. I find the direct approach more useful than the two-stage one, although, in principle, the answers should be the same.

The plan of the rest of this article is as follows: section 2 discusses several questions that could be investigated in an econometric production function study but that are usually assumed away in the standard productivity or technical change measurement study. Section 3 reports the results of a cross-sectional production function study that was designed to investigate some of these questions. It is found, in particular, that the estimated production function differs substantially from what one would infer on the basis of the factor shares approach, both in the relative weight that it assigns to different inputs and in its finding of economies of scale. Moreover, evidence is found to support the view that education is a relevant quality dimension of the labor variable. Section 4 brings together results from a number of previous studies on input change, indicating a substantially larger growth in inputs over time when they are measured in more relevant units and with more regard for quality change. Section 5 brings together the results of the previous two sections and shows that in this particular case this approach leads to an almost complete accounting of the sources of output growth in United States agriculture during 1940–60, leaving no 'unexplained' residual to be identified with unidentified 'technical changes'.

2 Why estimate production functions?

Econometric production function studies can be used to investigate the appropriate algebraic form for the assumed aggregate production function, the number and type of variables that should be included in the list of inputs and the appropriate way of measuring them, and what numerical values should be assigned to the coefficients to be attached to each of these variables. Even though these questions are basic to any attempt to allocate the observed growth in output to its various 'causes', they are usually assumed away rather than investigated in most of the studies measuring productivity and technical change.

The choice of a particular algebraic form for the production function is associated with the question of ease of substitution between different inputs. It is a question concerning the curvature of the isoquants. The elasticity of substitution is usually assumed to be either infinite, as illustrated by Kendrick's (1961) arithmetic total input indexes, or unity, as is the case when the Cobb–Douglas function or a geometric input index with constant or shifting weights is used.[8] It has been suggested recently that the elasticity of substitution should also be estimated rather than assumed beforehand. This suggestion leads to the use of an exponentially weighted harmonic average of input indexes[9] in estimating the 'residual' to be attributed to technical change. Which form of the production function is the most appropriate one is an empirical question and could be settled by testing different forms on the same set of input-output data. Unfortunately, the appropriate 'curvature' is probably too fine a question to be settled on the aggregate level. Our data are just not good enough to discriminate among these various alternatives. Fortunately, however, this does not matter much so far as growth accounting is concerned. Using the conventional input measures but different index number formulas leaves us with pretty much the same large 'residual'. The curvature question is, of course, of much greater importance for the theory of functional income distribution. But that is another matter.

The main candidates for addition to the conventional list of inputs are research and development capital, education of the labor force, and 'external' inputs such as research and extension activities of the government and other firms, and other non-market priced services such as the provision of transportation and communication facilities.[10] Ideally, we should investigate the relevance and actual numerical importance of such variables before we introduce them into our growth accounts. Most of the previous work on education, including my own, simply imputes part of the observed productivity increase to changes in education, using cross-sectional income-by-education tabulations as the source of its weighting scheme (see, e.g. Denison, 1962,

8. As in Solow (1957).

9. See Arrow, Chenery, Minhas and Solow (1961).

10. It does not really matter whether we treat some of these as variables to be added to the production function or as variables that modify the 'quality' of already included inputs. Thus, one could talk equally well about the services of educational capital or about education as an aspect of labor-force quality. Which is more convenient will depend on the form the data come in and on the interactions with the other variables that we are willing to assume (or assume away).

and Griliches, 1960). Many difficult questions are raised by the use of income-by-education data, and one cannot rule out the possibility that the observed associations may be in large part spurious. Moreover, until detailed tabulations of the 1960 Census become available, the only available income-by-education data are for the United States as a whole. One may not doubt the proposition that education is an important source of growth in the economy as a whole, and still not be convinced that it is very important in agriculture. By introducing the education of the labor force as a separate variable in an econometric production function study, it becomes possible to *estimate* rather than to *assume* its coefficient.

To measure that part of output change that results from a change in the level of a particular input, we must weight it by its respective production function coefficient. Assuming linear, or linear in the logarithms, production functions, constant returns to scale, and competitive equilibrium (at least in the weight-base period), these coefficients can be approximated by input market prices or their relative shares in total costs. But if, as has been alleged to be true for agriculture, a sector is in continuous disequilibrium, a weighting scheme based on factor shares will be incorrect for productivity comparisons. Many agricultural economists have argued for years that the marginal product of labor in agriculture is substantially below the going wage rate for hired labor and that the marginal product of capital is substantially above the conventional bank or mortgage rates (see, e.g. Schultz, 1947, and the literature cited there). They have been supported by the historically observed large outflow of labor from, and inflow of capital into, agriculture. A statistically estimated production function provides an alternative and conceptually more appropriate system of weights for compiling inputs into a 'total' input index. For a sector such as agriculture, where different inputs have had very different time trends, the conventional productivity estimates are quite sensitive even to a small shift in weights.

All the conventional productivity indexes assume constant returns to scale. So do also many of the estimated production functions. On the other hand, most of the cost curves and much of the programming and budgeting literature imply the existence of substantial economies of scale, both in agriculture and in industry. This whole area, however, is quite controversial. From my point of view, it is not very important whether the economies of scale go on indefinitely or the cost curve finally turns up. The interesting question is whether there were and are *some* additional economies to be had at existing scale levels. Unfortunately, the fitting of standard Cobb–Douglas type of production functions, such as I will report on below, is not very well suited to answering this question, since it assumes that the production function is homogeneous in all inputs. It may provide an answer whether the function is homogeneous of degree more or less than 1, and this may be interesting and valuable by itself, but it may miss many aspects of what we usually think of as sources of economies of scale, such as indivisibilities and disproportionalities by its assumption of homogeneity. To study the subject of economies of scale adequately will require the use of a production function that is not homogeneous over at least some range of the inputs.

3 The results of one cross-sectional study

I have recently investigated some of the problems discussed in the previous section by estimating an aggregate agricultural production function based on 1949 data for sixty-eight regions of the United States. This study is described in some detail elsewhere, including the presentation of alternative estimates and of detailed reservations (see Griliches, 1963). For the purposes of this paper only its main results are reproduced in Table 1.[11] The estimates summarized in this table indicate that (a) education as measured is a statistically significant variable with a coefficient that is not very different from the coefficient of the man-years-worked variable. Thus, it turns out that it would not have been very wrong to 'inflate' the labor variable by the computed 'quality' (education) per man index before estimating the production function. This makes it much easier to apply such a framework

11. These estimates are based on the fitting of a Cobb–Douglas type of equation. Several alternative forms of the production function were also tried but did not lead to any appreciable improvement in the results. A function allowing for cross-sectional differences in the mix of output (crops versus livestock) did fit the data somewhat better. But since the aggregate output mix has not changed much over time, these results are not reported here. For details see the above-cited papers.

to time series, since one can adjust the labor variable beforehand for quality change rather than carry education along as a separate variable.[12] (b) The estimated coefficients differ significantly from what has been assumed to be true on the basis of factor-shares data and equilibrium assumptions. Table 2 compares the coefficients as estimated in this study with the official estimates based on 1947–9 factor-shares data. The production function estimate of the labor coefficient is relatively smaller and the machinery coefficient is relatively larger than the official factor-shares-based estimates of these same coefficients. (c) There is evidence of substantial economies of scale.

These findings, particularly the last two, if accepted, will account for a substantial fraction of the conventionally measured productivity increases. Before we accept them, however, we should look for other pieces of evidence that would either confirm or contradict these findings. The somewhat smaller coefficient for labor found in the production function study relative to the one estimated officially on the basis of factor shares implies that the marginal product of labor was less than was assumed in the estimation of factor shares.[13] This is most likely due either to an overestimate of the amount of work actually performed by family labor or to what is almost the same thing, an overvaluing of this (family) work by the application of the hired wage rate, a rate that is heavily affected by the price necessary to attract sufficient labor into the industry at peakload (harvest) time. It is hard to 'prove' that the marginal product of labor in agriculture was below the estimated farm wage rate, but a variety of evidence points in this direction. Most of the cross-sectional agricultural production-function studies based on individual farm data seem to arrive at estimates of the marginal product of labor that are

12. Both the fit and the coefficients of the equation remain practically unchanged if we substitute labor times education for the two separate variables.

13. Whether this difference is significant can be tested by computing a predicted output for each observation in the sample using the official factor-shares as coefficients of the production function. The R^2 of the factor-shares equation is 0·89 against 0·98 for the estimated production function. The differences in the coefficients between (U17) and the factor-shares equation account for about 81 per cent of the residual variance left over from the latter equation, indicating that they are 'highly significant'.

Table 1 Aggregate Agricultural Production Function, United States, 1949, Sixty-Eight Regions*

Regression	Coefficients							R^2	Sum of coefficients (excluding E)
	X_1 (Livestock expense)	X_2 (Other current expense)	X_3 (Machinery)	X_4 (Land)	X_5 (Buildings)	X_6 (Man-years)	E (Education)		
(U17)	0·169 (0·023)	0·121 (0·032)	0·359 (0·048)	0·170 (0·033)	0·094 (0·044)	0·449 (0·072)	—	0·977	1·362
(R6)	0·140 (0·025)	0·111 (0·031)	0·325 (0·049)	0·167 (0·032)	0·085 (0·042)	0·524 (0·076)	0·431 (0·181)	0·979	1·352

* All variables are logarithms of original values unless otherwise specified: units = averages per commercial farm. Numbers in parentheses are the calculated standard errors of the respective coefficients.

X_1, log of (purchases of livestock and feed and interest on livestock investment).

X_2, log of (purchases of seed and plants, fertilizer and lime, and cost of irrigation water purchased).

X_3, log of (purchases of gasoline and other petroleum fuel, repairs of tractors and other machinery, machine hire, and depreciation and interest on machinery investment).

X_4, log of interest on value of land.

X_5, log of (building depreciation and interest).

X_6, log of average full-time equivalent number of workers per commercial farm.

E, logarithm of the average education of the rural farm population weighted by total US income by education class weights; not per commercial farm but per man.

Dependent variable – log of value of farm production per commercial farm.

Table 2 Alternative Input Weighting Schemes, United States Agriculture

| Source of weights | Inputs | | | | | |
| | Labor | Real estate | Power and machinery | Feed, seed, and livestock | Fertilizer, seed, and lime | Other |
	1	2	3	4	5	6
USDA, 1947–9 adjusted*	0·40	0·13	0·14	0·18	0·03	0·12
Cross-sectional aggregate production function, 1949†	0·33	0·19	0·26	0·13‡	0·02§	0·07§

* From R. A. Loomis, 'Production Inputs of US Agriculture', *The Farm Cost Situation*, May 1960, p. 3. Adjusted by substituting the estimated total value of purchased feed, seed, and livestock in 1947–9, instead of the 'value added by the non-farm sector' concept, and recomputing the weights accordingly.

† From Table 1, eq. (*U17*), computed by summing the relevant coefficients and dividing through by this sum.

‡ The measure used in the cross-sectional regression does not include 'seed' here but in the 'Other' category. Luckily, this is a minor item.

§ The cross-sectional measure included fertilizer and lime in the 'Other' category. The coefficient of 'other current expense' in the regression is distributed between the two categories proportionately to the USDA weights.

substantially below the comparable hired wage rate in the locality. Out of forty-three production-function estimates compiled by J. G. Elterich,[14] estimated from data for the years 1950–53 and for the states of Alabama, Illinois, Indiana, Iowa, Kentucky, Michigan and Montana, nineteen estimated the marginal product of labor (at the geometric mean of the sample) at less than $70 a month, seven of the estimates were between $70 and $100, eight between $100 and $150, and nine were over $150 a month.[15] The average monthly hired wage rate during these years was around $137 to $151, and higher than that in Illinois, Iowa and Michigan, where most of these studies were made. This finding is also consistent with the very large and continuing migration out of agriculture. In 1950, the year after our study, out-migration was equal to 5·2 per cent of the farm population and has averaged about 3·5 per cent per year since.

The higher estimate of the coefficient of power and machinery inputs is also supported by the finding of relatively high marginal returns to machinery investment in many farm production functions, programming, and budgeting studies. From 1950 to 1961, the ratio of machinery inputs to labor inputs increased by 74 per cent.[16] In the same period farm wages rose only by 9 per cent relative to machinery prices,[17] implying a substantial decline in the 'share' of labor in agriculture.[18] This, of course, could be explained by an elasticity of substitution that was larger than

14. Joachim G. Elterich, private communication, Michigan State University, East Lansing, 12 June 1961.

15. See also Heady and Dillon (1961), Table 17.1.

16. See US Department of Agriculture (1961), Table 22.

17. From the US Department of Agriculture, *The Balance Sheet of Agriculture 1961* (Agricultural Information Bulletin No. 247), and *Agricultural Prices, 1961 Annual Summary*, PR 1 – 3(62). Strictly speaking, we want here not a price index of machinery but a price index of machinery services. This would involve us in a multiplication of these indexes by interest and depreciation rates. Since interest rates rose on balance over this period and the expected life of machinery is unlikely to have lengthened much (given the observed rate of obsolescence), such an adjustment would have resulted in a relatively higher price index of machinery services and hence an even lower increase in the relative price of labor. All this, of course, does not allow for any quality changes in machinery or for that matter, in labor.

18. For additional evidence on the decline of the 'share' of labor in agriculture see Ruttan and Stout (1960).

unity, but it would have to be larger than eight, and evidence presented elsewhere implies a much smaller elasticity of substitution (see Griliches, 1963). One could, of course, assert that technical change during this period has been non-neutral and especially labor-saving, but there is little independent evidence for such a hypothesis. It is my belief that these facts can be best rationalized by the recognition that during most of the 1947–9 period agriculture was not in equilibrium, and hence that the 'base-period' factor shares did not reflect well, if at all, the underlying production conditions.

Perhaps the most controversial finding is the one of economies of scale. Our equations indicate that a 10 per cent proportional increase of all inputs in agriculture leads to a 13 per cent or more increase in output. This finding is subject, however, to several possible sources of bias.[19] In particular, the use of values rather than quantities for some of our variables could lead to biased estimates of the coefficients and their sum. But this, if anything, should have biased our estimates downward.[20] More serious is the objection that the form of the equation does not allow for the most interesting source of economies of scale – indivisibilities and the non-homogeneity of the production function. An experiment with a somewhat more complicated form of the production function did not lead, however, to superior results.

The finding of substantial economies of scale is consistent with much scattered information about United States agriculture. In the previously cited survey of micro-farm production function estimates – of forty-three surveyed production-function estimates, thirty-six had coefficients whose sum exceeded 1·0, and in twenty-three equations it exceeded 1·1. Similarly, much of the farm budgeting (including linear programming) literature implies that farmers could earn a higher rate of return on their owned

19. For a discussion of statistical sources of bias see Griliches (1957) and Hoch (1958).

20. The direction of the bias will depend, among other things, on whether the demand elasticity for the particular input is larger or smaller than unity. In the case of the Cobb–Douglas function all these elasticities should be larger than unity, biasing thereby the coefficients of values downward and hence also their sum. This is supported by the finding of a substantially lower coefficient of labor if labor 'expense' is used instead of man-years in the regression.

capital and labor if their farms were larger. Perhaps the most striking confirmation of the hypothesis of economies of scale is to be found in what has happened to the distribution of farms by size of output since 1950.[21] Table 3 presents the distribution of commercial farms by size of sales (in 1954 prices) in 1950 and 1959, and illustrates clearly the rapid growth in the relative number of larger farms and the decline of the smallest size class. This by itself, however, need not indicate the presence of economies of scale, because a growth in total productivity could increase the output of each farm without really changing its 'scale' (as measured by inputs) but still move some farms into the next size class (as measured by sales). Column (4) in Table 3 makes an outside estimate of the amount of shifting in the size distribution that could be due to a 21 per cent 'neutral' increase in total input productivity[22] on the assumption that all farms benefited from this increase equally and that farms are distributed uniformly within each size class. Both of these assumptions would lead to an overestimate of the 'technical change effect', but even as computed, column (4) is about midway between columns (2) and (3) of Table 3, underestimating substantially the relative decline of small farms and the relative increase of the larger ones. Thus there must have been a 'real' shift toward larger-scale farms since 1950, such as would have been predicted by the estimated production function.

The most important test of an estimated production function is not how well it fits the data it was derived from but rather whether and how well it can 'predict' and interpret subsequent behavior. The computed production function together with the assumption of some degree of economizing behavior 'predicts' a decline in labor inputs, a rise in machinery inputs, and an expansion in the share of larger scale farms. All of these 'predictions' were borne out by subsequent events. Moreover, the existence of economies of scale would imply a rise in the relative price of the most specialized input to agriculture, the one having the lowest supply elasticity. Land fits this description best and its price rose 83 per cent from 1947–9 to date, while during the

21. For a discussion of this test see Stigler (1958).
22. As estimated by the US Department of Agriculture for the period 1950–59.

Table 3 Percentage Distribution of Commercial Farms, by
Size Classes, United States, 1950 and 1959

| Economic class | Value of sales per farm (in 1954 dollars) | Percentage distribution of commercial farms by economic class | | |
| | | 1950* | 1959 (actual)† | 1959 ('predicted')‡ |
	1	2	3	4
I	25,000 and over	3·0	8·9	6·2
II	10,000–24,999	11·0	21·5	12·8
III	5000– 9999	20·8	25·0	25·6
IV	2500– 4999	25·4	23·6	22·0
V and VI	250– 2499	39·8	21·0	33·4

* From Jackson V. McElveen, *Family Farms in a Changing Economy*
(US Department of Agriculture Information Bulletin no. 171 [March,
1957]).

† From Karl A. Fox, 'Commercial agriculture: perspectives and prospects', *Farming, Farmers and Markets for Farm Goods* ('CED Supplementary Paper' no. 15, Committee for Economic Development, 1962,
Table 15).

‡ Computed on the assumption that the USDA's 21 per cent estimated
increase in total productivity since 1950 would have increased each unit's
output in the 1950 distribution by 21 per cent. The resulting shift from one
class to the next is estimated on the assumption of a uniform distribution
within each class interval. Thus, for example, a 21 per cent increase of
output for farms in classes V and VI would have moved all farms that were
above $2066 ($2500/1·21) into the above $2500 class. Using more detailed
information from the 1950 *Census of Agriculture* (vol. II, ch. ix), we know
that 14·7 per cent of farms were in the $1500–$2499 class in 1950. Assuming
a uniform distribution over this (smaller) class implies that about 6·4 per
cent of all farms were to be found in the $2066–$2499 interval, which are
then shifted to the row above. Similar calculations were performed for the
other classes. Even with the use of more detailed class data, the assumption
of uniformity biases these 'predicted' shifts upward, since the actual distribution is very much skewed to the right. The mean of the $10,000–$24,999
class is, for example, around $14,500 instead of the assumed $17,500.

same period, the index of prices paid by farmers for all (other)
production items including labor rose only 26 per cent.[23]

23. Here again we would like to have the price of land services rather
than the price of farm real estate. Scattered data (of dubious quality) on the
ratio of gross cash rent of value (for entire farms rented wholly for cash) for
the North Central states indicate that this ratio has not declined from 1950

4 Measuring input change

This section brings together the results of several studies of input quality change and other aspects of input measurement. These will be summarized in the form of a series of adjustments performed on the official input series for United States agriculture. The rationale and evidence for these various adjustments is discussed at greater length in the original papers (see Fettig, 1963; and Griliches, 1960 and 1963a). The magnitude of these adjustments is illustrated in Table 4, which presents the official input and output indexes for United States agriculture for 1940 and 1960 (1947–9 = 100) and the adjusted series for the same dates. The following major adjustments were made: [24]

Labor-force quality

Our production results indicate that we can proceed directly to multiply the man-years figures by our index of education per man (since the coefficients of these two variables are almost the same). This index was computed by weighting each school-year-completed class by the average 1950 income of all United States males (twenty-five years and over) in this class. The resulting measure of education is almost proportional to mean school years completed, except for a non-zero weight for 'no education'

---

to date (see various issues of *Farm Cost Situation*). The rise in land prices could, of course, also be due to the capitalization of the effects of the government support programs. All that is claimed in the text is that the subsequent events do not contradict the implications of the production function study. An additional bit of evidence in the same direction is the continued rise in the percentage of total farm purchases made for 'farm enlargement' purposes, from about 22 per cent of all sales in 1950 to 45 per cent of the total in 1960 (see various issues of *Farm Real Estate Situation*).

24. A relatively minor adjustment had to be performed on the official output and input series to make them comparable to the concepts used in estimating the cross-sectional production function. There, the output measure includes sales to other farmers and the input measures include purchases from other farms. The official time series exclude, however, interfarm sales of feed, seed and livestock from their definition of aggregate output and include only the 'value added by the non-farm sector' in their input measure. The adjustment consisted of adding the difference between the two feed, seed and livestock concepts (in 1947–9 prices) to both output and input. This had very little effect on the final results.

and a somewhat higher relative weight for college education. Its derivation is given in Table 5.

The adjustment itself amounts to about 5 per cent between 1940 and 1947–9 and 8 per cent from 1947–9 to 1960. The total increase in the 'quality' of the agriculture labor force between 1940 and 1960 was about 14·4 per cent. Additional adjustments were calculated for changes in the age, sex and race distribution of the agricultural labor force based on 1950 income data by these characteristics. The resulting adjustments were quite small and hence were not included in the final analysis.

Bias in the deflators

Since most machinery input estimates are based on cumulated deflated expenditure series, if the deflators are poor, so will also be the resulting 'constant price' capital estimates. All the deflators used for these purposes in agriculture are based on USDA-collected price statistics and are components or a recombination of components of the Prices Paid Index. While most machinery price indexes do poorly as far as quality change is concerned, the Prices Paid by Farmers Index has been especially affected by the official USDA insistence that it wants an index of 'unit values' rather than of prices and the consequent practices of relatively loose commodity specification, pricing the quality or brand 'most commonly bought by farmers', and pricing items with all 'the customarily bought attachments'. As the result of these practices, the USDA indexes have drifted upward over time relative to other similar price indexes (Table 6). The adjustments performed to counteract some of these biases consisted of computing for each of the machinery categories the drift of the USDA price index relative to a more tightly specified price index (the CPI new-automobile price index in the case of automobiles, and the comparable WPI components for motor trucks and other farm machinery). These relative indexes were averaged using 1947–9 gross investment in the respective categories as weights, and the resulting index was used to inflate the USDA estimates of farm gross investment in motor vehicles and farm machinery from 1947 to date. But even these more tightly specified price indexes do not take satisfactory account of the changes that have occurred in more complicated pieces of machinery such

Table 4 Various Input and Output Indexes, United States Agriculture, 1940 and 1960
(1947–9 = 100)

	Official* 1940 1	1960 2	Adjusted 1940 3	1960 4
Inputs:				
Labor	122	62	115	67[a]
Real estate	98	106	98	106
Power and machinery	58	142	54	152[b]
			66	181[c]
Feed, seed, livestock	63	149	71	157[d]
Fertilizer and lime	48	192	48	207[e]
Other	93	138	89	144[f]
Output:	82	129	81	132[g]
No. of commercial farms[h]	120	71		

* US Department of Agriculture, *Changes in Farm Production and Efficiency*.

(a) For a detailed description of this and subsequent adjustments see Griliches (1960). The official labor-input index was multiplied by an index of formal schooling per man in agriculture. See Table 5 for the derivation of this index.

(b) Adjusted for estimate bias in the deflators. Past investment expenditures were inflated upward from 1947 on the basis of the discrepancy between the USDA indexes and other indexes covering the same items. Automobile purchases were adjusted upward by the discrepancy between the CPI and USDA automobile price indexes, all the other price indexes were adjusted by the discrepancy between comparable components of the WPI and USDA indexes, except that for tractors. Fettig's (op. cit.) index of new tractor prices was used from 1950 on. See Table 6 for details. The resulting gross investment series were then substituted in the USDA formula for computing the capital stock series. This adjustment affects only about half of this index. The operation and maintenance components were left unadjusted.

(c) Based on a gross (fifteen-year moving sum of past investment) concept of capital. Computed from the previously described adjusted investment series. Again, the operation and maintenance component was unaffected by this adjustment.

(d) Total value of purchased feed, seed and livestock in 1947–9 prices from *Farm Income Situation* (deflated separately by the prices paid indexes for feed, seed and livestock from *Agricultural Prices*), substituted for the 'value added by the non-farm sector' concept.

(e) The fertilizer and lime index was increased by 8 per cent from 1947–9 to 1960 to reflect improvements in quality (mainly the shift to nitrogen) not caught by the unweighted plant nutrient measure (see Griliches, 1960, for details of this adjustment).

(f) The same 8 per cent increase was applied also to the 'other inputs' category on the assumption that quality improvements in this category have been *at least* of the same order of magnitudes as those not caught by an already quality-oriented measure of fertilizer inputs. This adjustment was distributed equally between the pre- and post-1947–9 periods.

(g) Adjusted by adding to it the purchased feed, seed and livestock index weighted by the *difference* in 1947–9 between all purchased feed, seed and livestock and the 'value added by the non-farm sector' concept.

(h) See notes to Table 3 for sources. Interpolated for 1940, 1947–9 and 1960 by changes in the number of all farms (from *Farm Income Situation*, July 1962, Table 8H).

as automobiles or tractors.[25] In the case of tractors, estimates were available of the marginal 'price' of horsepower and the differential value of diesel versus gasoline engines.[26] These estimates were used to construct an alternative tractor price index that was then substituted for the USDA deflators. The resulting adjusted gross investment series was then used to recompute the capital stock series using the same procedures as in the official series.[27] These adjustments amounted to about 12 per cent between 1947–9 and 1960 in the gross investment series, and to somewhat less for the resulting capital series. None of these adjustments, except for the tractor price index, are as far reaching as I have previously advocated in my work on automobile price indexes.[28] Even the tractor adjustment is relatively small (10 per

25. For additional discussion of this problem see Griliches (1963a).

26. These estimates are based on the coefficients of cross-sectional regressions of tractor prices on their respective specifications. See Fettig (1963), for details.

27. The above adjustments affect only about half of the machinery and power index – the services of the stock of capital component. The other half – maintenance and operation of machinery – was left unadjusted. These series should probably have been also adjusted upward, but not enough data are available to do this adequately. The 1955 Farm Expenditure Survey revealed a 26 per cent underestimate of this category of expenditures, but this finding has not been apparently incorporated in the subsequent revisions.

28. See Griliches (1961; 1963a). The adjustments suggested in these papers were not applied to this case, in the belief that most of the 'qualities' measured there had little relevance for automobiles as a production good (rather than as a consumption good) in agriculture.

Table 5 Index of Educational Level of Rural Farm Males

School years completed	Rural farm population: males, 25 years old and over, by school years completed			1950 mean income of males (25 years old and over), by school years completed, all US
	1940	1950	1960	
	1	2	3	4
None	5·2%	3·6%	2·7%	$1378
Grade school 1–4	18·3	15·8	11·1	1699
Grade school 5–7	24·7	23·5	20·2	2164
Grade school 8	29·7	26·9	26·8	2676
High school 1–3	10·8	12·8	14·4	3096
High school 4	6·2	10·7	17·9	3784
College 1–3	2·5	3·1	4·4	4449
College 4 or more	1·2	1·8	2·6	6318
Not reported	1·4	1·8	—	2471*
'Mean weighted income'	$2502	$2644	$2860	
Index, 1950 = 100	94·6	100	108·2	

* Computed from *1950 Census of Population, Education,* Series PE, no. 5B, p. 108, using Houthakker's midpoints (II. S. Houthakker, 'Education and income', *Review of Economics and Statistics,* vol. 41, 1959).

Source: col. 1, *1940 Census of Population,* vol. 4; col. 2, *1950 Census of Population,* vol. 2, Pt 1; col. 3, *1960 Census of Population,* PC(S1)-20 (4 June 1962), Table 76; col. 4, computed by averaging for the appropriate age groups the mean incomes given by Houthakker, op. cit. Table 1. The weights for these averages were taken from *1950 Census of Population, Education,* Series PE, no. 5B, pp. 42–3.

cent since 1950), because the USDA priced tractors within relatively narrow horsepower classes, preventing its index from drifting too far.

Differences in the concept of capital services

The official USDA machinery input series (excluding maintenance and operation) is proportional to its estimate of the value of the stock of machinery on farms in constant prices. It is based on a declining balance depreciation formula using relatively high rates of depreciation (about 16·5 per cent per year on the average). It can be shown that these rates are too high, even as approximations to market rates of depreciation. But what is even more important is that for productivity comparisons we want an estimate of the *current flow of services* rather than of the market

Table 6 USDA Prices Paid by Farmers Indexes for Machinery and Motor Vehicles Relative to Other Comparable Indexes, 1947 to 1960
(1947–9 = 100)

Year	Auto-mobiles (USDA/CPI)	Motor trucks (USDA/WPI)	Tractors* (USDA/WPI and Fettig)	Other farm machinery (USDA/WPI)	Average† (1947 = 100)
1947	97·7	99·1	96·9	97·4	100
1948	101·4	101·3	98·6	99·1	102
1949	100·7	99·9	103·9	103·5	105
1950	100·8	100·5	103·8	102·7	105
1951	101·4	98·9	104·1	101·1	104
1952	103·3	101·4	103·1	104·3	106
1953	103·9	102·1	103·0	104·8	106
1954	105·5	104·0	102·9	104·7	107
1955	114·4	101·8	107·0	103·7	108
1956	114·9	100·2	106·0	103·1	107
1957	116·8	104·1	111·0	104·1	110
1958	116·1	103·7	107·6	104·0	109
1959	120·1	105·7	113·3	104·8	111
1960	118·3	108·4	114·2	105·2	112

* Fettig's index linked to the WPI in 1950; 1951 and 1952 (not computed by Fettig) interpolated by the WPI.

† Weighted average, with weights based on the 1947–9 rate of gross investment in these different categories (0·102 for automobiles, 0·160 for motor trucks, 0·264 for tractors, and 0·474 for other farm machinery).

Source: USDA unpublished component indexes used to deflate gross farm investment. They are components or a recombination of components of the USDA prices paid by farmers index. The CPI and WPI figures are taken from various BLS releases (in particular Bulletin nos. 1295 and 1256). Fettig's index is taken from his unpublished dissertation (op. cit.). From among the various indexes computed by him, I am using the chain-linked one, with weights based on his linear single-year regressions.

value of all present and future services. The relevant measure would be approximated, in a perfect market, by the rental price per machine-hour times machine-hours used. The rental price would be a 'constant' price, which would not only adjust for fluctuations in the general price level but also for relative price changes between new and used equipment (obsolescence) induced by the availability of superior machines. This rental would change over time only if the physical flow of services were to deteriorate due to wear and tear (and age) or change its character of use due to the appearance of new and different machines. It would not change just because the appearance of new machines makes the use of the old ones less profitable and leads to capital losses by the owners of the old machines. Nor would it change if new machines are made more durable at a higher price but still provide the same *annual* service flow.

The current official measures assume that the flow of annual services from a machine falls by more than half in the first four years of its life, and that very little of it is left (about 16 per cent) after ten years. What little evidence we have on the mortality and use of machines with age indicates that this is not the case. An alternative, somewhat extreme, assumption is that the flow of services does not decline with age.[29] This would imply the use of 'gross' instead of 'net' measures of capital in productivity comparisons.[30] I have accordingly substituted a fifteen-year moving sum of past gross investment for the USDA measure. The new measure assumes that services are constant during the first fifteen years of life of a machine and fall to zero thereafter.[31] The two measures move closely together over long periods of time, but they can have very different time trends in the intermediate run.

29. Repair costs, if adequately measured, may rise with age and should be subtracted from the flow-of-services estimate. The available repair and maintenance statistics do not allow, however, for any growth due to the aging of the stock of capital equipment.

30. For additional discussion of some of these issues see Griliches (1966).

31. The available data indicate that the *average* life expectancy of farm machinery is about eighteen years, and that, at least for tractors, fifteen-year-old machines are still working as many as 88 per cent of the average hours worked by new machines. These figures are based on unpublished tabulations from the 1956 National Farm Machinery Survey. Thus, the assumption of fifteen years of relatively constant service does not appear to be extreme.

In our particular case, this adjustment leads to a substantially lower estimate of the growth of the stock of machinery on farms between 1940 and 1947–9, and a much higher estimate of the subsequent (to 1960) growth in this stock.[32]

Other quality changes

The main adjustment here is to the fertilizer and lime index. The USDA measure does not use tons as its measure, converting them into units of effective ingredients (plant nutrients). It does, however, simply add up the three main ingredients (nitrogen, phosphoric acid and potash), ignoring differences in their relative value. Since the consumption of the most expensive and important ingredient – nitrogen – has grown more over time, the resulting official fertilizer input series underestimate the actual growth in 'effective' fertilizer consumption. The adjustment for this bias consists of substituting a weighted plant nutrient measure for the unweighted one used by the USDA, the weights having been derived from a cross-sectional study of prices of different fertilizer mixtures in 1955. The adjusted index rises by about 8 per cent more from 1947–9 to 1960. A similar adjustment, distributed equally between the pre- and post-1947–9 period, is applied to the 'other inputs' category on the assumption that quality improvements in this category have been *at least* of the same order of magnitude as those not caught by an already quality-oriented measure of fertilizer inputs.

5 Summary

The adjusted input series together with the estimated production function weights are brought together in Table 7 to account for the growth in aggregate agricultural output between 1940 and 1960, and to compute the 'residual', unexplained increase in productivity. For comparison purposes, several intermediate measures are computed, using the official input series and weighting schemes. Also, since the estimated production function is of the Cobb–Douglas form, implying the use of geometric rather than arithmetic input indexes, the final estimates are presented on a geometric index base. As can be seen from this table,

32. Again, this adjustment affects only the 'depreciation and interest' component (about half) of the machinery and power input index.

Table 7 US Agriculture: Various Total Input and Total Input Productivity Indexes, 1940 and 1960 (1947–9 = 100)

	Total input indexes		Total productivity indexes		
	1940	1960	1940	1960	1960 (1940 = 100)
Arithmetic indexes:					
1. Official	97	102	85	126	148
2. Official adjusted to cross-sectional output concept	95	109	85	121	142
3. 'Corrected' input series (for quality change and deflator bias) and official (adjusted) weights	91	114	89	116	130
4. 'Corrected' input series (for quality change *and* capital stock concept); adjusted official weights	93	118	89	112	129
5. 'Corrected' inputs (as in 4); production function weights ($U17$), excluding scale effect (sum = 1·0)	90	124	90	106	118
Geometric indexes:					
6. Official adjusted for output concept	91	100	89	131	148
7. 'Corrected' inputs (as in 4), adjusted official weights	90	112	90	118	132
8. 'Corrected' inputs, production function weights, excluding scale effect	88	114	93	116	125
9. 'Corrected' inputs, production function weights, including scale effect (sum = 1·36)			104	98	94

Source:
1. From US Department of Agriculture, *Changes in Farm Production and Efficiency.*
2. Weights: row 1 of Table 2; input series: cols. 1 and 2 of Table 4, except for 'Feed, seed and livestock' and 'Output', where cols. 3 and 4 were used.

3. Weights: same as in 2; Input series: cols. 3 and 4 of Table 4; first row for power and machinery.

4. Same as 3, except that the second set of estimates was used for power and machinery.

5. Weights: row 2 of Table 2; input series – same as 4. Geometric indexes of productivity computed as antilog of

$$y - \sum a_i x_i,$$

where y is the log of output, x's are the logarithms of the input series and the a_i's are the respective weights.

6. Same sources as 2.

7. Same sources as 4.

8. Same sources as 5.

9. Productivity indexes equal antilog of

$$y - \sum a_i x_i - k(\sum a_i x_i - n),$$

where k is the excess of the sum of the coefficients in the original production function over unity (in this case $k = 0.36$) and n is the logarithm of the index of number of farms. The a_is used here still sum to unity. The above expression is equivalent to antilog of

$$(y - n) - \sum a_i (1 + k)(x_i - n).$$

there is little difference between the original official indexes and those adjusted for the different concept of output. Nor is there much difference between the arithmetic and geometric indexes for comparable combinations. Substituting 'corrected' input series into the official weighting scheme reduces the estimated productivity increase by about one third.[33] Using the estimated production function weights, without allowing, however, for scale effects, leads to another one-fourth to one-sixth in reduction in the original productivity growth estimates. If one allows, in addition, for economies of scale at the cross-sectionally estimated rate, they account for all (and somewhat more) of what is left.[34]

The production-function study results together with the estimates of input quality change account for all (if not more than

33. Almost all of this is accounted for by 'quality-change' adjustments. The adjustment for differences in capital concept affects only the timing of the estimated 'residual', shifting part of the estimated productivity increases from the post- to the pre-1947–9 period.

34. Keeping the same relative weights but using 1·2 as the sum of the coefficients (instead of the estimated 1·36) would have just about accounted for all of the measured productivity increases without leaving an embarrassment of a negative, albeit small, residual. Given the estimated variance-covariance matrix of the coefficients of the production function 1·36 is not significantly different from 1·2 at conventional significance levels.

all) of the observed increases in agricultural productivity.[35] They imply that (very) roughly about one-third of the observed productivity increases are due to improvements in the quality of inputs (among which the rise in education per worker plays an important part), about a quarter or so is due to a move toward the elimination of relative disequilibria due to the overpricing of labor (in particular of family labor) and the underpricing of capital services by the conventional market measures, and that the rest is due to the expansion that occurred in the scale of the average farm enterprise.[36] This 'complete' accounting for the observed productivity increases does not mean that there were no meaningful increases in agricultural productivity over this period. It means, rather, that we may have succeeded in providing an explanation for what were previously unexplained increases in farm output.

While the particular sources of measured productivity growth will be different in other sectors of the economy, the methodology presented in this paper is quite general and could be applied to quite different industries or sectors. Such breakdowns of measured productivity growth into its source components should

35. Throughout this investigation, we have tried to adjust inputs for quality change, but we have made no comparable adjustments in the output series. This has been based on the assumption that, in agriculture, the problem of quality change has been much less serious for output measures than for input measures. This assumption has not been investigated, however. The whole question of measuring aggregate agricultural output probably should be also reopened. Once this is done, it is quite likely that we would find that the growth in output has also been underestimated.

36. A possible source of bias may be hidden in our assumption that the various effects are additive. To take advantage of economies of scale, one may have to use a qualitatively different set of inputs. In the cross-sectional study, we did not allow for differences in the quality of inputs used in different areas (except in the case of labor). If larger-scale farms use inputs of higher quality, some of the measured returns to scale may actually be due to quality change. If this is the case, applying the cross-sectional returns to scale estimate to quality-adjusted input time series may involve us in some double counting. Also, some of these economies of scale may be external in the sense that they reflect the lower cost of purchasing certain services in larger bundles. If, for example, larger horsepower tractors are sold at a lower price per horsepower unit and horsepower is the important dimension that enters into the production function, then some of the apparent economies of scale are really external and specific to particular inputs.

prove helpful in forecasting future rates of output growth and in any future attempts to manipulate these rates.

Since this approach goes after particular numbers it is somewhat *ad hoc* and does not lend itself well to an incorporation into a standard 'growth model'. It does suggest, however, that the concept of 'embodiment' of technical change should be extended to all inputs rather than just to capital (as in Solow, 1962) or just to labor (as in Denison, 1962) that the possibility of disequilibrium and its consequences should play a much larger role in such models, and that some of the more interesting questions raised by such models may eventually resolve themselves into questions concerning the determinants of the demands for and supply of different 'qualities' and the feedback mechanism, if any, operating through the changing pattern of 'quality' prices.

References

ARROW, K. J., CHENERY, H. B., MINHAS, B. S., and SOLOW, R. M. (1961), 'Capital-labor substitution and economic efficiency', *Rev. Econ. Stat.*, vol. 43.

DENISON, E. F. (1962), *The Sources of Economic Growth in the United States and the Alternatives before Us*, Committee for Economic Development.

FABRICANT, S. (1959), 'Basic facts on productivity change', *Occasional Paper, National Bureau of Economic Research*, no. 63.

FETTIG, L. P. (1963), 'Price indexes for new farm tractors in the postwar period', Unpublished Ph.D. dissertation, University of Chicago.

GRILICHES, Z. (1957), 'Specification bias in estimates of production functions', *J. farm Econ.*, vol. 39.

GRILICHES, Z. (1960), 'Measuring inputs in agriculture: a critical survey', *J. farm Econ.*, vol. 42.

GRILICHES, Z. (1961), 'Hedonic price indexes for automobiles: an econometric analysis of quality change', in United States Congress, *Government Price Statistics Hearings*, Joint Economic Committee, pp. 173–96; reprinted as *The Price Statistics of the Federal Government*, National Bureau of Economic Research, general series, no. 73.

GRILICHES, Z. (1963), 'Estimates of the aggregate agricultural production function from cross-sectional data', *J. farm Econ.*, vol. 45.

GRILICHES, Z. (1963a), 'Notes on the measurement of price and quality changes', *National Bureau of Economic Research Conference on Models of Income Determination*, Princeton University Press.

GRILICHES, Z. (1963b), 'Capital stock investment functions: some problems of concept and measurement', in D. Patinkin (ed.), *Measurement in Economics*, Stanford University Press.

HEADY, E. O., and DILLON, J. S. (1961), *Agricultural Production Functions*, Iowa State University Press.

HOCH, I. (1958), 'Simultaneous equation bias in the context of the Cobb–Douglas production function', *Econometrica*, vol. 24.

KENDRICK, J. W. (1961), *Productivity Trends in the United States*, Princeton University Press.

RUTTAN, V. W., and STOUT, T. T. (1960), 'Regional differences in factor shares in American agriculture: 1925–57', *J. farm Econ.*, vol. 42.

SCHULTZ, T. W. (1947), 'How efficient is American agriculture?', *J. farm Econ.*, vol. 39.

SOLOW, R. M. (1957), 'Technical change and the aggregate production function', *Rev. Econ. Stat.*, vol. 39. [Reprinted as Reading 16 of the present selection.]

SOLOW, R. M. (1962), 'Technical progress, capital formation and economic growth', *Amer. econ. Rev.*, vol. 52.

STIGLER, G. J. (1958), 'The economies of scale', *J. Law Econ.*, vol. 1.

UNITED STATES DEPARTMENT OF AGRICULTURE (1961), *Changes in Farm Production and Efficiency*, Statistical Bulletin, no. 233, July.

Part Five
International Aspects of Technological Change

The papers in the previous sections have not explicitly taken into account the division of the world into nation-states and the fact that, in the past and in the present, sizeable differences in the levels of technology have existed among them. One of the most decisive aspects of the contemporary world is that countries differ significantly in the degree of technological sophistication at their control. Moreover, much of world history could – and should – be written around the consequences – military, political, demographic and social as well as economic – of these differences. The papers in this final section attempt to explore some of the implications of such differences.

The first paper, by Ames and Rosenberg, poses the following question: what are the implications for the rate of economic growth of individual economies of the fact that, at any moment of time, some countries may be identified as more or less 'advanced' than others? A view which has been widely held is that technological 'leadership' involves serious penalties and technological 'followership' significant advantages. This paper attempts to sort out, within a formal analytical framework, the possible meanings and implications of such arguments and to relate them to the larger questions of industrial growth.

The paper by Vernon opens up a dimension which has, until recently, received very little consideration. The theory of comparative advantage has attempted to account for the composition of trade flows among countries in terms of differences in resource endowments and factor prices. Vernon enlarges the traditional framework by considering the impact of the introduction of new products. He is particularly concerned with the fact that, in the course of their introduction and diffusion, new products go through a distinct series of stages with respect both to their design and their production. The requirements of these successive stages in the 'maturing' of a product influence the changing geographic location of its production in ways which are not apparent within the more traditional theory.

Baldwin's paper explores the ways in which the technological constraints underlying the production function for different commodities may affect the path and rate of economic development of newly-settled regions. His model examines the hypothetical development of two regions, initially sparsely populated, which begin their growth by concentrating upon the production and export of a primary commodity. Baldwin shows how this development may be decisively influenced by the basic characteristics of the technology shaping the production function for each commodity: the presence of economies of scale which influence the optimum size of the productive unit, the ease with which factors of production may be substituted for each other, the skills (managerial as well as technical) required of the labor force, differences in the factor intensity of the productive process, etc. Differences in these initial conditions may exert important influences by the ways in which they shape forms of economic and social organization, the distribution of income, the composition of demand for commodities and services, and the acquisition of entrepreneurial skills. Because of these kinds of relationships, a careful study of technological conditions may prove to be a highly fruitful way of advancing our understanding of the development process.

Finally, the very existence of sharp differences in technological achievements among countries may be regarded as a cause for optimism. Such differences hold out the prospect that significant economic development may be achieved through the borrowing, by less-developed countries, of techniques which already exist in more advanced countries. Such borrowings, however, are by no means simple, and the problems posed have often been drastically underestimated in the past. In the last paper in this section, Solo addresses himself to this range of problems by asking what factors influence a country's capacity to assimilate a more advanced technology. Some technologies, to begin with, are highly specific to a particular environment and may not be directly transferable – as is sometimes the case in agriculture. More generally, the transfer process involves the ability to recognize possibilities for transfer and an awareness of the sort of adaptations which may be preconditions for success. These adaptations may include alterations in the technology in order to apply it in a different environment or restructuring of the environment itself to provide more hospitable circumstances for the functioning of the new technology. Success in all this will depend in an important degree, as Solo emphasizes, upon a society's intellectual skills at many levels, or, as he calls it, upon its structure of cognition.

19 E. Ames and N. Rosenberg

Changing Technological Leadership and Industrial Growth[1]

E. Ames and N. Rosenberg, 'Changing technological leadership and industrial growth', *Economic Journal*, March 1963, pp. 13–31.

Although this science contains indeed a number of correct and very excellent precepts, there are, nevertheless, so many others, and these either injurious or superfluous, mingled with the former, that it is almost quite as difficult to effect a severance of the true from the false as it is to extract a Diana or a Minerva from a rough block of marble (Descartes).

This paper explores afresh the thesis that there is (or has been) a net penalty incurred by countries which have been innovators. A considerable literature explores this 'penalty for taking the lead' or 'handicap of the early start'.[2] Despite several controversies, in which the thesis has never been properly proved, nor even adequately formulated, it continues to be stated as a self-evident truth.[3] Since the thesis deals with important issues, we think it analytically worthwhile to state formally and in rather general terms the conditions under which the several forms of the thesis might be valid. This procedure serves to illuminate several aspects of the growth process, and to pose new questions for examiners.

Writers who have claimed there is a penalty in taking the lead have really only discussed one or two isolated phenomena

1. The authors are grateful to two of their colleagues, June Flanders and Jonathan R. T. Hughes, for useful comments on an earlier draft.

2. The subject has most recently been examined and the earlier literature partially reviewed in Kindleberger (1961). The most important earlier references are: Veblen (1966), Jervis (1947), Frankel (1955), Gordon (1956), Frankel (1956); and Svennilson (1954).

3. e.g. Rostow (1960, p. 70). On the other hand, Mandelbaum (1955, p. 3) discards the thesis on a rather casual basis. Mandelbaum appears to believe that once a country has been left behind it becomes increasingly difficult even to make a start.

relevant to this broader question of whether or not such a penalty exists. A simple formalization of the problem will place the entire issue in better perspective and make possible a critical examination of its implications.

We insist at the outset that even if (a) there are *some* penalties for taking the lead, it does not necessarily follow that (b) on balance, late starters are better off than early starters. Writers, however, who start off with the initial – rather innocuous – assertion often slip, unknowingly, into the latter proposition, which, we shall argue, is extremely difficult to defend.

Let us suppose two countries, Eastland and Westland, identically endowed at the time of Noah with population and resources. Suppose that Eastland remains in an agrarian, underdeveloped state until the year 1900; at that time it is in exactly the condition Westland was in 1700. Westland, however, began to industrialize in 1700, so that by 1900 it is an urban, factory society. We propose to discuss the thesis, developed by writers from Veblen through Kindleberger, that Eastland can develop more rapidly and/or to a higher level after 1900 than can Westland. In Kindleberger's (1961, p. 282) words, 'there may be a penalty in the early start, if institutions adapt themselves to a given technology, and if static patterns of capital replacement develop as habits'.[4]

It is certainly a fact that the countries whose industries have grown fastest in the past one hundred years are not those which grew most rapidly in the preceding century. The leading industrial countries of the Middle Ages – Brabant, Lombardy, Venetia – have never regained their former position in the world, any more than Egypt has regained rule over the grain trade. Certain special circumstances have intervened – e.g., the United Kingdom began running out of cheap coal about 1900[5] –

4. The concluding sentence of Habakkuk's (1962) highly interesting new book, *American and British Technology in the Nineteenth Century* reads: 'Such lags as there were in the adoption of new methods in British industry can be adequately explained by economic circumstances, by the complexity of her industrial structure and the slow growth of her output, and *ultimately by her early and long-sustained start as an industrial power*' (italics added).

5. Even the adequacy of a country's resource endowment, after all, depends on what industries use it, and hence on world prices and other economic facts. Partly, however, resource endowment depends upon

but it is an open question whether such factors are exclusively responsible for the facts. It is natural to inquire whether any economic theory of 'leadership' in industry can be formulated. Is the historical fact that a displaced leader does not regain its primacy an accident? Would it be natural to expect leadership to rotate among different countries?

The problem involves two distinct concepts: a country's output and a country's technology. Underdeveloped countries are, as a matter of fact, poor (they have small outputs *per capita*) and backward (they use technology which others have abandoned). It is perhaps possible for a country with a backward technology to be wealthy, or for a poor country to be technologically advanced, but these possibilities are seldom realized. The connection between the levels of output and of technology must therefore be distinguished conceptually in dealing with real situations in which they are closely related.

With regard to either output or technology, it is necessary to distinguish levels, amounts of change and rates of change. The fact that a country today is changing rapidly implies nothing, in principle, about its level of development either today or at some date in the future. A failure to distinguish between levels and rates of change may lead to interesting welfare results. If a country has a choice (in some sense) between developing for a century at a slow rate up to a level L_1, or else stagnating for half a century and then developing at a rapid rate to a higher level L_2, then its decision is based upon the familiar balancing of present and future satisfactions, with this peculiarity, that several generations are involved.

The importance of the late-comer theses may be illustrated by what is to us a pair of *reductio ad absurdum* propositions, which are sometimes almost seriously advanced:[6]

---

technology; this is a branch of knowledge which increases in a way still largely unknown, but certainly not wholly dependent upon economic events. See Ames (1961).

6. In this discussion, as elsewhere in the paper, we assume that all countries have the same resource endowments. Obviously the discussion is completely irrelevant if the differences between pairs of countries are explainable by differences in resource endowment. Thus, to us both the stocks of plant and technology must be variables, rather than resources, although for some purposes they can certainly be treated as resources. Our practice, in this

1. It is sometimes alleged that the reason industrial output in Germany and Japan has grown more rapidly, and has tended to become cheaper than output in the United States and the United Kingdom is that bombing attacks during the Second World War destroyed a mass of obsolete equipment in the former countries, giving these 'late comers' an advantage by paving the way for more modern techniques. If this argument were valid, a feasible form of US foreign economic assistance would then be the systematic bombing of the cities of our economically developed allies, in order to further their economic development.[7]

2. If an underdeveloped ally, envious of Japan or fearful of Communist China, should wish to industrialize, the United States should urge upon it the slogan, 'mañana!' For the later it begins, and the more backward it is when it begins, the sooner and the farther it can outstrip its competitors. 'Industrialize tomorrow,

---

paper, of enclosing resource endowment in the *ceteris paribus* pound does not necessarily reflect our own view of the importance of this factor. We merely find it here a convenient assumption, since we are attempting to meet on its own grounds a form of analysis which relegates resource endowment to an insignificant role. There has been a strong tendency in recent years to downgrade the independent significance of natural resources in the development process. Thus, the papers presented to the Conference on Natural Resources and Economic Growth, held at Ann Arbor, Michigan, 7–9 April 1960 [reprinted in *Natural Resources and Economic Growth*, J. J. Spengler (ed.)] approach the subject from a wide variety of interests, but stress two themes: (a) the powerful impact of classical economic theory (particularly the Ricardian–Malthusian variant) has led to a vast exaggeration of the relative importance of natural resources and the constraints which they impose upon an economy's development; (b) the relative importance of natural resources is a declining function of development itself. As Kindleberger (1961) puts it, 'It may be taken for granted that some minimum of resources is necessary for economic growth, that, other things being equal, more resources are better than fewer, and that the more a country grows the less it needs resources, since it gains capacity to substitute labor and especially capital for them' (p. 172).

7. Not only are people found who argue that economies may benefit from mass destruction; one author appears to argue that even armies do. 'The disaster which the British suffered at Dunkirk in 1940 served one useful purpose. It swept the slate clean, for the British Army had lost all its equipment in the evacuation. It taught them what would and what would not do, in attempting to defend their Island against the Luftwaffe and perhaps the Wehrmacht, should actual invasion come. Out of the bitter experience came improved designs for planes and tanks, and enormously

industrialize yesterday, but never industrialize today', is a para-phrase of a Red Queen, but it should perhaps be a US slogan.[8]

We suspect that such policies as these would somehow be un-sympathetically received by their intended beneficiaries. On the other hand, the absurdity of a mis-stated theory should not imply the incorrectness of a properly stated one. It is useful to see under what conditions a 'theory of the late starter' might be correct.

Suppose we were able to quantify the concept of a 'state of technology' so that we could locate three different states, A, B and C on a map, in such a way that we could measure the dis-tances A–B, B–C and A–C. It would be convenient if all these points were on a single line; but if they were, and technological change were thus one-dimensional, all the interesting features of latecomer theses would automatically be disproved, as later dis-cussion will show. For expository purposes we assume a two-dimensional technology which can be represented diagram-matically. Any movement to the north or east in our diagram represents improved technology. An origin is assumed to the south-west. The three states, A, B and C may then be represented as follows:

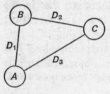

Figure 1

---

stepped-up requirements' (Hall 1959, p. 387). Habakkuk has suggested that differences in economic performance in the years after the Second World War may be explained, in part, by considerations of ideological and emo-tional reinforcement, which, in turn, are a consequence of the war. 'In our own day, the need to restore economies shattered by war has fired Europeans with a fervour for the task – a fervour which, together with the disruption of old routines by the war, goes some way to explain why, since 1945, France and Germany have grown more rapidly than countries like Britain and the USA, which have not had to meet the challenge of rebuilding large parts of their industrial capacity from scratch' (Habakkuk, 1962, p. 219).

8. Campbell (1960), in discussing the Soviet attempt at rapid industrial-ization, states that: 'Their task was not complicated by the presence of

We now say that an economic activity at A is 'underdeveloped', one at B is 'developed but obsolete' and one at C is 'developed and progressive'. As an empirical matter, we shall assert that a country at C produces more than one at B, and a country at B produces more than one at A. This assertion is not, however, a logical necessity. Now consider two economic entities, one at A and one at B. When will that at A move to C while that at B will not? Secondly, assuming that both will eventually get to C, under what conditions (if any) will that starting from A (the underdeveloped entity) reach C before that starting from B (the developed entity)?

Let us provisionally assume that two economies possess the attributes of two firms. This assumption will make it possible to talk about some aspects of the thesis. But a part of the literature on 'the late start' associates the problem with factors outside any single firm, so that the comparison of a pair of firms does not exhaust it. It will later become necessary, then, to consider how the corporate structure of two economies (one at A and one at B) may complicate the problem.

An enterprise at an arbitrary point P on the map shown above will move to a point P' if it profits from doing so; that is, we associate with an arbitrary point P a revenue Γ_p and a cost c_p, so that profit $\Pi_p = \Gamma_p - c_p$ at point P. The selection of a particular technology means that $\Pi_p > \Pi'_p$, so that technology P is preferred to any other technology P'. If P is optimal, then it is optimal both in respect to 'more advanced' and 'more backward' technologies in the neighbourhood of P.

Now we return to A, the 'underdeveloped', and B, the 'developed but obsolete' firms, and consider their movement to a 'developed and progressive' point C. Assume the two firms are purely competitive, so that prices may be taken as given. Then we may associate with the three points cost functions $C_A(x)$, $C_B(x)$ and $C_C(x)$, where x represents output. In terms of conventional thinking about technology, we assume that

$$C_A(x) > C_B(x) > C_C(x)$$

---

existing plants and an already familiar technology, and they could build a modern industry from the ground up' (p. 165).

for every relevant level of output x. Thus if a firm is confronted with given, perfectly elastic demand curves, it selects among the profit functions

$$\Pi_A(x), \Pi_B(x) = \Pi_A(x) + \{C_A(x) - C_B(x)\},$$
$$\text{and } \Pi_C(x) = \Pi_B(x) + \{C_B(x) - C_C(x)\}$$
$$= \Pi_A(x) + \{C_A(x) - C_C(x)\}.$$

On this assumption, the cost (and hence the profit) of producing at technology C is the same, regardless of whether the producer has always been at C, whether he is now at C but was formerly at B, or whether he is now at C but was formerly at A. In this case the inequalities

$$C_A(x) > C_B(x) > C_C(x)$$

will ensure that all producers will adopt technology C (unless, of course, there is a technology C' which is superior to C).[9]

A portion of the 'late-comer' literature is based upon propositions about how the movements to C from A and B will take place. They are all based upon advantages which A is alleged to have over B.

A 'weak' statement of this sort says: A and B will both move to C. But the time which A takes to reach C will be less than the time which B spent in getting from A's condition to C. (This statement does not imply that A will reach C first, but only that it will travel faster on the 'journey' to C than B.)[10]

In part, the 'weak' statement depends on the assertion that the late comer need not repeat the mistakes of its predecessors. The United States built an extensive canal system, which it replaced by railroads, and it is replacing its railroads by motor roads. The late comer will certainly avoid the canals, and probably the railroads. It is for this reason that in our diagram we

9. Gerschenkron (1952, p. 6) argues that apart from 'existing institutions', such as serfdom or political disunification, the more backward is a country, the more it gains from development. Nove (1961, p. 300) asserts that 'A further very important "advantage" in the growth race is backwardness itself . . .'.

10. If a country's welfare depends (at any time) upon the level, rather than the rate of change in its output or consumption, statements of this sort have nothing to do with the welfare implications of particular development patterns.

show a route from A to C which does not pass through B. But this formulation as yet does not explain why the late comer would use only highways, while the United States would retain the (developed but obsolete) railroad. It is necessary to explain why the late comer may reach an advanced technology which the developed but obsolete firm will not adopt.[11]

The late comer may well be able to avoid a further cost incurred by the pioneer. An 'early starter' may proliferate a wide variety of standards and specifications which later become difficult to change. It is in the nature of pioneering that exploration takes place on many fronts. A diversity of standards and specifications is often created which is, in effect, a 'residue' deposited upon the structure of the economy. This unnecessary diversity may in turn hamper the achievement of economies of large-scale production in, e.g. industry B which produces an intermediate (or capital) good which is an input for industry A.

11. This 'weak' statement does not explicitly consider the possibility that even 'false starts' can yield some advantage to the early comer; e.g. the *Great Eastern* was a spectacular commercial failure, but British industry may have learned something useful from building her. The late comer may not have to 'produce' new knowledge and new techniques, but he does have to produce the skills required to take advantage of existing technology. Whereas exploration and pioneering generate costs which are avoidable by the late comer, they also generate many other forces which are essential to further economic growth. Indeed, one may argue that industrialization is, in large measure, a learning process. The capital stock of any economy includes the acquired skills – technical, managerial, professional – in short, the 'intangible capital' embodied in its living population. This 'stock' of intangible capital was accumulated, in large part, as part of the process of exploration and pioneering in the country taking the lead. (See Schultz (1959, pp. 109–117); Schultz (1961, pp. 1–17); Goode (1959, pp. 147–55); Kuznets (1955, pp. 3–28); Hirschman (1958), especially ch. 1. Furthermore, such skills, once acquired, constitute a major portion of an economy's capital stock. Astonishment at the rapid rates of post-war recovery in Germany and Japan is due, in part, to persistent failure to give appropriate recognition to the importance of intangible 'human capital'. A demolished Hamburg in 1945 was in no sense reduced to the equivalent of a Bombay or a Calcutta. [As related miscalculation, even the massive RAF bombing raids upon Hamburg in the summer of 1943 had a much more limited impact in reducing the city's contribution to German war-time production than had been estimated at the time. See the various reports by the US Strategic Bombing Survey, European War, especially the overall report (1945).]

Thus, in the production of railway locomotives in the United States in the mid-nineteenth century, makers of locomotives found themselves producing to a wide range of specifications for different railroad companies, each of which laid down precise standards of workmanship (many of which were determined by historical 'accidents') to suit its own needs.[12] The result was a considerable reduction in the economies of large-scale production which a more highly standardized product would have permitted.

One real advantage of the late starter may be the clear perception of the advantages of standardization, and therefore the early adoption of uniform standards. The pioneering country, on the other hand, may be able to adopt uniform standards, but due to technological interrelatedness only at a substantial cost, at later stages of its development.[13]

A 'moderate' late-comer statement may also be found: if today A and B both start towards C, A (the less-developed country) will reach C first. In other words, there will be a change in leadership. This conclusion is apparently reached in one of two ways. One way is to assert that as a country develops, its rate of development slows down. If A is sufficiently backward, its rate of growth can be so great that it can overtake B.

This statement contains a logical weakness. Suppose that there

12. Fitch (1880), points out that the early standardization of equipment by each railroad made it difficult for locomotive manufacturers to standardize their output at a later date, since the adoption of standardization by manufacturers would cause a period of mixed standards among the users of their products, and hence high repair and maintenance costs. See also Kindleberger (1961), pp. 284–91. In somewhat analogous fashion, Svennilson (1954) attributes the backwardness of the British steel industry and its failure to achieve an optimum size plant in the inter-war years, in part, to the failure to achieve more standardization in the steel-using industries. Svennilson argues (p. 125) that plants were small and costs were high in the steel industry because sellers were unable to influence buyers, who insisted on very specialized qualities of steel. He admits that mass-produced low-cost steel might have led buyers to revise their policies; but he fails, in our opinion, to ask a reasonable question: why did new and integrated firms not enter the industry to displace the inefficient plants already existing?

13. Habakkuk (1962, p. 185), in attempting to account for British failure to exploit technological possibilities more rapidly in the nineteenth century, suggests that 'because labour was abundant, the proposals for uniformity in the dimensions and fitting of machinery, put forward by Whitworth, were very slow to be adopted'. See Whitworth (1841).

is a decreasing relation between the level of a country's development and its rate of change. Then any country at a given 'distance', technologically, from C, will be developing at a given rate. By the time A has reached a level of technology no farther from C than B's starting-point, it will have slowed down to B's rate of development. Therefore, B will necessarily reach C first, as it had a shorter distance to travel.[14]

The defect is not *necessarily* fatal. For suppose that a very drastic change in technology occurs in 1900. Then it may be that by 1950 B's technology is completely obsolete with reference to what is needed for the year 2000. Then the distance which B must travel to reach C will be at least as great as that which A must travel to reach C. (In our diagram, $D_2 \geqslant D_3$.)[15]

But this argument depends upon some very special propositions about the nature and sequence of technological change. One must postulate sharp discontinuities and disjunctions, such that the newly emerging technologies at any time are of such a nature that the currently most technologically advanced economies possess no special advantages (either in skills, experience, scientific knowledge and adaptability) in their development. If, however, as Usher (1960) has persuasively argued, technological change must be understood as a continuous process of cumulative synthesis emerging out of a perception of deficiencies in existing techniques and knowledge, then there is at least a strong presumption that the (unknown) technology of the year 2000 will be 'closer' to that of the United States in the year 1950 than to that of Ghana in the year 1950. '. . . the history of technology can be much more adequately presented from the point of view of continuously emergent novelty than from the romantic concept

14. Soviet literature gets around this difficulty by asserting that if A is socialist and B is capitalist, then B will slow down while A will not. Therefore A can 'overtake and surpass' B. Some such *deus ex machina* is needed to meet this difficulty. But even here, a late-comer thesis could exist in an assertion that Communist China will overtake the USSR.

15. Imagine that in 2000 technology is based on completely new principles. Then it will be as easy to teach these principles to Ghanaians as to US factory workers of 1950. Moreover, Ghanaian plant and equipment can be as readily converted to the new technology as US plant and equipment. Then this argument holds. If, however, the new technology of 2000 makes both the Ghanaian and US economies of 1950 more obsolete, then the argument does not hold.

of occasional innovation at widely spaced intervals' (pp. 109–10).[16]

In addition to the 'weak' and 'moderate' statements, a 'strong' statement is also found. This statement asserts that A will move to C, while B will not move at all.[17] On this view, A will certainly overtake B. It is worth considering this possibility (apart from the fact that it is mentioned in the literature) because it brings into direct discussion a point which is interesting and perhaps important. We return to the three cost functions $C_A(x)$, $C_B(x)$ and $C_C(x)$. All of these are functions of output; and it is natural to associate their respective differences,

$$d_{AB}(x) = C_A(x) - C_B(x),$$
$$d_{AC}(x) = C_A(x) - C_B(x)$$

and $d_{BC}(x) = C_B(x) - C_C(x)$

with the 'distance', or magnitude of the technological changes from A to B, from A to C, and from B to C, respectively. Here, obviously, $d_{AC} = d_{AB} + d_{BC}$.[18] It apparently makes no difference (as the problem has so far been formulated) whether the underdeveloped country (A) retraces the developed-but-obsolete country's route $A - B - C$, or 'takes the short cut $A - C$'; both countries will ultimately end up at C.

Suppose, however, that the development process involves both a change in technology and a change in scale. This possibility is suggested by what Samuelson (1947) has called the 'Generalized LeChatelier Principle'. This principle asserts that if the number of restraints on a cost function is reduced the function itself can only remain constant or decrease at any level of output. Thus a firm at B has all the technology available at A, and more besides; one at C has all that available at B and more besides. Marginal cost at each level of output at B is thus not more than at A, and at C not more than at B. Output at B is thus not less than at A, and at C not less than at B. Replace 'not more than'

16. The *locus classicus* for Usher's analysis is his *A History of Mechanical Inventions*, 2nd edition, 1954. See also Usher (1955). A recent comparison of Usher and Schumpeter may be found in Ruttan (1959).

17. Gerschenkron (1952), p. 7, argues that the backward country will adopt techniques at least as modern as the advanced country. He thus adopts a stand between 'moderate' and 'strong' on our scale.

18. This assertion differs, of course, from the assumed conditions of the weak and moderate statements.

by 'less than' in the foregoing sentences, and assume that the cost of moving from a technology T to another technology T' is an increasing function of output at T, but does not depend on T'.[19] Call this transition cost $M(x)$, then we find that a firm at A selling at price p has a choice between locating at A, producing x_A units with cost function $C_A(x)$; locating at B, paying $M(x_A)$ to move, and then selling at p subject to cost function $C_B(x)$; or locating at C, paying $M(x_A)$ to move, and selling at p, subject to cost function $C_C(x)$. Whatever $M(x)$, if B is sufficiently close to A, the firm may find it profitable to move to a more distant technology C, where the savings in cost $[C_A(x) - C_C(x)]$ are greater, but not to move to a nearer technology B, where the savings in cost $[C_A(x) - C_B(x)]$ are less. In other words, it will make large, but not small, technological changes.[20]

On the other hand, a firm at B may confront a higher transition cost $[M(x_B) > M(x_A)$, since $x_B > x_A]$ than the firm at A, while the cost reduction $[C_C(x) - C_B(x)]$ it achieves is less than that achieved by the former. That is, the fact that C is closer to B than to A may offset the greater cost reduction resulting because output at C is greater than that at B. A situation is then possible in which the 'underdeveloped' firm at A would move to C, while the 'developed but obsolete' firm at B would not.[21]

The concept of transition cost is flexible enough to cover a variety of situations. Consider four agricultural economies: in A there is neither fertilizer nor contour ploughing; in B there is

19. One reason it was hard to introduce steel in Britain was that it competed with iron in many uses, so that considerable and lengthy tests of its physical and economic properties were required throughout all industry. Once innovators had incurred the transition costs of testing, however, imitators could rapidly follow (Clapham, 1952, p. 56).

20. This statement implicitly assumes that there exists an answer to the following dilemma: granted a sufficiently long period, any transition cost, however large, can be made to pay for itself. This statement, in turn, states that the accumulated income is infinite. The problem of evaluating income streams over an infinite future is common to all growth theory, and is not satisfactorily solved to date. See Ramsey (1928). We are indebted to our colleague, Professor Stanley Reiter, for these observations.

21. Cipolla (1952, pp. 178 *et seq.*) argues that guild restrictions in seventeenth-century Italy produced what we would call 'infinite' transition costs, which prevented the Italian textile industry from adopting the improved methods of its British, Dutch and French competitors and caused its disappearance.

fertilizer but no contour ploughing; in C there is contour ploughing but no fertilizer; and in D there is contour ploughing and
fertilizer. D is 'the economy of the future', and A is an 'underdeveloped economy'. Assume that a country without contour
ploughing suffers erosion, and that an investment in terracing
must take place before contour ploughing can be instituted. In
this case, C would move to D without a transition cost, while A
would not. Here a developed but obsolete economy has an advantage over the late comer. However, B has no such advantage
over A. In this sense there is a transition cost from A or B to
C or D, but no transition cost from C to D. In this sense, transition costs depend upon the entire path of a country's past
history, or (to put it another way) the direction from which a
particular technology (in this example, contour ploughing plus
fertilizer) is approached.[22, 23]

The transition costs $M(x)$ have been assumed to be an increasing function of output, but independent of technology.[24] On
this basis we have argued that a more-developed economy will
fail to make a change to an advanced technology, while a less-
developed economy will make the change. But transitional costs
may depend on the level of technology from which the change is
initiated. If it is true that the more advanced the technology
being taught workers, the more expensive the education, then
the higher the level to which an economy is moving, the greater
the transition costs; if it is true that the more advanced a technology is, the easier it is to retrain the labour force, then the

22. A former colleague of the authors was employed teaching the Lebanese
to grow cedars. The problem was that destruction of the forests at an earlier
date had led to erosion, and hence transition costs. The soil-exhausting
techniques of tobacco and cotton cultivation in the American south might
also be cited.

23. Richardson (1960, p. 114), makes essentially this point, in a diffuse
reference to Frankel. Habakkuk (1962, p. 218) cited Richardson's point in
an explanation of the loss of British supremacy in textiles. These various
references do not clearly distinguish between cases where individual firms
are burdened by the past and cases where the past impedes entry, by making
it hard for new firms to fit into the interstices among existing industries.

24. This assumption means, in effect, that if technological change
involves retraining labour, it does not matter to what new skill the labour is
retrained; if factories must be retooled or demolished, it does not matter
what the nature of the new equipment or building is.

higher the level from which an economy starts, the less the transition costs. If the more advanced the existing technology is, the greater are the retooling and rebuilding costs, then the late-comer thesis is strengthened. If new technology is capital-saving compared to older (for given levels of output) all late-comer theses are weakened.

One part of the 'late-comer' thesis is that the economies now developed but obsolete will remain so. This is not necessarily implied by the discussion of transition costs thus far presented. If it is true that a firm at A will move to C, while one at B will not, we can imagine the following sequence of technologies:

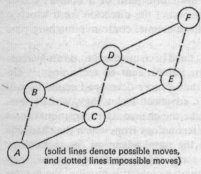

(solid lines denote possible moves, and dotted lines impossible moves)

Figure 2

Here a firm at A would move to C and then to E, while a firm at B would move to D and then to F.[25] These moves would imply a 'leap frog' game in which first one and then the other would assume technological 'leadership'. The process would exist because large technological changes are profitable, while smaller ones are not. But there is no necessary reason why this process should come to an end.

25. The literature on 'turnpike theorems' may turn out to be relevant to this discussion (see Dorfman, Samuelson and Solow, 1958, chs. 11 and 12). These theorems are based upon the dual concepts of rates of growth and composition of capital. Formally, they assert that rates of growth may be increased if a non-optimal capital-mix is altered; and it makes no difference if one type is increased, or another destroyed, in the alteration process. This subject is still in an unsatisfactory state, and to elaborate on it would require a separate treatment.

If, however, transition costs rise sufficiently fast with output, then at each stage the firm must contemplate a larger leap than it made the last time.[26] If knowledge increases at an even pace, the firm must wait longer at each stage than the last to accumulate the knowledge necessary to make the (larger) next leap. At some point, then, each firm will find itself expecting to have to accumulate knowledge for an indefinitely long period before it can afford to make another leap. At this point we would expect technological progress to stop for ever – even though at the end of the period an indefinitely large jump would occur. Thus, one firm or the other might establish a permanent lead.

From this point of view, one can draw some of the dubious historical examples so dear to economists. The Russian iron industry of 1750 was the largest in the world.[27] It was based on a charcoal technology. It was displaced in the 1790s by the British, who changed from charcoal to coke as fuel. Now in the twentieth century the Soviets, by use of sintering, oxygen blast and so on, would replace the West by adopting on a larger scale technology which is one step ahead of that in use elsewhere. Likewise the Indians and Chinese, whose technology in 1000 surpassed that of Western Europe, fell behind in the eighteenth century, and would presumably be in line to regain a lead in the $(20+x)$th century without in the meanwhile having passed through our own stage of development. If, however, the higher the level of technology, the greater the obstacles to its change, then either the Asiatic countries will never regain the lead or if they do regain it they will never lose it again. Clearly the difficulty with this type of argument is that it is not easy to ascertain which

26. The assumption that transition costs are a rising function of output is consistent with Frankel's 'technological interrelatedness', as we understand the concept. Technological interrelatedness means that single components of a productive process cannot be replaced on a 'one-at-a-time' basis. 'Interrelatedness has the effect of requiring the new method, if it is to be introduced, to earn higher profits than would otherwise be necessary. Let the difference between required profits with interrelatedness and required profits without it be called the "profit gap". The size of this gap may be taken, in any particular case, as a measure of the handicap imposed on innovation by interrelatedness' (Frankel, 1955, pp. 306–7).

27. Bowden, Karpovich and Usher (1937, p. 301). British *output* in the eighteenth century (ibid., p. 385) may be compared to Russian *exports*, as given in Liubomirov (1947).

frog leaps last. Whether Western Europe, or the USSR, or China or India leaps last makes presumably a great deal of difference, but there is no *a priori* way of deciding the last (and therefore the best) leap.

Let us now return to the distinction we have made between a firm and an economy. An economy may progress by having new firms replace old firms. Inflexibility of existing firms need be no obstacle to the economy. Thus if cost and revenue functions are not 'real', but subjective constructions by enterprises, then as 'middle age' advances, any enterprise, however successful it may have been, will see increasing obstacles to further innovations.[28] Here $M(x)$ is an increasing function of time, as well as merely of output.[29] But to the extent that these obstacles are 'imaginary' it is natural to ask why new firms are not formed.

28. In contrast, Gerschenkron (1952, p. 6) treats the beginning of industrialization as if it were in part the removal of a real or imaginary transition cost.

29. There may exist real impediments to change. Thus, England's alleged 'stagnation', supposedly beginning around the turn of the present century, may simply reflect the fact that as coal reserves have been depleted, the cost of producing at a constant rate of output has risen over time. Such a situation would affect firms of all ages in the coal industry. Although output per man-year figures (for all persons employed in mining) are subject to some obvious limitations, they are still of considerable interest. Output per person employed in British mines was reported to be at a yearly average of 319 tons for the period 1879–83, 287 tons for 1894–8, 257 tons for 1909–13 and 221 tons for 1922–5 (Committee on Industry and Trade, 1928, pp. 435, 439). One need not share the somewhat apocalyptic concern of Jevons in *The Coal Question* (1865) over the eventual exhaustion of coal resources to point to the rising costs of coal extraction and its importance to the competitive position and rate of growth of the British economy (see Gordon, 1956, p. 650). Gordon also points out that 'to some extent the "stagnation" of the British economy is an appearance created by comparisons with Germany and the United States' and that 'German per capita output has not as yet surpassed that of Britain despite a resource picture which, if anything, favors the former' (*loc. cit.*; see also Taylor (1961) and Phelps Brown and S. J. Handfield-Jones, 1952). The 'resource exhaustion' argument has no logical connection with the 'penalty of taking the lead' argument, since resources are exhausted simply as a function of their rate of utilization. Such exhaustion hardly constitutes a part of the penalty for technological leadership as such. The superior resource position of the late industrializers (with an identical *initial* endowment of resources, by assumption) at any particular point in time merely reflects their earlier failure to utilize their resource base.

For these firms, the transition cost would be low, and consequently the economy would adopt new technology by the simple device of forming new firms (or, alternatively, by existing multiple-product firms adding to their product lines). 'Freedom of entry', then, is a way to eliminate transition costs due solely to the immobility of older firms.[30] Even where the obstacles are 'real' rather than imaginary for older firms, i.e. where there are transition costs (due to, say, technological interrelatedness) for firms with existing plants, the freedom of new firms to enter the industry with new plants embodying the new technology will guarantee technological progressiveness for the industry as a whole. Moreover, occupational immobility within a single generation[31] does not imply that younger people must take jobs in

30. The failure to deal with at least the possibility of new entrants seriously vitiates much of Frankel's interesting discussion of technical interrelatedness (Frankel, 1955, pp. 296–319). Frankel deals with situations where it is impossible to replace single components of a productive sequence separately because of specifications imposed by the other machines in the existing productive sequence. 'Confronted with this situation, the enterprise would compare the new and old methods not on a component or machine basis but on an entire plant basis since only by replacing existing plant in toto could it utilize the new machine' (p. 302). But one cannot jump from the firm to the industry, since new firms are not subject to the constraints which the past presumably imposes upon the present *via* technical interrelatedness. To complete the argument it would be necessary to demonstrate why *new* firms do not enter the industry (in Schumpeterian fashion) when superior techniques are available and are not being employed by existing firms. Frankel does not address himself to this highly important aspect of his problem. Indeed, in discussing whether an industry will adopt a new innovation, he inserts in a footnote (p. 304), quite gratuitously, the 'assumption' that there is no entry of new firms: 'This statement assumes that, for whatever reason, entry of new firms is blocked.' But in making this assumption Frankel is, in effect, conceding that the analysis which he developed at the level of the individual firm is not transferable to an entire industry. In attempting to restate the crux of his argument in reply to Gordon's strictures, Frankel appears to us to reduce it to the following propositions: (a) for any particular period of time, all existing capital goods constitute a free gift from the past; (b) however, some capital goods are 'freer' gifts than others. It is better to have a capital stock inheritance from the past which is free of interrelatedness ('no strings attached') than to have an inheritance characterized by interrelatedness. Interrelatedness constitutes a constraint upon freedom of choice (Gordon (1956) and Frankel, *op. cit.*, pp. 646–57). We find little with which to disagree in Frankel's 'Reply'.

31. In the United States, for instance, it has been difficult to persuade

their fathers' industries. In this case technologically progressive firms would have young employees and 'mature' firms older employees.

If obstacles to entry were the same for an economy in all stages of its development they would be irrelevant to the late-comer thesis. But if they increase as a country develops they contribute to a sluggishness of early starters compared with late comers. For example, early English automobile manufacturers were newly formed firms, but they found it hard to introduce mass production because of the entrenched position of other machinery firms and of their own employees (Habakkuk, 1962, pp. 202–3). Likewise, the 'late-coming' continental industry in effect invented investment banking, while 'early starting' England never adopted it (Gerschenkron, 1952, pp. 13 *et seq*.). It is therefore sometimes argued that early starters develop rigidities which affect not only firms but also freedom of entry. The point is that in certain situations external economies appear to exist, but in fact cannot be utilized by newcomers.[32] These situations appear

---

coal-miners and railroad workers to change their occupation. This difficulty seems partly due to the fact that these workers often live in small towns, where alternative employment does not exist. To seek a job means also to move to a new community. The children of these workers, however, may be attracted to jobs elsewhere.

32. An example is the case of Kindleberger's Tunnel. English coal-mines own freight cars, and railroads own the tunnels. Railroads would not profit by adopting larger-sized coal cars (which have less dead weight per unit of capacity), and in fact would have to enlarge the tunnels. The coal-mines will not adopt larger cars, since they would not go through the tunnels. Therefore a potential economy is not achieved because it is external to both industries: 'wholesale destruction would have helped, and the British economy has been worse off, rather than better, with the free gift of the past. Capital investment may be more readily made good in a developed economy than the blocks to technical change from external economies overcome.' But his reasoning is not entirely convincing: 'wholesale destruction might have enabled the country to make a fresh start. And a fresh start was needed. No nation normally has the option of destroying or scrapping existing equipment and starting again, in the same way as a developing country, or one emerging from a destructive war.' But why, indeed, not? In the very next sentence Kindleberger states, as if in explanation: 'Private owners clinging to their privileges inhibited the railroads from acting as if the old cars, or freight cars without brakes, had no value' (Kindleberger, 1961, p. 288). On this basis, it appears that Kindleberger's analysis reduces simply

to depend upon alleged complexity and interrelatedness in early starting economies (the authors know nothing about obstacles to entry in Afghanistan, India, Ghana, etc., but suspect they may be at least as great as those in the United Kingdom).

Suppose an economy in which each firm produces a single product, and in which a given set of consumer goods is produced. Development consists in introducing new intermediate goods, including equipment. It is possible that every time a new type of equipment is introduced an old one is removed from the market, but there is no reason why this situation should exist. Indeed, the development process seems to have increased 'specialization', in the sense that more names of commodities, occupations, etc., exist in a developed than in an underdeveloped economy. Despite the increased variety of output, costs of finished goods do not rise. In some sense, then, the firm in a developed economy buys a wider variety of inputs, and hence, on the average, a smaller number of units of each kind of input per unit of output.[33]

Specialization in a developed economy is different from that in an underdeveloped economy. The two deserve different names, but do not have them. In the developed economy the specialized firm produces articles which have only a single purpose; the many purposes served in such an economy make it probable that another firm exists which produces articles which either are

---

to irrational or non-maximizing behaviour on the part of the capitalists. Kindleberger had earlier (p. 286) cited the fact 'that in the United States 15,000 miles of track was converted from broad to narrow gauge in two days in 1886'.

33. If this is a reasonable view, and if (reasoning from general equilibrium theory) each industry buys from each other industry, the number of kinds of inter-industry relations in an economy of n industries is $n(n-1)$, which varies with n^2. Thus 'complexity' of the economy increased with the square of the number of commodities produced, that is, with the level of development. If the firm has a specialized service concerned with buying inputs, and if the 'span of attention' of the individual buyer is such that he cannot keep track of more than some number of different inputs, then the overhead costs of a firm will necessarily rise as the economy develops. These overhead costs are externally induced, and do not vary with output of the firm. In this sense the development of the economy imposes an external diseconomy on each firm. For the treatment of some related issues see Melman (1951).

similar except in some small detail[34] or which are made using very much the same labour skills and equipment. In contrast, the specialization in an underdeveloped economy takes place in an environment where many fewer kinds of commodity[35] are produced. It is not obvious that specialization in the sense of multiplicity of commodity names or skill names makes for less occupational mobility than specialization in the sense of a small number of rather disparate commodity names or skills.

Suppose, however, that it were the case that the early stages of economic development involve concentration on a relatively small number of industries, and that only in later stages does there occur the proliferation of specialties which characterizes the highly developed economy. In this case, if it is also true that the highly differentiated economy tends to have relative immobility of resources, then underdeveloped countries could grow more rapidly than the developed countries. Nothing in this possibility, however, suggests that late comers would not be as affected by retardation as they reached the stage of increased complexity as early developers, and it is not easy to make even a 'moderate' or 'strong' proposition for this argument.[36]

The foregoing discussion does not touch upon a major aspect of the 'late-comer' thesis: the rate of growth is a fraction, the numerator of which is the change in output and the denominator the total output in a period; and part of the literature states that for any given output increase (resulting, for instance, from a

34. J. M. Clark has mentioned jurisdictional disputes between the makers of roofing and of siding for Quonset huts.

35. Lange (1945) has shown that articles whose prices vary proportionately over time may be treated as a single commodity.

36. We do not propose to discuss Habakkuk's view (1961 ch. 6, *passim*) that innovation is a result rather than a cause of economic growth.
The contrary view is explained adequately in Schumpeter's *Theory of Economic Development*. Sociologically, Habakkuk's attitude on another point, the existence of an industrial plant in Britain, 'which in many branches was adequate to the demands made upon it' (p. 212), resembles the comment attributed by Americans to British employers and labour (pp. 198–9) that 'the working-man does not need more than so many shillings a week'.

foreign-aid programme of given size) the lower the output base is, the higher the growth rate. This preoccupation with the denominator breeds neglect of the numerator: the smaller the base level of output, the harder it is to attain any given numerator. Our discussion has been mainly concerned with the connection between levels and changes of output and technology, but implicitly it makes a further point: a moderate or strong late-comer thesis must assert that the retardation which affected the early starters either will not affect the late comer at all or will only affect it after it has overtaken the early starters.

Finally, we may examine arguments sometimes advanced from 'economic sociology'. Economic growth is almost always associated with changes in the composition of aggregate demand and therefore of total output. (The relative importance within GNP of $C+I+G+E_x$ changes. So also does the composition of each of these categories.) Successful economic growth therefore involves, *inter alia*, the maintenance of a high degree of flexibility of resource use. One measure of success in the operation of market forces is thus the responsiveness of resources to the changes dictated by changes in the composition of aggregate demand. An economy which finds it increasingly difficult to accommodate itself to these changes will necessarily encounter increasing difficulties in sustaining its economic growth. It may be argued, then, that movement along any historical growth path involves the introduction of rigidities and resistances and thereby reduces an economy's capacity to accommodate itself to further change.

This is essentially Svennilson's explanation of European stagnation in the inter-war years. Western European countries, in varying degrees, experienced difficulty in bringing about the transformation of domestic resource use required by the drastically altered composition of post-war demand. But Svennilson's analysis states further that a 'mature' economy is less viable and more resistant to change than a less-developed economy.

There is no doubt that, in the oldest and most advanced industrial countries, transformation of the economy in accordance with the new trends in technology and demand meets with strong resistance from the accumulated stock of capital, from the traditional special skill of labour

– in fact, from the whole organization of society. . . . On the other hand, resistance is much weaker in a society which offers a virgin soil for the development of manufacturing industries (Svennilson, 1954, p. 206).[37]

The resistances to resource reallocation in inter-war Western Europe were, indeed, real enough. But Svennilson's thesis requires more for its support than the trivial statement that we do not live in a frictionless world. It presumes that the obstacles to resource mobility are an increasing function of an economy's maturity. But this extremely important proposition, when appropriately expressed in a *comparative* sense, is not at all obvious. Even if there was considerable resource immobility in British industry during the inter-war years, it is not clear that these immobilities were greater than those encountered in the Brazilian (or Bulgarian or Indian) economies during the same period. On the contrary, we should be inclined to argue that industrial countries, even the 'oldest and most advanced', possess, *on balance*, far greater versatility, flexibility, and capacity to accommodate to change than do predominantly agricultural, low-income economies which presumably offer 'a virgin soil for the development of manufacturing industries'.

Here again it is important to maintain the distinction between factors which are alleged to be responsible for an eventual retardation of growth in industrial economies and factors demon-

37. See also ibid., pp. 9–10, 34–40, 44–52. Cf. Schumpeter's brilliant analysis of the atrophy of capitalist institutions in *Capitalism, Socialism and Democracy*, Pt II. John Jewkes has argued that advanced capitalist economies develop social and political resistances to change, although they need not succumb to them; nor does he share Svennilson's view on the comparison of the mature with the immature economy: 'an advanced industrial country will have created classes and institutions, vested interests each of which, it is arguable, will seek, in the last resort by political action, to maintain its own position. The entrepreneur may be powerful enough to force the State to give him protection, the trade union may seek to maintain the same earnings in the same industry for an undiminished number of workers. Of course, the community as a whole cannot, by staking its claim, determine its economic standing (unless it can find some benefactor in the international field). But the efforts of groups to do so may be successful and, by delaying necessary readjustments, may undermine the position of an advanced industrial country. Some such cause, probably, lay at the root of the British industrial decline between the wars. Of course, industrial retardation of this kind is not inevitable' (Jewkes, 1951, pp. 10–11). Cf. also Lewis (1949); Lewis (1957); Robinson (1954).

strating the existence of penalties for taking the lead. For the problems involved are not those resulting from early technological leadership as such, but are, in some sense, an outgrowth of the growth process itself. If retardation is inevitable in a mature economy, it is, presumably, inevitable for the followers as well as for the leaders.

Conclusions

The late-comer thesis can be formulated in three variants. The *weak* thesis asserts that the late comers will pass through any sequence of development more rapidly than early starters. The *moderate* thesis asserts that late comers will ultimately reach higher levels of development than early starters, even though the latter do not cease developing. The *strong* thesis states that late comers will surpass early starters, partly because the latter will cease to develop. Each of these theses implies preceding ones in the list, but not later ones.

The *weak* thesis depends ultimately on the propositions that late comers can avoid past mistakes of early starters; that late comers will make no more current mistakes than late starters; and that mistakes do not contain useful experience (in some sense) for those who make them.

The *moderate* thesis is based in part on the assertion that early starters are subject to retardation which does not affect late comers. This is possible if a continuing technological change affects the economies of early starters more adversely than those of late comers (in terms of obsolescence of skills and plant).

The *strong* thesis is based upon transition costs: if the cost of moving from a lower to a higher technology is an increasing function of the level of technology already reached the rate of development will slow down as a country develops. This is true because changes sufficiently large to be worthwhile can be made only at increasingly infrequent intervals (and in the limit can never be made). Thus if a late comer 'makes the last change' it attains a permanent leadership.

The validity of these theses depends upon a variety of empirical problems:

1. Granted that late comers have certain advantages, they also

have certain disadvantages. It is moot whether on balance they have an advantage in terms of any of these theses.

2. Granted that late comers may initially grow at a rapid rate, it is moot whether they will escape the retardation afflicting early starters when they reach comparable states of development.

3. Granted that in fact some late comers have tended to catch up with the British, it is moot whether their success is due to economic rather than to sociological or 'irrational' factors.

4. Many of the arguments used to defend the theses apply only to economies into which new firms may not enter; the moot question of whether obstacles to entry increase as economies develop – for economic or non-economic reasons – is therefore of major importance.

5. The definition of interrelatedness given in the literature states that a firm has interrelated assets (processes) when it cannot replace one without simultaneously replacing others. But the word interrelatedness is also used in the literature on late comers in the context of a group of industries. For this concept no adequate definition is given; and it is not clear why (a) interrelatedness should increase as a country develops, or (b) obstacles to entry should increase as interrelatedness increases. Yet the late-comer theses typically make assertions of this sort.

The results of our argument were in large part reached by the explicit assumption that technologies of different countries could be placed in ordinal, or even cardinal, relationships to each other. This assumption, it is clear, makes it possible to achieve results which are more precise than would otherwise be attained. There exists, of course, no measure of the sort assumed. To this extent, our results lack precision. However, even a cursory examination of the literature reveals constant use of notions of this sort, so that at least we have brought some dirty linen into the open, where it belongs. Those who object to this expository device will, of course, be careful to avoid using similar concepts themselves. We shall be interested to see how they will cope with interesting and important problems such as those raised by the late-comer thesis.

Even though no such measure of technology now exists, we can describe some of its properties, using problems arising in the late-comer thesis. A measure of technology, for instance, cannot be a movement along a single line. If it were, the (real) possibility of short-cuts by late comers would have to be discounted. We use a two-dimensional technology in our diagrams; if present growth rates are substantially affected by the course of past development a multi-dimensional measure of technology may be necessary.

Such a concept might provide a bridge between Schumpeter's innovation, which represents discrete change, and Usher's technology, which represents continuous change. Economic discussion of technological change is hampered by the fact that both views can be defended, while no test is available to show how either may be related to real economies. This discussion has been weakened by the absence of such a test; if it existed, many other economic problems could also be handled more readily.

All published late-comer theses have logical defects, but these may often be repaired. What remains is a series of empirical conjectures, the verification of which depends upon the truth of certain hypotheses about the relation between changes in output and changes in technology. Even if no late-comer thesis were a logical necessity (and we feel this is the case), some such thesis might well be true as historical fact. But the available factual discussion goes little beyond tantalizing suggestions, and indeed can hardly do so until the development of adequate theory and measures of technological change.

References

AMES, E. (1961), 'Research, invention, development and innovation', *Amer. econ. Rev.*, June, pp. 370–81.

BOWDEN, W., KARPOVICH, M., and USHER, A. P. (1937), *An Economic History of Europe Since 1750*, American Book.

CAMPBELL, R. (1960), *Soviet Economic Growth*, Riverside Press.

CIPOLLA, M. (1952), 'The decline of Italy'. *Econ. hist. Rev.*, vol. 5, second series.

CLAPHAM, J. H. (1952), *An Economic History of Modern Britain*, vol. 2, Cambridge University Press..

COMMITTEE ON INDUSTRY AND TRADE (1928), *Survey of Metal Industries*, HMSO.

DORFMAN, R., SAMUELSON, P., and SOLOW, R. (1958), *Linear Programming and Economic Analysis*, McGraw-Hill.

FITCH, C. H. (1880), *Tenth Census of the US*, vol. 2, Washington, DC.

FRANKEL, M. (1955). 'Obsolescence and technical change in a maturing economy', *Amer. econ. Rev.*, June, pp. 296–319.

FRANKEL, M. (1956), 'Reply', *Amer. econ. Rev.*, September, pp. 652–56.

GERSCHENKRON, A. (1952), 'Economic backwardness in historical perspective', in B. F. Hoselitz (ed.), *The Progress of Underdeveloped Areas*, University of Chicago Press.

GOODE, R. B. (1959), 'Adding to the stock of physical and human capital', *Amer. econ. Rev.*, *Pap. Proc.*, May, pp. 147–55.

GORDON, D. (1956), 'Obsolescence and technological change: comment', *Amer. econ. Rev.*, September, pp. 646–52.

HABAKKUK, H. J. (1962), *American and British Technology in the Nineteenth Century*, Cambridge University Press.

HALL, C. R. (1954), *History of American Industrial Science*, Library Publishers.

HIRSCHMAN, A. (1958), *The Strategy of Economic Development*, Yale University Press.

JERVIS, F. R. J. (1947), 'The handicap of Britain's early start', *Manchester School*, vol. 15, pp. 112–22.

JEVONS, W. S. (1865), *The Coal Question*, Macmillan.

JEWKES, J. (1951), 'The growth of world industry', *Oxf. econ. Pap.*, new series, February, pp. 1-15.

KINDLEBERGER, C. P. (1961), 'Obsolescence and technical change', *Oxf. Univ. Inst. Stat. Bull.*, August, pp. 281–97.

KUZNETS, S. (1955), 'Problems in comparison of economic trends', in S. Kuznets *et al.* (eds.), *Economic Growth: Brazil, India and Japan*, Duke University Press.

LANGE, O. (1945), *Price Flexibility and Employment*, Principia Press.

LEWIS, W. A. (1949), *Economic Survey 1919–1939*, Harper & Row.

LEWIS, W. A. (1957), 'International competition in manufactures, *Amer. econ. Rev. Pap. Proc.*, May, pp. 578–87.

LIUBOMIROV, P. G. (1947), *Studies in the History of Russian Industry*, Moscow.

MANDELBAUM, K. (1955), *The Industrialization of Backward Areas*, 2nd edn, Kelley & Millman.

MELMAN, S. (1951), 'The rise of administrative overhead in the manufacturing industries of the United States, 1899–1947', *Oxf. econ. Pap.*, new series, February, pp. 62–93.

NOVE, A. (1961), *The Soviet Economy*, Praeger.

PHELPS BROWN, E. H., and HANDFIELD-JONES, S. J. (1952), 'The climacteric of the 1890s', *Oxf. econ. Pap.*, October, pp. 266–307.

RAMSEY, F. P. (1928), 'A mathematical theory of saving', *Econ. J.*, vol. 38, pp. 543–59.

RICHARDSON, G. B. (1960), *Information and Investment*, Oxford University Press.

ROBINSON, E. A. G. (1954), 'Changing structure of the British economy', *Econ. J.*, September, pp. 443-61.

ROSTOW, W. W. (1960), *The Stages of Economic Growth*, Cambridge University Press.

RUTTAN, V. (1959), 'Usher and Schumpeter on invention, innovation and technological change', *Quart. J. Econ.*, November, pp. 596–606. [Reprinted as Reading 3 of the present selection.]

SAMUELSON, P. A. (1947), *Foundations of Economic Analysis*, Harvard University Press.

SCHULTZ, T. (1959), 'Investment in man; an economist's view', *Soc. Sci. Rev.*, June, pp. 109–17.

SCHULTZ, T. (1961), 'Investment in human capital', *Amer. econ. Rev.*, March, pp. 1–17.

SVENNILSON, I. (1954), *Growth and Stagnation in the European Economy*, United Nations Economic Commission for Europe, Geneva.

TAYLOR, A. J. (1961), 'Labour productivity and technological innovation in the British coal industry, 1850–1914', *Econ. hist. Rev.*, August, pp. 48–70.

USHER, A. P. (1955), 'Technical change and capital formation', in *Capital Formation and Economic Growth*, National Bureau Committee for Economic Research, pp. 523–50. [Reprinted as Reading 2 of the present selection.]

USHER, A. P. (1960), 'Industrialization of modern Britain', *Technology and Culture*, Spring, vol. 1, no. 2, pp. 109–27.

US STRATEGIC BOMBING SURVEY, EUROPEAN WAR (1945), *Report*, US Government.

VEBLEN, T. (1915), *Imperial Germany and the Industrial Revolution*, Macmillan.

WHITWORTH, J. (1841), *On a Uniform System of Screw Threads*, Institute of Civil Engineers.

20 R. Vernon

International Investment and International Trade in the Product Cycle[1]

R. Vernon, 'International investment and international trade in the product cycle', *Quarterly Journal of Economics*, May 1966, pp. 190–207.

Anyone who has sought to understand the shifts in international trade and international investment over the past twenty years has chafed from time to time under an acute sense of the inadequacy of the available analytical tools. While the comparative cost concept and other basic concepts have rarely failed to provide some help, they have usually carried the analyst only a very little way toward adequate understanding. For the most part, it has been necessary to formulate new concepts in order to explore issues such as the strengths and limitations of import substitution in the development process, the implications of common market arrangements for trade and investment, the underlying reasons for the Leontief paradox, and other critical issues of the day.

As theorists have groped for some more efficient tools, there has been a flowering in international trade and capital theory. But the very proliferation of theory has increased the urgency of the search for unifying concepts. It is doubtful that we shall find many propositions that can match the simplicity, power, and universality of application of the theory of comparative advantage and the international equilibrating mechanism; but unless the search for better tools goes on, the usefulness of economic theory for the solution of problems in international trade and capital movements will probably decline.

The present paper deals with one promising line of generalization and synthesis which seems to me to have been somewhat neglected by the main stream of trade theory. It puts less em-

1. The preparation of this article was financed in part by a grant from the Ford Foundation to the Harvard Business School to support a study of the implications of United States foreign direct investment. This paper is a by-product of the hypothesis-building stage of the study.

phasis upon comparative cost doctrine and more upon the timing of innovation, the effects of scale economies, and the roles of ignorance and uncertainty in influencing trade patterns. It is an approach with respectable sponsorship, deriving bits and pieces of its inspiration from the writings of such persons as Williams, Kindleberger, MacDougall, Hoffmeyer and Burenstam-Linder.[2]

Emphases of this sort seem first to have appeared when economists were searching for an explanation of what looked like a persistent, structural shortage of dollars in the world. When the shortage proved ephemeral in the late 1950s, many of the ideas which the shortage had stimulated were tossed overboard as prima facie wrong.[3] Nevertheless, one cannot be exposed to the main currents of international trade for very long without feeling that any theory which neglected the roles of innovation, scale, ignorance and uncertainty would be incomplete.

Location of new products

We begin with the assumption that the enterprises in any one of the advanced countries of the world are not distinguishably different from those in any other advanced country, in terms of their access to scientific knowledge and their capacity to comprehend scientific principles.[4] All of them, we may safely assume, can secure access to the knowledge that exists in the physical, chemical and biological sciences. These sciences at times may be difficult, but they are rarely occult.

It is a mistake to assume, however, that equal access to scientific principles in all the advanced countries means equal probability of the application of these principles in the generation of new products. There is ordinarily a large gap between the knowledge of a scientific principle and the embodiment of the principle in a marketable product. An entrepreneur usually has

2. Williams (1947), Kindleberger (1950), Hoffmeyer (1958), MacDougall (1957), Burenstam-Linder (1961).

3. The best summary of the state of trade theory that has come to my attention in recent years is Bhagwati (1964). Bhagwati refers obliquely to some of the theories which concern us here; but they receive much less attention than I think they deserve.

4. Some of the account that follows will be found in greatly truncated form in Vernon (1962). The elaboration here owes a good deal to the perceptive work of Hirsch (1965).

to intervene to accept the risks involved in testing whether the gap can be bridged.

If all entrepreneurs, wherever located, could be presumed to be equally conscious of and equally responsive to all entrepreneurial opportunities, wherever they arose, the classical view of the dominant role of price in resource allocation might be highly relevant. There is good reason to believe, however, that the entrepreneur's consciousness of and responsiveness to opportunity are a function of ease of communication; and further, that ease of communication is a function of geographical proximity.[5] Accordingly, we abandon the powerful simplifying notion that knowledge is a universal free good, and introduce it as an independent variable in the decision to trade or to invest.

The fact that the search for knowledge is an inseparable part of the decision-making process and that relative ease of access to knowledge can profoundly affect the outcome are now reasonably well established through empirical research.[6] One implication of that fact is that producers in any market are more likely to be aware of the possibility of introducing new products in that market than producers located elsewhere would be.

The United States market offers certain unique kinds of opportunities to those who are in a position to be aware of them.

First, the United States market consists of consumers with an average income which is higher (except for a few anomalies like Kuwait) than that in any other national market – twice as high as that of Western Europe, for instance. Wherever there was a chance to offer a new product responsive to wants at high levels of income, this chance would presumably first be apparent to someone in a position to observe the United States market.

Second, the United States market is characterized by high unit labor costs and relatively unrationed capital compared with practically all other markets. This is a fact which conditions the demand for both consumer goods and industrial products. In the case of consumer goods, for instance, the high cost of laundresses

5. Note Kindleberger's (1962, p. 15 *passim*) reference to the 'horizon' of the decision-maker, and the view that he can only be rational within that horizon.

6. See, for instance, Cyert and March (1963), esp. ch. 6; and Aharoni (1966).

contributes to the origins of the drip-dry shirt and the home washing machine. In the case of industrial goods, high labor cost leads to the early development and use of the conveyor belt, the fork-lift truck and the automatic control system. It seems to follow that wherever there was a chance successfully to sell a new product responsive to the need to conserve labor, this chance would be apparent first to those in a position to observe the United States market.

Assume, then, that entrepreneurs in the United States are first aware of opportunities to satisfy new wants associated with high income levels or high unit labor costs. Assume further that the evidence of an unfilled need and the hope of some kind of monopoly windfall for the early starter both are sufficiently strong to justify the initial investment that is usually involved in converting an abstract idea into a marketable product. Here we have a reason for expecting a consistently higher rate of expenditure on product development to be undertaken by United States producers than by producers in other countries, at least in lines which promise to substitute capital for labor or which promise to satisfy high-income wants. Therefore, if United States firms spend more than their foreign counterparts on new product development (often misleadingly labeled 'research'), this may be due not to some obscure sociological drive for innovation but to more effective communication between the potential market and the potential supplier of the market. This sort of explanation is consistent with the pioneer appearance in the United States (conflicting claims of the Soviet Union notwithstanding) of the sewing machine, the typewriter, the tractor, etc.

At this point in the exposition, it is important once more to emphasize that the discussion so far relates only to innovation in certain kinds of products, namely to those associated with high income and those which substitute capital for labor. Our hypothesis says nothing about industrial innovation in general; this is a larger subject than we have tackled here. There are very few countries that have failed to introduce at least a few products; and there are some, such as Germany and Japan, which have been responsible for a considerable number of such introductions. Germany's outstanding successes in the development and use of plastics may have been due, for instance, to a traditional concern with her lack of a raw materials base, and a recognition that a market might exist in Germany for synthetic substitutes.[7]

Our hypothesis asserts that United States producers are likely to be the first to spy an opportunity for high-income or labor-saving new products.[8] But it goes on to assert that the first producing facilities for such products will be located in the United States. This is not a self-evident proposition. Under the calculus of least cost, production need not automatically take place at a location close to the market, unless the product can be produced and delivered from that location at lowest cost. Besides, now that most major United States companies control facilities situated in one or more locations outside of the United States, the possibility of considering a non-United States location is even more plausible than it might once have been.

Of course, if prospective producers were to make their locational choices on the basis of least-cost considerations, the United States would not always be ruled out. The costs of international transport and United States import duties, for instance, might be so high as to argue for such a location. My guess is, however, that the early producers of a new product intended for the United States market are attracted to a United States location by forces which are far stronger than relative factor-cost and transport considerations. For the reasoning on this point, one has to take a long detour away from comparative cost analysis into areas which fall under the rubrics of communication and external economies.

By now, a considerable amount of empirical work has been done on the factors affecting the location of industry.[9] Many of these studies try to explain observed locational patterns in conventional cost-minimizing terms, by implicit or explicit reference

7. See two excellent studies: Freeman (1963) and Hufbauer (1965). A number of links in the Hufbauer arguments are remarkably similar to some in this paper; but he was not aware of my writings nor I of his until after both had been completed.

8. There is a kind of first-cousin relationship between this simple notion and the 'entrained want' concept defined by Barnett (1953), p. 148. Hirschman (1958, p. 68) also finds the concept helpful in his effort to explain certain aspects of economic development.

9. For a summary of such work, together with a useful bibliography, see Meyer (1963).

to labor cost and transportation cost. But some explicitly introduce problems of communication and external economies as powerful locational forces. These factors were given special emphasis in the analyses which were a part of the New York Metropolitan Region Study of the 1950s. At the risk of oversimplifying, I shall try to summarize what these studies suggested.[10]

In the early stages of introduction of a new product, producers were usually confronted with a number of critical, albeit transitory, conditions. For one thing, the product itself may be quite unstandardized for a time; its inputs, its processing, and its final specifications may cover a wide range. Contrast the great variety of automobiles produced and marketed before 1910 with the thoroughly standardized product of the 1930s, or the variegated radio designs of the 1920s with the uniform models of the 1930s. The unstandardized nature of the design at this early stage carries with it a number of locational implications.

First, producers at this stage are particularly concerned with the degree of freedom they have in changing their inputs. Of course, the cost of the inputs is also relevant. But as long as the nature of these inputs cannot be fixed in advance with assurance, the calculation of cost must take into account the general need for flexibility in any locational choice.[11]

Second, the price elasticity of demand for the output of individual firms is comparatively low. This follows from the high degree of production differentiation, or the existence of monopoly in the early stages.[12] One result is, of course, that small cost differences count less in the calculations of the entrepreneur than they are likely to count later on.

Third, the need for swift and effective communication on the

10. The points that follow are dealt with at length in the following publications: Vernon (1960), pp. 38–85; Hall (ed.) (1959), pp. 3–18, 19; Lichtenberg (1960), pp. 31–70.

11. This is, of course, a familiar point elaborated in Stigler (1939, pp. 305 ff.).

12. Hufbauer (1965), suggests that the low price elasticity of demand in the first stage may be due simply to the fact that the first market may be a 'captive market' unresponsive to price changes; but that later, in order to expand the use of the new product, other markets may be brought in which are more price responsive.

part of the producer with customers, suppliers, and even competitors is especially high at this stage. This is a corollary of the fact that a considerable amount of uncertainty remains regarding the ultimate dimensions of the market, the efforts of rivals to preempt that market, the specifications of the inputs needed for production, and the specifications of the products likely to be most successful in the effort.

All of these considerations tend to argue for a location in which communication between the market and the executives directly concerned with the new product is swift and easy, and in which a wide variety of potential types of input that might be needed by the production unit are easily come by. In brief, the producer who sees a market for some new product in the United States may be led to select a United States location for production on the basis of national locational considerations which extend well beyond simple factor cost analysis plus transport considerations.

The maturing product[13]

As the demand for a product expands, a certain degree of standardization usually takes place. This is not to say that efforts at product differentiation come to an end. On the contrary; such efforts may even intensify, as competitors try to avoid the full brunt of price competition. Moreover, variety may appear as a result of specialization. Radios, for instance, ultimately acquired such specialized forms as clock radios, automobile radios, portable radios, and so on. Nevertheless, though the subcategories may multiply and the efforts at product differentiation increase, a growing acceptance of certain general standards seems to be typical.

Once again, the change has locational implications. First of all, the need for flexibility declines. A commitment to some set of product standards opens up technical possibilities for achieving economies of scale through mass output, and encourages long-term commitments to some given process and some fixed set of facilities. Second, concern about production cost begins to take the place of concern about product characteristics. Even if in-

13. Both Hirsch (1965) and Freeman (1963) make use of a three-stage product classification of the sort used here.

creased price competition is not yet present, the reduction of the uncertainties surrounding the operation enhances the usefulness of cost projections and increases the attention devoted to cost.

The empirical studies to which I referred earlier suggest that, at this stage in an industry's development, there is likely to be considerable shift in the location of production facilities at least as far as internal United States locations are concerned. The empirical materials on international locational shifts simply have not yet been analyzed sufficiently to tell us very much. A little speculation, however, indicates some hypotheses worth testing.

Picture an industry engaged in the manufacture of the high-income or labor-saving products that are the focus of our discussion. Assume that the industry has begun to settle down in the United States to some degree of large-scale production. Although the first mass market may be located in the United States, some demand for the product begins almost at once to appear elsewhere. For instance, although heavy fork-lift trucks in general may have a comparatively small market in Spain because of the relative cheapness of unskilled labor in that country, some limited demand for the product will appear there almost as soon as the existence of the product is known.

If the product has a high income elasticity of demand or if it is a satisfactory substitute for high-cost labor, the demand in time will begin to grow quite rapidly in relatively advanced countries such as those of Western Europe. Once the market expands in such an advanced country, entrepreneurs will begin to ask themselves whether the time has come to take the risk of setting up a local producing facility.[14]

How long does it take to reach this stage? An adequate answer must surely be a complex one. Producers located in the United States, weighing the wisdom of setting up a new production facility in the importing country, will feel obliged to balance a number of complex considerations. As long as the marginal

14. Posner (1961, p. 323 *et seq.*), presents a stimulating model purporting to explain such familiar trade phenomena as the exchange of machine tools between the United Kingdom and Germany. In the process he offers some particularly helpful notions concerning the size of the 'imitation lag' in the responses of competing nations.

production cost plus the transport cost of the goods exported from the United States is lower than the average cost of prospective production in the market of import, United States producers will presumably prefer to avoid an investment. But that calculation depends on the producer's ability to project the cost of production in a market in which factor costs and the appropriate technology differ from those at home.

Now and again, the locational force which determined some particular overseas investment is so simple and so powerful that one has little difficulty in identifying it. Otis Elevator's early proliferation of production facilities abroad was quite patently a function of the high cost of shipping assembled elevator cabins to distant locations and the limited scale advantages involved in manufacturing elevator cabins at a single location (see Phelps, 1963, p. 4). Singer's decision to invest in Scotland as early as 1867 was also based on considerations of a sort sympathetic with our hypothesis (see Dunning, 1958, p. 18).[15] It is not unlikely that the overseas demand for its highly standardized product was already sufficiently large at that time to exhaust the obvious scale advantages of manufacturing in a single location, especially if that location was one of high labor cost.

In an area as complex and 'imperfect' as international trade and investment, however, one ought not anticipate that any hypothesis will have more than a limited explanatory power. United States airplane manufacturers surely respond to many 'noneconomic' locational forces, such as the desire to play safe in problems of military security. Producers in the United States who have a protected patent position overseas presumably take that fact into account in deciding whether or when to produce abroad. And other producers often are motivated by considerations too complex to reconstruct readily, such as the fortuitous timing of a threat of new competition in the country of import, the level of tariff protection anticipated for the future, the political situation in the country of prospective investment and so on.

We arrive, then, at the stage at which United States producers have come around to the establishment of production units in the advanced countries. Now a new group of forces are set in train.

15. The Dunning book is filled with observations that lend casual support to the main hypotheses of this paper.

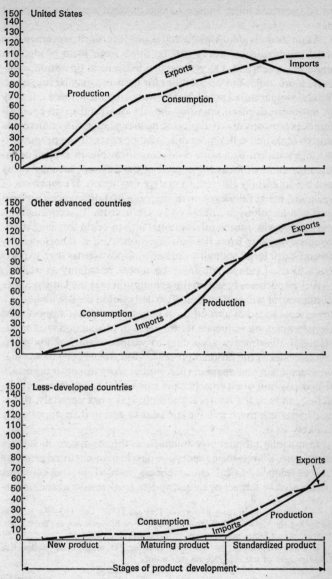

150 **United States**
140
130
120
110
100
90
80
70
60
50
40
30
20
10
0

Production

Exports

Consumption

Imports

150 **Other advanced countries**
140
130
120
110
100
90
80
70
60
50
40
30
20
10
0

Exports

Consumption

Imports

Production

150 **Less-developed countries**
140
130
120
110
100
90
80
70
60
50
40
30
20
10
0

Exports

Consumption

Imports

Production

| New product | Maturing product | Standardized product |

Stages of product development

Figure 1

In an idealized form, Figure 1 suggests what may be anticipated next.

As far as individual United States producers are concerned, the local markets thenceforth will be filled from local production units set up abroad. Once these facilities are in operation, however, more ambitious possibilities for their use may be suggested. When comparing a United States producing facility and a facility in another advanced country, the obvious production-cost differences between the rival producing areas are usually differences due to scale and differences due to labor costs. If the producer is an international firm with producing locations in several countries, its costs of financing capital at the different locations may not be sufficiently different to matter very much. If economies of scale are being fully exploited, the principal differences between any two locations are likely to be labor costs.[16] Accordingly, it may prove wise for the international firm to begin servicing third-country markets from the new location. And if labor cost differences are large enough to offset transport costs, then exports back to the United States may become a possibility as well.

Any hypotheses based on the assumption that the United States entrepreneur will react rationally when offered the possibility of a lower-cost location abroad is, of course, somewhat suspect. The decision-making sequence that is used in connection with international investments, according to various empirical studies, is not a model of the rational process.[17] But there is one theme that emerges again and again in such studies. Any threat to the established position of an enterprise is a powerful galvanizing force to action; in fact, if I interpret the empirical work correctly, threat in general is a more reliable stimulus to action than opportunity is likely to be.

In the international investment field, threats appear in various forms once a large-scale export business in manufactured products has developed. Local entrepreneurs located in the countries which are the targets of these exports grow restive at the oppor-

16. Note the interesting finding of Kreinin (1965, pp. 130–39). Kreinin finds that the higher cost of labor in the United States is not explained by a higher rate of labor productivity in this country.

17. Aharoni (1966) provides an excellent summary and exhaustive bibliography of the evidence on this point.

tunities they are missing. Local governments concerned with generating employment or promoting growth or balancing their trade accounts begin thinking of ways and means to replace the imports. An international investment by the exporter, therefore, becomes a prudent means of forestalling the loss of a market. In this case, the yield on the investment is seen largely as the avoidance of a loss of income to the system.

The notion that a threat to the *status quo* is a powerful galvanizing force for international investment also seems to explain what happens after the initial investment. Once such an investment is made by a United States producer, other major producers in the United States sometimes see it as a threat to the *status quo*. They see themselves as losing position relative to the investing company, with vague intimations of further losses to come. Their 'share of the market' is imperiled, viewing 'share of the market' in global terms. At the same time, their ability to estimate the production-cost structure of their competitors, operating far away in an unfamiliar foreign area, is impaired; this is a particularly unsettling state because it conjures up the possibility of a return flow of products to the United States and a new source of price competition, based on cost differences of unknown magnitude. The uncertainty can be reduced by emulating the pathfinding investor and by investing in the same area; this may not be an optimizing investment pattern and it may be costly, but it is least disturbing to the *status quo*.

Pieces of this hypothetical pattern are subject to empirical tests of a sort. So far, at any rate, the empirical tests have been reassuring. The office machinery industry, for instance, has seen repeatedly the phenomenon of the introduction of a new product in the United States, followed by United States exports,[18] followed still later by United States imports. (We have still to test whether the timing of the commencement of overseas production by United States subsidiaries fits into the expected pattern.) In the electrical and electronic products industry, those elements in the pattern which can be measured show up nicely (see Hirsch, 1965). A broader effort is now under way to test the United States trade patterns of a group of products with high income elasticities;

18. Reported in US Senate, Interstate and Foreign Commerce Committee, *Hearings on Foreign Commerce*, 1960, pp. 130–39.

and, here too, the preliminary results are encouraging.[19] On a much more general basis, it is reassuring for our hypotheses to observe that the foreign manufacturing subsidiaries of United States firms have been increasing their exports to third countries.

It will have occurred to the reader by now that the pattern envisaged here also may shed some light on the Leontief paradox (see Leontief, 1953 and 1956). Leontief, it will be recalled, seemed to confound comparative cost theory by establishing the fact that the ratio of capital to labor in United States exports was lower, not higher, than the like ratio in the United States production which had been displaced by competitive imports. The hypothesis suggested in this paper would have the United States exporting high-income and labor-saving products in the early stages of their existence, and importing them later on.[20] In the early stages, the value-added contribution of industries engaged in producing these items probably contains an unusually high proportion of labor cost. This is not so much because the labor is particularly skilled, as is so often suggested. More likely, it is due to a quite different phenomenon. At this stage, the standardization of the manufacturing process has not gotten very far; that is to come later, when the volume of output is high enough and the degree of uncertainty low enough to justify investment in relatively inflexible, capital-intensive facilities. As a result, the production process relies relatively heavily on labor inputs at a time when the United States commands an export position; and the process relies more heavily on capital at a time when imports become important.

This, of course, is an hypothesis which has not yet been subjected to any really rigorous test. But it does open up a line of

19. These are to appear in a forthcoming doctoral thesis at the Harvard Business School by Louis T. Wells, tentatively entitled 'International trade and business policy'.

20. Of course, if there were some systematic trend in the inputs of new products – for example, if the new products which appeared in the 1960s were more capital-intensive than the new products which appeared in the 1950s – then the tendencies suggested by our hypotheses might be swamped by such a trend. As long as we do not posit offsetting systematic patterns of this sort, however, the Leontief findings and the hypotheses offered here seem consistent.

inquiry into the structure of United States trade which is well worth pursuing.

The standardized product

Figure 1, the reader will have observed, carries a panel which suggests that, at an advanced stage in the standardization of some products, the less-developed countries may offer competitive advantages as a production location.

This is a bold projection, which seems on first blush to be wholly at variance with the Heckscher–Ohlin theorem. According to that theorem, one presumably ought to anticipate that the exports of the less-developed countries would tend to be relatively labor-intensive products.

One of the difficulties with the theorem, however, is that it leaves marketing considerations out of account. One reason for the omission is evident. As long as knowledge is regarded as a free good, instantaneously available, and as long as individual producers are regarded as atomistic contributors to the total supply, marketing problems cannot be expected to find much of a place in economic theory. In projecting the patterns of export from less-developed areas, however, we cannot afford to disregard the fact that information comes at a cost; and that entrepreneurs are not readily disposed to pay the price of investigating overseas markets of unknown dimensions and unknown promise. Neither are they eager to venture into situations which they know will demand a constant flow of reliable marketing information from remote sources.

If we can assume that highly standardized products tend to have a well-articulated, easily accessible international market and to sell largely on the basis of price (an assumption inherent in the definition), then it follows that such products will not pose the problem of market information quite so acutely for the less-developed countries. This establishes a necessary if not a sufficient condition for investment in such industries.

Of course, foreign investors seeking an optimum location for a captive facility may not have to concern themselves too much with questions of market information; presumably, they are thoroughly familiar with the marketing end of the business and are looking for a low-cost captive source of supply. In that case, the low cost of labor may be the initial attraction drawing the investor to less-developed areas. But other limitations in such areas, according to our hypothesis, will bias such captive operations toward the production of standardized items. The reasons in this case turn on the part played in the production process by external economies. Manufacturing processes which receive significant inputs from the local economy, such as skilled labor, repairmen, reliable power, spare parts, industrial materials processed according to exacting specification, and so on, are less appropriate to the less-developed areas than those that do not have such requirements. Unhappily, most industrial processes require one or another ingredient of this difficult sort. My guess is, however, that the industries which produce a standardized product are in the best position to avoid the problem, by producing on a vertically-integrated self-sustaining basis.

In speculating about future industrial exports from the less-developed areas, therefore, we are led to think of products with a fairly clear-cut set of economic characteristics.[21] Their production function is such as to require significant inputs of labor; otherwise there is no reason to expect a lower production cost in less-developed countries. At the same time, they are products with a high price elasticity of demand for the output of individual firms; otherwise, there is no strong incentive to take the risks of pioneering with production in a new area. In addition, products whose production process did not rely heavily upon external economies would be more obvious candidates than those which required a more elaborate industrial environment. The implications of remoteness also would be critical; products which could be precisely described by standardized specifications and which could be produced for inventory without fear of obsolescence would be more relevant than those which had less precise specifications and which could not easily be ordered from remote locations. Moreover, high-value items capable of absorbing significant freight costs would be more likely to appear than bulky items low in value by weight. Standardized textile products are, of course, the illustration par excellence of the sort of product that meets the criteria. But other products come to

21. The concepts sketched out here are presented in more detail in Vernon (1963).

mind such as crude steel, simple fertilizers, newsprint, and so on.

Speculation of this sort draws some support from various interregional experiences in industrial location. In the United States, for example, the 'export' industries which moved to the low-wage south in search of lower costs tended to be industries which had no great need for a sophisticated industrial environment and which produced fairly standardized products. In the textile industry, it was the grey goods, cotton sheetings and men's shirt plants that went south; producers of high-style dresses or other unstandardized items were far more reluctant to move. In the electronics industry, it was the mass producers of tubes, resistors and other standardized high-volume components that showed the greatest disposition to move south; custom-built and research-oriented production remained closer to markets and to the main industrial complexes. A similar pattern could be discerned in printing and in chemicals production.[22]

In other countries, a like pattern is suggested by the impressionistic evidence. The underdeveloped south of Italy and the laggard north of Britain and Ireland both seem to be attracting industry with standardized output and self-sufficient process.[23]

Once we begin to look for relevant evidence of such investment patterns in the less-developed countries proper, however, only the barest shreds of corroboratory information can be found. One would have difficulty in thinking of many cases in which manufacturers of standardized products in the more advanced countries had made significant investments in the less-developed countries with a view of exporting such products from those countries. To be sure, other types of foreign investment are not uncommon in the less-developed countries, such as investments

22. This conclusion derives largely from the industry studies conducted in connection with the New York Metropolitan Region study. There have been some excellent more general analyses of shifts in industrial location among the regions of the United States. See, e.g. Fuchs (1962). Unfortunately, however, none has been designed, so far as I know, to test hypotheses relating locational shifts to product characteristics such as price elasticity of demand and degree of standardization.

23. This statement, too, is based on only impressionistic materials. Among the more suggestive, illustrative of the best of the available evidence, see Toothill (1962).

in import-replacing industries which were made in the face of a threat of import restriction. But there are only a few export-oriented cases similar to that of Taiwan's foreign-owned electronics plants and Argentina's new producing facility, set up to manufacture and export standard sorting equipment for computers.

If we look to foreign trade patterns, rather than foreign investment patterns, to learn something about the competitive advantage of the less-developed countries, the possibility that they are an attractive locus for the output of standardized products gains slightly more support. The Taiwanese and Japanese trade performances are perhaps the most telling ones in support of the projected pattern; both countries have managed to develop significant overseas markets for standardized manufactured products. According to one major study of the subject (a study stimulated by the Leontief paradox), Japanese exports are more capital-intensive than is the Japanese production which is displaced by imports (see Tatemoto and Ichimura, 1959); this is what one might expect if the hypothetical patterns suggested by Figure 1 were operational. Apart from these cases, however, all that one sees are a few provocative successes such as some sporadic sales of newsprint from Pakistan, the successful export of sewing machines from India, and so on. Even in these cases, one cannot be sure that they are consistent with the hypothesis unless he has done a good deal more empirical investigation.

The reason why so few relevant cases come to mind may be that the process has not yet advanced far enough. Or it may be that such factors as extensive export constraints and overvalued exchange rates are combining to prevent the investment and exports that otherwise would occur.

If there is one respect in which this discussion may deviate from classical expectations, it is in the view that the overall scarcity of capital in the less-developed countries will not prevent investment in facilities for the production of standardized products.

There are two reasons why capital costs may not prove a barrier to such investment.

First, according to our hypotheses, the investment will occur in industries which require some significant labor inputs in the production process; but they will be concentrated in that subsector of the industry which produces highly standardized products capable of self-contained production establishments. The net of these specifications is indeterminate so far as capital-intensiveness is concerned. A standardized textile item may be more or less capital-intensive than a plant for unstandardized petro-chemicals.

Besides, even if the capital requirements for a particular plant are heavy, the cost of the capital need not prove a bar. The assumption that capital costs come high in the less-developed countries requires a number of fundamental qualifications. The reality, to the extent that it is known, is more complex.

One reason for this complexity is the role played by the international investor. Producers of chemical fertilizers, when considering whether to invest in a given country, may be less concerned with the going rate for capital in that country than with their opportunity costs as they see such costs. For such investors the alternatives to be weighed are not the full range of possibilities calling for capital but only a very restricted range of alternatives, such as the possibilities offered by chemical fertilizer investment elsewhere. The relevant capital cost for a chemical fertilizer plant, therefore, may be fairly low if the investor is an international entrepreneur.

Moreover, the assumption that finance capital is scarce and that interest rates are high in a less-developed country may prove inapplicable to the class of investors who concern us here.[24] The capital markets of the less-developed countries typically consist of a series of water-tight, insulated, submarkets in which wholly different rates prevail and between which arbitrage opportunities are limited. In some countries, the going figures may vary from 5 to 40 per cent, on grounds which seem to have little relation to issuer risk or term of loan. (In some economies, where inflation is endemic, interest rates which in effect represent a negative real cost are not uncommon.)

These internal differences in interest rates may be due to a number of factors: the fact that funds generated inside the firm

24. See Rosen (1958) who finds that in the period studied from 1937 to 1953, 'there was no serious shortage of capital for the largest firms in India'. Papanek makes a similar finding for Pakistan for the period from 1950 to 1964 in a book about to be published.

usually are exposed to a different yield test than external borrowings; the fact that government loans are often floated by mandatory levies on banks and other intermediaries; and the fact that funds borrowed by governments from international sources are often re-loaned in domestic markets at rates which are linked closely to the international borrowing rate, however irrelevant that may be. Moreover, one has to reckon with the fact that public international lenders tend to lend at near-uniform rates, irrespective of the identity of the borrower and the going interest rate in his country. Access to capital on the part of underdeveloped countries, therefore, becomes a direct function of the country's capacity to propose plausible projects to public international lenders. If a project can plausibly be shown to 'pay its own way' in balance-of-payment and output terms at 'reasonable' interest rates, the largest single obstacle to obtaining capital at such rates has usually been overcome.

Accordingly, one may say that from the entrepreneur's viewpoint certain systematic and predictable 'imperfections' of the capital markets may reduce or eliminate the capital-shortage handicap which is characteristic of the less-developed countries; and, further, that as a result of the reduction or elimination such countries may find themselves in a position to compete effectively in the export of certain standardized capital-intensive goods. This is not the statement of another paradox; it is not the same as to say that the capital-poor countries will develop capital-intensive economies. All we are concerned with here is a modest fraction of the industry of such countries, which in turn is a minor fraction of their total economic activity. It may be that the anomalies such industries represent are systematic enough to be included in our normal expectations regarding conditions in the less-developed countries.

Like the other observations which have preceded, these views about the likely patterns of exports by the less-developed countries are attempts to relax some of the constraints imposed by purer and simpler models. Here and there, the hypotheses take on plausibility because they jibe with the record of past events. But, for the most part, they are still speculative in nature, having been subjected to tests of a very low order of rigorousness. What is needed, obviously, is continued probing to determine whether the 'imperfections' stressed so strongly in these pages deserve to be elevated out of the footnotes into the main text of economic theory.

References

AHARONI, Y. (1966), *The Foreign Investment Decision Process*, Division of Research, Harvard Business School.

BARNETT, H. G. (1953), *Innovation: The Basis of Cultural Change*, McGraw-Hill.

BHAGWATI, J. (1964), 'The pure theory of international trade', *Econ. J.*, vol. 74.

BURENSTAM-LINDER, S. (1961), *An Essay on Trade and Transformation*, Almqvist & Wicksells, Uppsala.

CYERT, R. M., and MARCH, J. G. (1963), *A Behavioural Theory of the Firm*, Prentice-Hall.

DUNNING, J. H. (1958), *American Investment in British Manufacturing Industry*, Allen & Unwin.

FREEMAN, C. (1963), 'The plastics industry: a comparative study of research and innovation', *Nat. Instit. Econ. Rev.*, no. 26.

FUCHS, V. R. (1962), *Changes in the Location of Manufacturing in the United States Since 1929*, Yale University Press.

HALL, M. (ed.) (1959), *Made in New York*, Harvard University Press.

HIRSCH, S. (1965), 'Location of industry and international competitiveness', Unpublished Doctoral Thesis, Harvard Business School.

HIRSCHMAN, A. O. (1958), *The Strategy of Economic Development*, Yale University Press.

HOFFMEYER, E. (1958), *Dollar Shortage*, North Holland Publishing Co.

HUFBAUER, G. C. (1965), *Synthetic Materials and the Theory of International Trade*, Duckworth.

KINDLEBERGER, C. P. (1950), *The Dollar Shortage*, Wiley.

KINDLEBERGER, C. P. (1962), *Foreign Trade and the National Economy*, Yale University Press.

KREININ, M. (1965), 'The Leontief scarce-factor paradox', *Amer. econ. Rev.*, vol. 55.

LEONTIEF, W. (1953), 'Domestic production and foreign trade: the American capital position re-examined', *Proc. Amer. Philos. Soc.*, vol. 97.

LEONTIEF, W. (1956), 'Factor proportions and the structure of American trade: further theoretical and empirical analysis', *Rev. Econ. Stat.*, vol. 38.

LICHTENBERG, R. M. (1960), *One-Tenth of a Nation*, Harvard University Press.

MACDOUGALL, D. (1957), *The World Dollar Problem*, Macmillan.

MEYER, J. (1963), 'Regional economics: a survey', *Amer. econ. Rev.*, vol. 53.

PHELPS, D. M. (1963), *Migration of Industry to South America*, McGraw-Hill.

POSNER, M. V. (1961), 'International trade and technical change', *Oxf. econ. Pap.*, vol. 13.

ROSEN, G. (1958), *Industrial Change in India*, Free Press.

STIGLER, G. F. (1939), 'Production and distribution in the short run', *J. polit. Econ.*, June.

TATEMOTO, M., and ICHIMURA, S. (1959), 'Factor proportions and foreign trade: the case of Japan', *Rev. Econ. Stat.*, vol. 41.

TOOTHILL, J. N. (1962), *Inquiry Into the Scottish Economy*, Scottish Council.

VERNON, R. (1960), *Metropolis, 1985*, Harvard University Press.

VERNON, R. (1962), 'The Trade Expansion Act in perspective', *Emerging Concepts in Marketing*, Proceedings of the American Marketing Assn, December, pp. 384–9.

VERNON, R. (1963), *Problems and Prospects in the Export of Manufactured Products from the Less Developed Countries*, UN Conference on Trade and Development, December, mimeographed.

WILLIAMS, J. H. (1947), 'The theory of international trade reconsidered', in *Postwar Monetary Plans and Other Essays*, Blackwell, ch. 2.

21 R. Baldwin

Patterns of Development in Newly Settled Regions

R. Baldwin, 'Patterns of development in newly settled regions', *Manchester School of Economic and Social Studies*, May 1956, pp. 161–79.

To aid in the formulation of effective development programmers, economists must seek to understand the reasons why certain parts of the 'backward' world have become enmeshed in what appears to be a vicious circle of poverty. Why is it that these particular regions failed to become economically developed?[1]

One economic relation which may be useful for answering some aspects of this question is the input–output variation among commodity production functions. Although everyone is aware that there are significant differences among commodities concerning the nature of the physical output possibilities from different quantities and combinations of the factors of production – as witness the frequent use of such terms as 'labor intensive' and 'capital intensive'. commodities – these engineering differences among production functions only infrequently have been made an operational part of economic theorizing. By far the most comprehensive use of these differences for economic analysis is the input–output studies initiated by Professor Leontief (1951; 1953). He has measured the average technical input coefficients employed in producing the outputs of various industries in the United States. These coefficients include both current or flow coefficients, i.e. the quantities of various products (measured in dollars) and the number of workers (or man-years) which are used to produce a dollar's worth of a particular commodity per year, and capital or stock coefficients, i.e. the quantity of capital (measured in dollars) used to produce a dollar's worth of a commodity per year. On the basis of these coefficients it is

1. See Haavelmo (1954), no. 3, pp. 1–6, for an excellent discussion of the objectives and possibilities of a theoretical approach to the problem of economic development.

possible to estimate the derived demands which would arise from various changes in the final bill of goods.

Table 1 presents a few of the Leontief labour and capital coefficients. The figures in Table 2 are rough estimates of the labour coefficients for a few agricultural commodities that are produced to an important degree under a plantation system.

Table 1 Labour and Capital Requirements for Selected United States Industries per $1000 of Output, 1947*†

	Capital (dollars)	Labor (man-years)
Agriculture and fisheries	2524·4	0·082
Textile mill products	493·6	0·110
Chemicals	592·7	0·049
Iron and steel	1026·3	0·077
Agriculture, mining and construction machinery	838·6	0·087
Motor vehicles	565·8	0·060
Coal, gas and electric power	2222·6	0·099
Railway transportation	3343·3	0·153
Trade	984·9	0·165
Communications	4645·4	0·163

* Leontief (1953).

† Two other important studies dealing with inter-industries differences among capital–output ratios in the United States are Creamer (1954) and Borenstein (1954).

These coefficients are, of course, average figures. Even within the United States intra-industry differences are important because of variations in the techniques employed within an industry, differences among factor price ratios, dissimilarities in the quantity of the labour supply and natural resource conditions, differences in managerial skill, etc. For similar reasons, one would expect a wide degree of intra-industry variability among different countries.[2] However, if new firms, using the same technological and managerial knowledge, were established in the

2. Information on some of the international differences in capital–output ratios has been collected by Grosse (1953), Bhatt (1954) and Mandelbaum (1947).

various industries and their means of production were secured from some common, perfectly competitive factor market, Professor Leontief's investigations do seem to indicate that there

Table 2 Selected Labour Coefficients in Tea, Rubber and Sugar

Tea* (1940 Colombo price)	
Ceylon	6·0–6·3 men per $1000 per year
Rubber† (1939 Singapore price)	
Malaya and FMS estates	2·6–4·0 men per $1000 per year
Ceylon (estate and total)	4·2–5·7 men per $1000 per year
Sugar‡ (cane)	
Cuba (1939 Cuban price)	2·1 men per $1000 per year
Hawaii (1939 New York price)	0·6 men per $1000 per year
Or (1939 Cuban price)	1·1 men per $1000 per year

* The sources for these figures are *The Census of Ceylon*, 1946, vol. 1, pt 1 (6·3 per $1000); *The Annual General Report for 1934 on the Economic, Social, and General Conditions of the Island, 1934–6* (6·0 per $1000); and V. D. Wickizer, *Coffee, Tea and Cocoa* (6·3 per $1000 on the basis of information on p. 162 and his yield figures). It is interesting to note that using the real labour coefficient for India in 1915 (*Report of the Production of Tea in India, 1915*, Calcutta, 1916) together with the 1940 Colombo price gives a coefficient of 8·9 per $1000.

† This information was obtained from P. T. Bauer, *The Rubber Industry*, pp. 266–7 (Malayan estates, July 1940 to June 1941 – 2·6 per $1000; FMS estates, 1933 – 2·6 per $1000; FMS estates, 1929 – 4·0 per $1000); *Census of Ceylon, 1946* (total exports 1946 – 5·7 per $1000); and *The Annual General Report* (4·2 per $1000).

‡ The sources are: US Cuban Sugar Council, *Sugar – Facts and Figures, 1952* (total production, 1950 – 2·7 per $1000); *Printed Reports of the 68th Annual Meeting of the Hawaiian Sugar Planters' Association, 1948* (total production of these companies, 1939 – 0·6 or 1·1 per $1000).

would be significant industry differences among the labour and capital coefficients employed in these firms.[3]

While the Leontief analysis assumes fixed production coefficients for each industry, this assumption will not be followed here, since the analysis will be conceptual rather than statistical. What

3. The coefficient of variation of the capital–output ratios in 1929 prices in thirty-seven manufacturing industries analysed by Creamer, op. cit., for 1948 was 30·7 per cent. However, the coefficient has declined steadily from 1900, when it was 66·9 per cent.

will be assumed is merely that there are significant engineering differences among some commodity production functions over their input–output range. These variations concern the manner in which returns to scale behave for different factor ratios and also the manner in which the marginal rates of factor substitution vary for different output levels and factor ratios.

This paper will utilize the concept of these production function differences in analysing the problem of differential rates of growth between newly settled regions. While technological conditions of production influence the pattern of growth in an economy at all stages of development, it appears that they can be particularly important in conditioning the potential for growth in newly settled regions. Consequently, the procedure to be followed will be to contrast the *hypothetical* development of two regions – both of which, initially, are assumed to be sparsely populated. The two areas are assumed to develop simultaneously within a given and constant state of technology and to draw their immigrants and capital from some common, populated region where all the inputs and outputs are represented.[4] The socio-political environment of this more populated region is assumed to be conducive to the development of the two sparsely populated regions.

Each region's economy is assumed to be small enough in its early stages of development to have no effect on the given hierachy of factor and commodity prices prevailing in the more developed, third region.[5] Furthermore, the two regions are equidistant from the older area and this distance is sufficiently great to make the costs of labour migration fairly substantial. It also is assumed that the economic development of each of the two sparsely populated regions begins in the export sector with the production of a primary commodity.

The differences between the two regions concern their natural resource conditions. One of the regions is assumed to possess a

4. The effects of improvements in technological knowledge will be discussed later. In a general sense, the exploitation of these new regions may be considered a technological change.

5. The labour supply is divided into a number of imperfectly competing groups.

soil and climate highly suitable for the initial cultivation of a plantation crop in contrast to the other area which is assumed to enjoy conditions most conducive to the initial production of a non-plantation type commodity such as wheat. However, in both regions there is assumed to be an abundant supply of mineral resources such as coal, iron-ore, ferro-alloys, oil, etc. These are not exploited immediately since, initially, they are at a prohibitive distance from the export ports.

The purpose of most of these assumptions is to minimize differences among the many other factors which can cause dissimilarities between the two regions in their patterns of development. The development model to be analysed can easily be compared with the differential growth patterns which might result by varying these initial assumptions. Some of the consequences of such other assumptions will be examined later. However, it seems that even with a wide range of possible initial conditions facing newly settled areas the effects of production function differences still emerge as an important (and neglected) determinant of development patterns.

Given the above conditions, the contention here will be that the extent to which the export sector induces the subsequent development of other sectors in the two economies depends to an important degree upon the technological nature of the production function of the export commodity (assuming there is only one major export item in each new region). For, given the price of the export commodity and the array of factor prices in the third region, this function will greatly affect subsequent development by initially influencing the nature of the labour and capital supply which flows into each region and the distribution of each economy's national income. It is from this framework that some of the many other important factors which determine the pattern of development will be introduced into the analysis.

Assume that the following conditions exist in one of the two regions. Factor and commodity prices in the populated area and the climate and soil of the new region indicate that the most profitable opportunity for initial development is the production of a plantation type commodity. Assume the production function for this particular commodity is such that for a wide range of labour–capital price ratios the most efficient organization for any level of production is on a relatively labour intensive basis.[6,7] In other words, efficient production of a dollar's worth of the commodity technologically tends to require a relatively large number of labourers to perform comparatively simple tasks. Beyond a certain proportion of capital to labour, the amount of capital which must be substituted for a given decrease in labour in order to maintain a given level of output is relatively large. Furthermore, there are significant increasing returns to scale in the cultivation and processing of the commodity. Consequently, comparatively large amounts of both capital and labour are necessary for the most efficient size of the production unit. A high level of managerial and technical skill also is needed to direct large plantations effectively.

Small, family-size farms are attractive for the very low income groups in the older region, but the independent entrance of these groups is prevented by the cost of migration and the initial capital outlay on even this type of small productive unit. Nor are these people able to borrow the funds in the capital market, since severe capital rationing tends to operate against these very low income groups.

Those establishing productive units migrate from middle and higher income groups of the developed region and either possess the necessary funds for migration or are able to borrow them in the capital market. In order to produce the commodity at the lowest possible costs, these entrepreneurs in turn create a demand in the older region for the labour of very low income groups (who are assumed to possess the requisite skill to perform the comparatively simple tasks involved in production or can be trained easily to perform them). Plantation owners or their agents seek out these low wage groups and finance their migration. And they protect this investment by attempting to tie the workers to the plantations for a certain number of years.

In the second sparsely populated region assume the following

6. Since information concerning the variability of capital coefficients among agricultural commodities is meagre, it will be assumed that optimal capital requirements per dollar of output are about the same for the two types of agricultural commodities discussed here.

7. For a general survey of the methods of production for a few plantation type products, see Wickizer (1951); Pim (1946); Bauer (1948); and Fay (1936).

conditions hold. Prices in the older area and the environment of the new region favour the development of a non-plantation agricultural commodity. The production function for this commodity differs from the plantation commodity in two respects. First, a family-size farm gives an efficient scale of production. In particular, large-scale production based on the intensive use of cheap, imported labour is not the best form of economic organization. Furthermore, the absolute amount of capital required is less for the optimum size of a production unit, and the level of managerial and technical skill need not be so high for a productive unit of the most efficient size. Secondly, the technological possibilities of capital intensification on the family-size farm are much greater. Varying the labour–capital price ratios over a wide range causes much more factor substitution in producing a given level of output than with the plantation crop.

As in the previous case, the very low income groups in the older region tend to be prevented from independent migration because of the costs of migration and the difficulty of borrowing funds. The level of knowledge and skill required for establishing a farm also rules out the migration of many from this labour group. Because of the assumed distance conditions, financing the movement of this type of labour is relatively costly. This fact and the wide range of alternative factor combinations prevent any extensive importation of cheap, unskilled labour by small-scale cultivators. Consequently, migrants flow from the income groups which can provide the necessary initial outlays on transportation and production. However, capital rationing also works against the latter group to some extent. The smaller scale of operations hampers the supervision of direct lending. Consequently, direct inter-regional lending is not as significant as in the plantation economy.

The agricultural development in both regions stimulates a simultaneous development of some supporting industries – such as transportation – which are directly linked with exporting the agricultural commodity (see Nurkse, 1954). The large amount of capital necessary for even a minimum amount of this type of social capital is supplied comparatively readily by foreign investors. Not only are these industries directly tied to the exchange-earning export industry and, consequently, are particularly attractive to foreign investors, but also they are organized on a large enough scale to take advantage of the established capital markets in the developed region.

However, the obstacles to the vigorous expansion of the plantation economy into a developed, higher *per capita* income economy are much greater than with the non-plantation economy. The relevant factors for an analysis of the development potential of the two regions from the stage already discussed can be grouped into demand and supply forces. First, in order to introduce domestic production of commodities for which an export advantage does not exist, there must be the basis of an internal demand for such products. And, secondly, given the demand, the natural resource situation and the supply of capital and labour must be adequate enough to meet foreign competition.

For both regions the composition of the family budget is assumed to depend upon the level of income.[8] At very low levels, the budget consists almost entirely of a few basic foodstuffs, clothing, household needs and shelter. As incomes rise, the food budget is diversified and, eventually, a smaller proportion of the budget consists of foodstuffs. Durable consumer good expenditures and savings increase in relative importance.

In the plantation economy at this initial stage a large part of the population is in the very low income brackets. Consequently, most of this group's effective demand consists of a few basic foods, simple clothing and other consumer durables, and minimum shelter needs. While production of the plantation crop requires large quantities of labour, this labour is not needed throughout the entire year. Consequently, workers lease small plots of land from the plantation owners (who also usually provide the capital) and supplement their income by growing part of their food requirements. During the idle period, the choice to a labourer of working more on the plantation and less for himself

8. Mandelbaum (1947) utilizes a budget approach in estimating the flow of demand in his hypothetical model of development for south-eastern Europe. He also employs capital, labour and commodity coefficients in computing the supply requirements for his programme. Because of a lack of data, this procedure so far has been only used in a rough fashion for the formulation of actual development plans in the backward countries.

does not exist. The alternative essentially is between leisure and working for himself. And since his plantation income is very low, his marginal utility for commodities is relatively high. Therefore, he is willing to devote much of his free time to growing part of his own food in a very socially inefficient manner. He drives his marginal productivity in this line down to nearly zero. The same phenomenon tends to take place with respect to part of his clothing, shelter and durable consumption goods needs. The family unit produces many of these items.[9]

Why do not the plantation workers break away from the plantation, produce the crop themselves, and raise their income level appreciably above the plantation wage? Some do break away. However, most of these unskilled, low income workers cannot save or borrow enough to start anything but a very small, low income yielding unit of production.

In attempting to expand from such small units, the cultivators are hampered by the technical constraints of the crop's production function. They must secure more labour, land and capital, i.e. expand horizontally, for efficient production. But this is very difficult. First, the initial income level on these farms is so low that their saving is almost insignificant. Nor are they able to borrow sufficient funds for a large scale unit. Secondly, it is difficult for this group to enlarge gradually its holdings of good land. The fertile land is cultivated by the plantation method and its owners are reluctant to sell or lease parcels of it. Large tracts must be taken at one time. But the small farmers cannot overcome this discontinuity. The best they can do is obtain isolated parcels of good land or more contiguous but relatively poor land. However, expanded production on this type of land is not very efficient. Thirdly, the level of knowledge and skill of these people is so low that they are not capable of supervising and controlling the greater amounts of capital and labour necessary for increased production. The supervision of the labour is particularly important. It is probably more difficult to direct the greater amount of non-family labour than the increased quantity of capital. Finally, the plantation class tends to develop a social antipathy towards this very low income group. It does not want the group

9. For a general discussion of some of the production and labour conditions in such backward areas, see Greaves (1935) and Moore (1951).

to move into the plantation class and erects social and economic barriers in the path of the group's expansion.

All of these factors tend to prevent this group from increasing the size of its farms and thus its income level. And, because production of the crop does not require the same amount of labour throughout the entire year, these small farm families (like the plantation workers) also grow part of the food they consume (or even a market supply) and produce many of the durable consumer goods they consume. Disguised unemployment tends to arise within this sector of the economy.

Small scale planters who employ some non-family labour are another important group in the economy. These individuals either break through the exclusively family-labour type of farming or initially possess sufficient funds to establish a small plantation. They resemble somewhat the middle income migrants in the other region. However, these planters face more difficulties in reaching the optimum productive unit. In the first place, they must accumulate much more capital to attain this level. And, as in the non-plantation area, the capital rationing barrier forces the planters to rely on current saving for most of their investment funds. Furthermore, given funds equivalent to the requirements for the smaller optimum size farm in the other region, a small-scale planter will not earn as high an income as his counterpart in the other region. The production unit is too small. Even if organized as efficiently as possible, he cannot use his managerial skill to full advantage. Merely directing production does not require his full time nor yield a very high income. Yet the only alternative to leisure is to perform the low productivity tasks of the hired help. Consequently, because of their low income level, these planters cannot expand their productive units as rapidly as the farmers of the non-plantation region. In addition, most of these small planters do not possess the high degree of managerial ability and technical skill required to expand the scale of operations in an optimum manner. Consequently, they tend to keep the amount of labour and land employed about the same and reinvest their savings in capital improvements which do little in lifting their level of income, because of the nature of the crop's production function.

A reasonably stable hierarchy of export producers emerges

within the economy. At one end stand the plantations employing large quantities of low wage labour. The other end of the scale is composed of many small, family cultivators who operate under a tenure system or perhaps own their land. The income level of these farmers is not much higher than the plantation wage, and the possibilities for expansion by these producers is not favourable. The small scale planter who combines family and hired labour lies between these two groups. While incomes among these planters are higher than the very small farmers, they are below the level achieved with a similar investment in the non-plantation economy.

Perhaps 70 per cent of the economy's income is spent on foodstuffs (Schultz, 1953). The remainder is devoted to services, consumer durables (of which expenditures on items other than simple clothing and household articles are a small percentage), and saving. The effective market demand for the higher class of consumer goods and services stems largely from the middle and high income groups, who are composed of large plantation owners, those performing the marketing services associated with the export item, and to some extent the small planters. A large number of these commodities are imported from the more developed region.

Why do not efficient domestic industries quickly develop and capture both the import markets and the domestic markets which are supplied in a socially inefficient manner?

Consider the opportunities in the fields of simple, mass consumed durables and luxury durables – many of which are imported. A major obstacle confronting prospective domestic manufacturers is the problem of training the labour force to the factory system. The large, low income labour supply possesses such a low level of education and skill that its costs of training represent a large, initial outlay. While there is always the alternative of recruiting skilled foreign labour, this too is expensive. At this early stage of development, the region cannot rely to any significant extent upon the voluntary migration of suitable labour. This labour migrates at its own expense only after the industrial sector has begun to expand vigorously and employment opportunities become well known. In addition, although the marginal productivity of the low income farm labour may be near zero, it is necessary to offer them a higher figure in order to induce them to move into urban factories. Both of these factors make it difficult to capture the import market.

They are particularly forceful with respect to luxury imports. Many of these items require a very high degree of labour skill. Conspicuous consumption also applies to some of these goods, and considerable outlays on advertising are necessary to overcome a preference for foreign commodities. Still another important factor with respect to some of these consumer durables is the internal and external economies involved in their production. The domestic market is too small to take advantage of these economies.

Most of these obstacles also apply to those consumption items produced on a household scale. However, another obstacle confronting more efficient domestic industry is one which prevents the importation of these items, namely, the high costs of internal transportation and the lack of other marketing facilities. In this region, the bulk of the population in the hinterland is so poor that the construction of transportation facilities (other than the minimum necessary for the export crop) proceeds very slowly. Governments cannot raise enough revenue from these people to build adequate facilities. The higher income groups are so spread out that they cannot support these facilities either. To obtain many of the commodities and services, which they desire, they travel to a few large cities where the marketing facilities for the export commodity are located. Outside of these central cities, few other trading cities spring up and, consequently, transportation facilities in the interior remain crude. Therefore, domestic manufacturers find it too expensive to tap the interior markets for mass consumption goods.

Two other factors on the supply side, which are relevant to this discussion, are the rate of saving and the supply of entrepreneurial labour. Because of the greater income inequality the proportion of saving to national income is likely to be higher in this region than in the non-plantation economy. However, a larger share of the saving flows back to the more developed area in the form of interest and dividend returns on foreign investments in the plantations and the auxiliary service industries. The foreign earnings which are retained tend to be employed for a further expansion of the export industry, since foreign investors prefer investments which are directly linked with the foreign exchange earning ability of the economy. Furthermore, foreign investments in industries producing for an internal demand are discouraged by the lack of an adequate market in addition to the other factors already enumerated. For the same reasons, large domestic savers also tend to employ their funds in the export and import trades or in such ventures as residential and business construction. But, because of the nature of the production function for the export crop, investment in this sector does little to improve the distribution of income; it merely enlarges the existing productive structure as more cheap labour is imported.[10] Nor does the investment in elaborate homes, office buildings, shops, etc. do much in inducing a better pattern of growth. With respect to entrepreneurship, the most obvious source of leadership for manufacturing – the large plantation owners – provides a meagre supply. This group, because of the unique non-pecuniary advantage of the plantation life, tends to develop a social antipathy towards occupations in the manufacturing field. And, the low income group possesses neither sufficient training nor the social and economic opportunities necessary to provide more than the occasionally successful entrepreneur.

All of these factors tend to restrain the economy from breaking out of its predominantly export-oriented nature.[11] As transportation facilities improve, the mineral resources are tapped, but this sector too becomes export-oriented. Domestic manufacturing industries based on these raw materials are blocked by the same obstacles previously mentioned. The only real possibility for exploiting the minerals is as raw material, export industries. And, because of the general lack of technological and entrepreneurial skill within the economy, many of the firms are owned and operated by foreigners. Although these industries may provide an important source of saving in the form of royalty payments and, depending on the quality of labour they require, may help

10. Lewis (1954) employs the assumption of an elastic labour supply in his interesting article.

11. For a discussion of some of the development obstacles in this kind of economy see Singer (1950) and Mosk (1951). Also on the general topic of the effect of foreign trade on newly settled areas see Myint (1954–5).

to improve the distribution of income, these effects will not be as favourable for growth as those that would result if the internal market were large enough to induce related domestic manufacturing industries.

When the development potential in the other region is analysed, a more optimistic outlook appears. The nature of the export crop's production function is an important reason for this view. As already mentioned, labour and capital requirements for an optimum size farm in this region are much smaller than for the plantation type commodity. The family unit gives an efficient scale of operations. As in the other region, the very low income families in the older regions are excluded from independent emigration by the relatively high costs of the movement. However, unlike the plantation region, the more wealthy individuals do not finance their passage, since very unskilled labour cannot be employed as effectively in this type of agriculture. Instead, most of the migrants come from the income groups which possess sufficient funds for migration. In this region there are relatively fewer individuals at both ends of the absolute income scale.

Unlike the plantation economy, as this region's export sector expands, the economy does not devote a large portion of its investment to securing and supporting a greater quantity of cheap, unskilled labour. Although many of the migrants to this region originally do not establish the most efficient size unit, the limitation is not so much labour, but rather the inability to secure sufficient capital. However, these migrants do not start, like those breaking away from the plantations, at such a low level of income that their saving is almost nothing.

In the early stages of development, these farmers also produce much of their food, clothing, shelter and simple durable consumer goods. But they are not blocked from optimum expansion as are most of the small planters and the family-size farmers in the other economy. Since the marginal productivity of labour and capital is higher in agriculture than in these activities, the farmers, by reinvesting their saving, increase the output of the cash commodity and curtail the family production of food and consumer durables. Moreover, as their income level increases, the family prefers to purchase more of its clothing, food, shelter, services and other consumer durables in the open market.

The more equitable distribution of income, which arises as the economy develops its export production, is more favourable for the induced development of domestic industry. A smaller proportion of the national budget is devoted to food expenditure. And the production of this food is undertaken on efficient, family-size farms. Furthermore, there is a relatively larger market demand for services and durable consumer goods. Profit opportunities arise in these lines of commodities. Initially, some of these goods are imported, while others are not consumed at all because of the high costs of transportation. But gradually trading centres spring up to answer the demands for medical, legal and personal services as well as to provide the marketing facilities for the imported commodities. All of this means investments in homes, offices, warehouses, roads, schools, hospitals, etc., which have a multiplier effect on the volume of trade (Duesenberry, 1950). As this development occurs, the mineral resources begin to be exploited. However, instead of merely becoming exports, these resources are also used to supply domestic manufacturers. For, the more favourable distribution of income and thus the relatively large demand for durables stimulates domestic manufacturing. Because of the relatively larger market demand for such items and the higher level of skill of the agricultural population, the problem of recruiting foreign labour or training domestic workers for manufacturing activity also is not as difficult as in the plantation area.

All of these factors and their interaction tend to induce a faster and a more balanced type of development. This economy has a better chance of climbing from its initial export orientation. Domestic industries spring up which, in turn, stimulated the further expansion of other domestic or export industries through external economies and the familiar multiplier-accelerator interactions.

In order to emphasize the role which technological differences among production functions can play in the process of economic development, a number of restrictive assumptions were made in the preceding analysis. When these are lifted, the factors stressed in the traditional explanations of differential growth patterns re-emerge to a more prominent position.

Firstly, there is the matter of the production functions themselves. In the above discussion, it was assumed that the production functions in the export industries of the two regions were such as to impose rigid constraints on the nature of the development process. While I believe this factor is and has been an important element in shaping actual development in several regions, this is not to say that it always plays an important role in the development process. For example, the production functions of some crops may be such that both plantation and small-scale production are equally efficient. And there may be wide possibilities of factor substitution with relatively slight changes in the factor price ratios. In these cases the engineering constraints of the production function will not be important in determining the character of development. It was also assumed that each region drew its productive means from a common equi-distant, purely competitive market. But obviously, if the array of factor and product prices differ in the older regions which initiate development in the two new regions and the distances to these new regions vary, the patterns of development will be affected accordingly. Differences among the factor supplying regions in the state of their technological knowledge, in their entrepreneurial spirit, in their tastes and in their social, economic and political ideas and institutions generally also will play an important role in determining the nature of development within the two regions. And, of course, the dissimilarities between the new regions with respect to their natural resource conditions are highly relevant. The effects on the preceding analysis which regional differences in the above factors can cause are fairly obvious.

Another condition which has been maintained in this discussion is the assumption of an unchanged state of technology. Probably, most of the technological knowledge actually introduced over the last 200 years has been of two types: (a) those changes which required more capital and less labour (or other resources) per unit of output than previously; and (b) those changes which required less of all factors (see Grosse and Duesenberry, 1952). How do these types of technological progress affect economic development?

Clearly, the development problem cannot be dismissed with the assertion that technological progress will guarantee successively higher levels of *per capita* income in an automatic fashion. Three major factors should be considered in analyzing the problem: demand, the supply of capital, and the nature and growth of the population. The first factor, demand, is extremely important for those agricultural exporting nations which are so large that changes in their output affect international prices. Price and income elasticities for many agricultural products are low in the higher *per capita*, agricultural importing regions. Consequently, part of the possible real income benefits of technological progress may be lost through an adverse movement in the terms of trade. Secondly, in order to achieve the maximum growth allowed by technological progress, the requisite capital must be forthcoming. But there is no reason to assume that the saving propensities of the public and business will adjust automatically to take advantage of the new technique. In low *per capita* countries this can be an especially serious problem. Rather similar barriers with respect to the nature of the labour supply also can prevent maximum growth. Shortages of particular kinds of labour and/or general lack of entrepreneurial ability are examples of this type of bottleneck. Finally, the growth of population in relation to the increase in income will determine what happens to *per capita* income.

But, of course, technological progress does operate in the direction of encouraging a more rapid rate of increase in national income. This is especially so if the progress in technological knowledge is such that less of all factors of production are required per unit of output. However, to the extent that technological progress is such that the relative position of each commodity in the scale of labour and capital coefficients remains roughly the same, technological progress can be handled in the model by interpreting much of the development behaviour of the two regions in relative rather than absolute terms. But a radical shift in the relative position of a commodity in the labour and capital coefficient hierarchy must be treated as an autonomous change, and the analysis must be modified to take into account this new engineering relationship.

Even with the many special assumptions in this analysis,

differences in the technological nature of production functions still, I think, emerge as an important factor determining actual patterns of economic development. Briefly, the argument is that the technological nature of the production function for the major commodities initially selected for commercial production influences the potentialities for further development in newly settled regions. In conjunction with market conditions in the more developed areas, these engineering constraints affect the nature of factor migration and the early distribution of income within a region. The latter factors, in turn, affect the stimuli for further economic development. While much more empirical and historical investigation is necessary to determine the extent of the technological restraints of various production functions, these differences can, I think, prove useful in contrasting actual historical development of some plantation-type economies in the world with those regions which at an early stage specialized on such commodities as livestock and grains. Furthermore, they must be carefully considered in the formulation of plans for future development.

References

BAUER, P. T. (1948), *The Rubber Industry*, Harvard University Press.

BHATT, V. H. (1954), 'Capital–output ratios of certain industries: a comparative study of certain countries', *Rev. Econ. Stat.*, vol. 36.

BORENSTEIN, I. (1954), 'Capital and outlay trends in mining industries, 1870–1948', *Occasional Paper*, National Bureau of Economic Research, no. 45.

CREAMER, D. (1954), 'Capital and output trends in manufacturing industries, 1880–1948', *Occasional Paper*, National Bureau of Economic Research, no. 41.

DUESENBERRY, J. S. (1950), 'Some aspects of the theory of economic development', *Explorations in Entrepreneurial History*, December.

FAY, C. R. (1936), 'The Plantation Economy', *Econ. J.*, vol. 46.

GREAVES, I. C. (1935), *Modern Production Among Backward Peoples*, Allen & Unwin.

GROSSE, R. N. (1953), 'The structure of capital', in W. W. Leontief *et al.*, *Studies in the Structure of the American Economy*, Oxford University Press.

GROSSE, R., and DUESENBERRY, J. S. (1952), *Technological Change and Dynamic Models*, prepared for the Input–Output Meeting of the Conference on Research in Income and Wealth, October.

HAAVELMO, T. (1954), *A Study in the Theory of Economic Evolution*, no. 3 in Contributions to Economic Analysis, North Holland Publishing Co.

LEONTIEF, W. W. (1951), *The Structure of the American Economy, 1919–1939*, Oxford University Press.

LEONTIEF, W. W. (1953), 'Domestic production and foreign trade: the American capital position re-examined', *Proc. Amer. philos. Soc.*, vol. 97.

LEONTIEF, W. W., *et al.* (1953), *Studies in the Structure of the American Economy*, Oxford University Press.

LEWIS, W. A. (1954), 'Economic development with unlimited supplies of labour', *Manchester School*, May.

MANDELBAUM, K. (1947), *The Industrialization of Backward Areas*, monograph no. 2, Oxford University Institute of Statistics.

MOORE, W. E. (1951), *Industrialization and Labor*, Russell.

MOSK, S. A. (1951), 'Latin America versus the United States', *Amer. econ. Rev., Pap. Proc.*, May.

MYINT, H. (1954–5), 'The gains from international trade and the backward countries', *Rev. econ. Stud.*, no. 58.

NURKSE, R. (1954), 'The problem of international investment today in the light of nineteenth-century experience', *Econ. J.*, vol. 64.

PIM, A. (1946), *Colonial Agricultural Production*, Oxford University Press.

SCHULTZ, T. W. (1953), *The Economic Organization of Agriculture*, McGraw-Hill.

SINGER, H. W. (1950), 'The distribution of gains between investing and borrowing countries', *Amer. econ. Rev., Pap. Proc.*, May.

WICKIZER, V. D. (1951), *Coffee, Tea and Cocoa*, Stanford University Press.

22 R. Solo

The Capacity to Assimilate an Advanced Technology

R. Solo, 'The capacity to assimilate an advanced technology', *American Economic Review, Papers and Proceedings*, May 1966, pp. 91–7.

We observe one set of societies where the norm of productivity is very low compared to that of another set. The former have been called 'developing', the latter 'advanced'. The complex of techniques, including the equipment in use and the organization of operations in the advanced which, presumably, accounts for their high productivity, will be termed 'advanced technology'. The question is, what determines the capacity of the developing society for incorporating this advanced technology into its own operations and, thereby, itself achieving high productivity?

It may simply be a question of financing investment. There need not be the resources to cover the high initial costs of reorganizing operations. Foreign exchange to purchase key producer durables may be lacking. Indeed the provision of investable resources through foreign loans or grants or through local savings to cover the costs of transformation seems hitherto to have been the principal concern of development economists and agencies who have taken their cue from post-Keynesian growth models. But simply to have the resources available for investment is not a guarantee of development. Resources must be matched by the opportunity to use them in transforming operations – an opportunity which will depend in part on the social capacity to assimilate advanced technology.

Nontransferable components of advanced technology

One reason why advanced technologies are not used in low-productivity economies may be because they are not usable there. They have evolved in and consequently are adapted to a social and a physical environment which differs significantly from that of developing societies. On account of these differences, their use in the developing society will sometimes be uneconomic and technically retrogressive.

The most evident difference in the contexts of technical operations between low- and high-productivity economies is in the physical environment. Characteristically, advanced technologies have developed in temperate climates while low productivity societies are found, for the most part, in tropical and subtropical zones. For that reason different vegetation, different fish and fowl, different animals flourish in each. The structure of the soils and the practices appropriate to soil conservation will differ. There are different crops, differently cultivated, with different problems in their preservation and processing. Different diseases attack men, animals and plants. When the diseases are the same, their vectors are likely to differ. Correspondingly, it is not possible to transfer the technologies and the sciences of agriculture, horticulture, animal husbandry, medicine and public health developed in temperate conditions directly to low-productivity societies in tropical and subtropical zones. As in the physical, similarly differences in the social and economic context of technical operations may bar the direct transference of advanced technologies.

Inasmuch as advanced technologies are not directly transferable, their assimilation requires social action and social competence in: (a) recognizing what can be transferred directly and what might be adapted for transference, then in (b) adapting advanced technologies for application in the low-productivity economy, or in (c) restructuring the context of operations to provide a more hospitable environment for the advanced technology.

These three – the capacity to recognize the feasibility of attempting directly to transfer or adapt advanced technology, the capacity to adapt technology to the physical, social and economic context, and the capacity to adapt social and economic conditions to the requisites of technology – together constitute the capacity to assimilate advanced technologies.

Adapting the advanced technologies

Consider the adaptation of advanced technologies to the context of operations in low-productivity economies – which can be understood as taking place at three different levels.

At the simplest level advanced technologies are adapted on the spot by adjusting or modifying machines which are in use else-where or processes and techniques which are practiced elsewhere to a particular need or circumstance. However, a machine or technique only expresses and partially embodies a corpus of knowledge. The mastery of that knowledge and its use to design the appropriate mechanisms and techniques can be understood also as an adaptation of advanced technology and one that con-veys a wider range and a greater power of assimilation than is possible through on the spot adjustments or modifications of already designed and practiced techniques.

Advanced technologies are based, in considerable part, on the knowledge of science. Hence, to exploit the corpus of information upon which advanced technology rests requires the competence of the scientist or the science-trained engineer who, in this instance, must also be deeply familiar with the circumstances and needs of a low-productivity society. But even the mastery of the knowledge of technology by those who also understand the cir-cumstances of the low-productivity economy need not suffice.

The knowledge underlying advanced technology is neither infinite nor static. It has itself been produced and is continuously being produced through experience and research. Since it has chiefly evolved under the circumstances of high-productivity societies in the effort to solve the problems of high-productivity societies, correspondingly it will be less applicable, less adequate to solve the problems of low-productivity societies or to design a technology suited to their circumstances.

Inasmuch as existing knowledge does not suffice, what must be acquired is the capacity to produce knowledge that does. It is the problem-solving, information-producing apparati that must be mastered, adapted and applied. Science itself must be develop-ment-oriented. To adapt and to focus the analytic concepts of science and its research method on those problems which arise at the nexus of need and circumstances in a developing society means to exploit the most dynamic component of advanced technology. It is to adapt advanced technology's very mechanism for adaptation and for further advance.

An instance of development-oriented science, familiar to this audience, is that of development economics. The conventional social technology for economic control and policy determination could not be directly transferred to low-productivity societies. And the accumulation of concept, theory and information – constituting economics – underlying that technology was likewise unsuited. Therefore, the research focus of our discipline was redirected and its problem-solving apparatus adapted so that it might produce the needed knowledge on the basis of which viable policies and controls for economic development can be designed. This suggests another fact. The capacity to adapt advanced technology to the requisites of low-productivity societies need not be wholly indigenous to those societies. Nor can it be entirely outside, since, to be effective, that capacity at each of its three levels must be integrated into the system of choice and power.

To recapitulate, with an illustration of the three levels of adaptation, take the hypothetical need to control insects on a tropical plantation. There might be equipment in use elsewhere which can be used for spraying insecticides, but it would have to be adapted on the spot to climate and terrain, to the shape of the infested plant, and to the locus of its infestation or to the skill of indigenous labor. Such on the spot adjustment of already designed equipment would be adaptation at the first level. It might then be necessary to compound and produce an appropriate insecticide which would require a knowledge of the chemistry of pesticides and of the habits and vulnerabilities of the insect to be controlled. To design this new component of technology by reference to existing knowledge would be a second-level adaptation. And if, as is frequently the case, not enough is yet known about the habits and vulnerabilities of the pest or the pesticidal efficacy of locally available materials or the possibilities of control through the use of the pest's own parasites, then a conceptual and analytic apparatus of science must be turned to search for the new information which is required. This development-orientation of science is the third level in the adaptation of advanced technology.

The greater the differences in the context of technical operations, the more difficult it will be to adapt a technology that has evolved in a high-productivity society to the needs and circumstances of a low-productivity society. This may account for the seemingly paradoxical emphasis on industrial development by predominantly agricultural low-productivity societies; since the ineradicable differences in the natural parameters of agriculture (and also the very deep differences in the social circumstances of agriculture) in low-productivity *vis-à-vis* high-productivity societies probably makes it more difficult to adapt advanced agricultural technologies than to adapt advanced industrial technologies for assimilation.

Whether or not it will be economic to use a particular technology will depend on the cost pattern of materials, of equipment, of energy sources, of the various skills of labor, and of access to markets. Differences in cost patterns will check the transference of advanced technologies. Particularly significant are cost differences which reflect the relative lack of skilled labor in low-productivity economies. An adaptation of advanced technology to this lack may call for the use of a less costly accumulation of producer's durables per unit of output, i.e. for 'low capital intensity' techniques, than is usual in high-productivity economies – or it may call for exactly the opposite. Thus Norbert Weiner, while working in India, observing the lack of ordinary industrial skills there, quite correctly deduced that this would make it impossible to transfer the technology practiced in high-productivity societies directly to India, and that the costs of training the masses to the requisite skills would be very great. He also observed that there existed in India a number of highly trained and very able mathematicians and scientists who might be mobilized to design and introduce fully automated industrial processes where the need for ordinary industrial skills would be minimal. Therefore, Weiner supposed that the opportunity costs of automation in India would be less than in, say, the United States and that it would be more economic to take the path of high automation in a low-productivity society such as India than it would be in a high-productivity society such as the United States. I am told that Keleki took the same position in regard to economic development in Poland. I doubt that a conclusive general case could be made that technology adapted for assimilation into low-productivity economies should be more or less capital intensive than the norm. Suffice that since the pattern of comparative costs is likely to be different, the transferred technology should in some way be reshaped. Perhaps the same could be said for many transfers of technology between high-productivity societies or even for transfers between sectors in the same economy.

Social and economic organization as a variable

An advanced technology must be adapted also to prevailing forms of social and economic organization. Or conversely, forms of social and economic organization might be changed to exploit the potential benefits of an advanced technology. For example, data processing and other computer-based techniques which constitute a spectacular component in the technology of advanced societies are significant more or less specifically as a support to the planning and control functions of very large organizations. Where decision-taking is decentralized and operations are carried on by large numbers of discrete entities in competitive markets, even as in the agricultural or in the small business sector of the American economy, these elements of technology are not significant. Therefore the assimilation of these components of advanced technology presumes the existence or at least the desirability of large organizations. The capacity to control a multitude of activities under centralized political direction or through the quasi-voluntary association of great corporate enterprise itself requires particular cultural proclivities.

Creating a hospitable context for advanced technology

Nor is it only the adaptation of technology to a particular context of operations which is at issue. The context may be adapted to the available technology. Thus new and complex techniques using concrete as a material in residential construction have been developed for certain low-productivity economies, because the lack of skilled craftsmen precluded the technology of construction as it is conventionally practiced in high-productivity economies. Alternatively, carpenters and bricklayers could have been trained. New and complex techniques of aerial survey for mapping and for the inventorying of natural and social resources have been developed more or less specifically for the low-productivity economies. Alternatively, roads and internal transportation facilities could have been built which would have made it possible to use the survey techniques conventional in high-productivity economies. Not only can advanced technology be adapted to the environment of the low-productivity economy; the environment of the low-productivity economy can be adapted to facilitate the transference of advanced technologies. Certainly efforts are made thus to reshape the context of operations, particularly in the development of the infrastructure in the building of roads, harbors, airfields, and in making available transport facilities, in providing protection against contagion, violence, theft, fraud, flood, hurricane, in offering the services of banks, and equity markets, etc. Especially relevant to the capacity to adapt technology or to adapt to advanced technologies are those science-based components of the infrastructure which enable decision-makers to test, measure, quantify the variables related to economic choice, or precisely to determine and precisely to specify the performance characteristics and requirements of materials, products, structures, or more accurately to forecast the context or consequences of decision and policy – thereby raising the capacity to control and to innovate in any sphere of technology.

The capacity to assimilate and the motivation to innovate

The environment may be hospitable to the incorporation of an advanced technology. The advanced technology may be transferable or there may be available in the developing society all the skills required to adapt it for transference. Yet it may fail to be incorporated into operations. Nor are such failures to realize upon opportunities for technological progress confined to low-productivity economies. They occur everywhere and are part of the story of nearly every ultimately successful innovation. For transformation to an advanced (or to any new) technology, there must not only be the capacity to transform, which has been the concern of this paper, but also there must be a gearing-together of the competence to evaluate the technical feasibility and economic benefits of change with the power to effectuate change and with the motivation to transform an existing organization of operations, including sometimes a structure of privilege.

The structure of cognition

A society's capacity to adapt itself to the requisites of advanced technology and to adapt the advanced technology to its own circumstances and objectives, as well as its capacity to innovate, will depend in part on the intellectual skills, the acquired knowledge and know-how, the problem-solving competencies – in a word, on the cognitions possessed by those who constitute that society.

The required structure of cognition might be conceived as a pyramid. At its base, pervading the whole society, would be simply a sense of the machine, of its logic, of its manipulability, of its limits, of its demands upon its caretakers, of its values in extending the individual's power, i.e. a cognition of mechanism. Merging with and emerging out of this cognition of mechanism, carrying it into vocation and practice, would be the cognitions implicit in mechanical and technical skills. A mass cognition of mechanism at its base, a middle mass of mechanical and technical skills occupying the center part of the pyramid, gives any society its capacity to respond to the signals of technical leadership and to adapt to and to adjust an advanced technology around the whole periphery of its operations.

Merging with and emerging out of the middle mass of technical skills is another sort of cognition which comprehends the interrelation of machines, materials, labor, and information in processes of producing goods and services. This cognition of process is necessary to set in motion, to control, or further to transform an advanced technology.

And at the apex of the pyramid, interacting with the cognition of process, are the cognitions of science, of a development-oriented science, and a science-based development engineering. These are necessary in order to 'apply' research method and the world accumulations of scientific knowledge to problems impeding technical advance, or to evaluate the feasibilities of transferring science-based technology or to adapt science-based technologies for assimilation.

This structure of cognitions is not the only precondition of development. For example, an enlightened political leadership or an aggressive entrepreneurship, as traditionally conceived, may also be needed. But the wisest, most benign political leadership and the slickest, quickest, wheeling-dealingest entrepreneurship will not produce a stream of technical transformations unless they act within this structure of cognition.

R. Solo 487

In Western Europe and the United States, the cognition of mechanism and the skills of mechanics and technicians and the cognition of process can be acquired – and for the most part are being acquired through daily observation – through apprenticeships, through *ad hoc* training or through managerial experience on the job – all outside the system of formal education. Formal education, inasmuch as it has had a functional objective, has produced the scientific and technical elites at the apex. But in low-productivity societies which have not yet crossed the threshold of industrialization, the mass cognition of mechanism, the skills of a middle mass of mechanics and technicians, and the essential cognition of process cannot be acquired spontaneously, outside the system of formal education. Rather the gigantic task of inculcating them needs to be planned and programmed. This suggests that a quite different policy and system of formal education is needed in developing societies than is traditional in the West.

Further Reading

The reader who is interested in exploring earlier treatments of the subject of technological change by economists should consult Adam Smith's *Wealth of Nations*, Karl Marx's *Capital*, and two books by Joseph Schumpeter: *The Theory of Economic Development*, Harvard University Press, 1934, and *Capitalism, Socialism, and Democracy*, Harper & Row, 1942. Among the books written by economic historians the starting point should be Abbott P. Usher, *A History of Mechanical Inventions*, Harvard University Press, 1954. This is a work of rare scholarship. A more detailed presentation of Usher's conceptual framework than is presented in Part I of the present volume will be found in the first four chapters of that book. An excellent, sweeping survey of the role of technological change in the economic development of Western Europe over the past two hundred years may be found in David Landes, *The Unbound Prometheus*, Cambridge University Press, 1969. An earlier, shorter version of this book appeared as chapter 5 in M. M. Postan and H. J. Habakkuk (eds.), *The Cambridge Economic History of Europe*, vol. VI: *The Industrial Revolution and After*, Cambridge University Press, 1965. In this earlier version Landes included an invaluable bibliography of over sixty pages (see Part 2, pp. 943–1007). For a provocative and stimulating comparative analysis of Anglo-American technological developments, see H. J. Habakkuk, *American and British Technology in the Nineteenth Century*, Cambridge University Press, 1962. For the American side alone, see W. Paul Strassmann, *Risk and Technological Innovation: American Manufacturing Methods During the Nineteenth Century*, Cornell University Press, 1959.

Two highly influential books written by economists in recent years are W. E. G. Salter, *Productivity and Technical Change*, Cambridge: University Department of Applied Economics, 1960, and Jacob Schmookler, *Invention and Economic Growth*, Harvard University Press, 1966. In addition, *The Rate and Direction of Inventive Activity*, Princeton University Press, 1962, contains a wide range of materials by a number of economists, including several of those whose work has been most influential in this area. Some attempts have recently been made to summarize, synthesize and evaluate the rapidly expanding body of analytical and empirical material on the subject of technological change. The most highly recommended among these are Richard R. Nelson, Merton J. Peck and Edward D. Kalachek, *Technology, Economic Growth and Public Policy*, Brookings, 1967; Edwin Mansfield, *The Economics of Technological Change*, Norton, 1968, and Murray Brown, *On the Theory and Measurement of Technological Change*, Cambridge University Press, 1968.

Beyond the books mentioned there is now a huge literature, mostly in the professional economic journals. No purpose would be achieved by a comprehensive bibliography of the subject. The following list is presented with the dual purpose of citing some of the most significant works and of indicating the range of approaches which have been taken toward this subject.

W. Adams and J. Dirlam, 'Big steel, invention and innovation', *Quart. J. Econ.*, May 1966.

E. Ames, 'Research, invention, development and innovation', *Amer. Econ. Rev.*, June 1961.

E. Ames and N. Rosenberg, 'The Enfield arsenal in theory and history', *Econ. J.*, December 1968.

K. Arrow, 'The economic implications of learning by doing', *Rev. econ. Stud.*, June 1962.

H. Barnett, *Innovations: The Basis of Cultural Change*, McGraw-Hill, 1953.

C. F. Carter and B. R. Williams, *Industry and Technical Progress*, Oxford University Press, 1957.

R. Day, 'Technological change and the sharecropper', *Amer. econ. Rev.*, June 1967.

E. F. Denison, *The Sources of Economic Growth in the United States and the Alternatives Before Us*, Allen & Unwin, 1962.

E. Domar, 'On the measurement of technological change', *Econ. J.*, December 1961.

R. Eckaus, 'Notes on invention and innovation in less developed countries', *Amer. econ. Rev., Pap. Proc.*, May 1966.

W. Fellner, 'Profit maximization, utility maximization and the rate and direction of innovation', *Amer. econ. Rev., Pap. Proc.*, May 1966.

A. Fishlow, 'Productivity and technological change in the railroad sector, 1840–1910', in D. Brady (ed.), *Output, Employment and Productivity in the United States after 1800*, Studies in Income and Wealth, no. 30, New York, 1966.

Z. Griliches, 'Hybrid corn: an exploration in the economics of technological change', *Economet.*, October 1957.

Z. Griliches and J. Schmookler, 'Inventing and maximizing', *Amer. econ. Rev.*, September 1963.

T. Hagerstrand, 'Quantitative techniques for analysis of the spread of information and technology', ch. 12 in C. A. Anderson and M. J. Bowman (eds.), *Education and Economic Development*, Aldine, 1965.

D. Hamberg, 'Size of firm, oligopoly and research: the evidence', *Canad. J. Econ. polit. Sci.*, February 1964.

H. Jerome, *Mechanization in industry*, National Bureau of Economic Research, 1934.

J. Jewkes, D. Sawers and R. Stillerman, *The Sources of Invention*, St Martin's Press, 1958.

D. Jorgenson, 'The embodiment hypothesis', *J. polit. Econ.*, February 1966.

D. Jorgenson and Z. Griliches, 'The explanation of productivity change', *Review of Economic Studies*, July 1967; reprinted in *Surv. curr. Bus.*, May, 1969, together with E. Denison, 'Some major issues in productivity analysis: an examination of estimates by Jorgenson and Griliches'.

J. Kendrick, *Productivity Trends in the United States*, Princeton University Press, 1961.

C. Kennedy, 'Induced bias in innovation and the theory of distribution', *Econ. J.*, September 1964.

B. Klein, 'A radical proposal for R and D', *Fortune*, May 1958.

K. Lancaster, 'Change and innovation in the technology of consumption' *Amer. econ. Rev.*, *Pap. Proc.*, May 1966.

F. Machlup, *An Economic Review of the Patent System*, study of the Subcommittee on Patents, Trademarks and Copyrights of the Senate Committee on the Judiciary, 85th Congr., 2nd sess., Washington; USGPO, 1958.

F. Machlup, *The Production and Distribution of Knowledge in the United States*, Princeton University Press, 1962.

G. S. Maddala and P. T. Knight, 'International diffusion of technical change – a case study of the oxygen steel-making process', *Econ. J.*, September 1967.

E. Mansfield, 'Entry, Gibrat's law, innovation and the growth of firms', *Amer. econ. Rev.*, December 1962.

E. Mansfield, 'Intrafirm rates of diffusion of an innovation', *Rev. Econ. Stat.*, November 1963.

E. Mansfield, 'The speed of response of firms to new techniques', *Quart. J. Econ.*, May 1963.

E. Mansfield, *Industrial Research and Technological Innovation*, Norton, 1968.

B. Massell, 'A disaggregated view of technical change', *J. polit. Econ.*, December 1961.

R. R. Nelson, 'The economics of invention: a survey of the literature' *J. Bus.*, April 1959.

R. R. Nelson, 'A "diffusion" model of international productivity differences in manufacturing industry', *Amer. econ. Rev.*, December 1968.

R. R. Nelson and E. S. Phelps, 'Investment in humans, technological diffusion and economic growth', *Amer. econ. Rev.*, *Pap. Proc.*, May 1966.

G. F. Ray, 'The diffusion of new technology: a study of ten processes in nine industries', *Nat. Inst. Econ. Rev.*, May 1969.

Report of the National Commission on Technology, Automation and Economic Progress, *Technology and the American Economy*, vol. 1, Washington, USGPO, 1966.

E. M. Rogers, *The Diffusion of Innovations*, Free Press, 1962.

N. Rosenberg, 'Technological change in the machine-tool industry, 1840–1910', *J. econ. Hist.*, December 1963.

N. Rosenberg, 'The direction of technological change: inducement mechanisms and focusing devices', *Econ. Devel. cult. Change*, October 1969.

V. Ruttan, 'Research on the economics of technological change in American agriculture', *J. farm Econ.*, November 1960.

P. A. Samuelson, 'A theory of induced innovation along Kennedy–Weisäcker lines', *Rev. Econ. Stat.*, November 1965.

F. H. Scherer, 'Firm size, market structure, opportunity and the output of patented inventions', *Amer. econ. Rev.*, December 1965.

J. Schmookler, 'The changing efficiency of the American economy, 1869–1938', *Rev. Econ. Stat.*, August 1952.

J. Schmookler, 'Bigness, fewness and research', *J. polit. Econ.*, December 1959.

J. Schmookler, 'Technological progress and the modern American corporation', in E. S. Mason (ed.), *The Corporation in Modern Society*, Harvard University Press, 1960.

R. A. Solow, 'Investment and technical progress', in K. Arrow, S. Karlin and P. Suppes (eds.), *Mathematical Methods in the Social Sciences, 1959*, Stanford University Press, 1960.

R. A. Solow, 'Technical progress, capital formation and economic growth', *Amer. econ. Rev.*, *Pap. Proc.*, May 1962.

A. B. Stafford, 'Is the rate of invention declining?' *Amer. J. Soc.*, May 1952.

G. Stigler, 'The division of labor is limited by the extent of the market', *J. polit. Econ.*, June 1951.

G. Stigler, 'The economics of information', *J. polit. Econ.*, June 1961.

W. P. Strassmann, 'Creative destruction and partial obsolescence in American economic development', *J. econ. Hist.*, September 1959.

W. P. Strassmann, *Technological Change and Economic Development*, Cornell University Press, Cornell, 1968.

P. Temin, 'Labor scarcity and the problem of American industrial efficiency', *J. econ. Hist.*, September 1966.

H. Villard, 'Competition, oligopoly and research', *J. polit. Econ.*, December 1958.

O. E. Williamson, 'Innovation and market structure', *J. polit. Econ.*, February 1965.

Acknowledgements

Permission to reproduce the Readings published in this volume is acknowledged from the following sources:

1 Royal Economic Society and William Fellner
2 National Bureau of Economic Research, Inc.
3 *Quarterly Journal of Economics*, Harvard University Press and Vernon W. Ruttan
4 *Economica* and M. Blaug
5 Economic History Association
6 University of Chicago Press and William Parker
7 University of Chicago Press and Richard Nelson
8 Princeton University Press
9 University of Chicago Press and Zvi Griliches
10 Royal Economic Society
11 American Association for the Advancement of Science and Zvi Griliches
12 John Wiley and Sons, Inc., Paul David and Henry Rosovsky
13 American Economic Association Review and Peter Temin
14 Econometric Society
15 *American Economic Association Review* and Moses Abramovitz
16 *Review of Economics and Statistics*
17 University of Chicago Press and Edward Denison
18 University of Chicago Press and Zvi Griliches
19 Royal Economic Society and Edward Ames
20 Harvard University Press and Raymond Vernon
21 The Manchester School and Robert Baldwin
22 American Economic Association and Robert A. Solo

Author Index

Dunning, J. H., 448

Eavenson, H. N., 280
Edwards, E. E., 256
Elterlich, J. G., 392
Ewell, R. H., 197

Fabricant, S., 354, 383
Falconer, J. I., 236, 238, 240, 243, 250, 261, 263, 271
Fay, C. R., 466
Fellner, W. J., 76, 91, 94, 95, 101, 102, 106, 108, 116, 144, 203, 204, 344, 347, 354, 357
Fishlow, A., 281
Fitch, C. H., 421
Fite, E. D., 237
Fogel, R. W., 276, 281
Foote, R. J., 187
Forbes, R. J., 144
Fox, H. G., 44, 45
Fox, K. A., 395
Francis, C., 285
Frankel, M., 413, 425, 427, 429
Freedman, P., 53
Freeman, C., 444, 446
Freeman, J. D., 81
Fuchs, V. R., 455

Gates, P. W., 236, 239, 246, 248, 261, 268
Gerschenkron, A., 229, 230, 419, 423, 428, 430
Gilfillan, S. C., 78, 117
Goldsmith, R. W., 327, 348, 349, 353, 357
Goode, R. B., 420
Gordon, D., 413, 428, 429
Gordon, M., 297
Greaves, I. C., 469
Green, H. A. J., 101, 204
Griliches, Z., 116, 180, 182, 209, 211, 212, 220, 291, 300, 318,

382, 384, 387, 388, 393, 396, 398, 399, 402
Gross, N., 225
Grosse, R., 462, 476
Grotewold, A., 224

Haavelmo, T., 461
Habakkuk, H. J., 97, 240, 243, 274, 275, 414, 417, 421, 425, 430, 432
Hague, D. C., 101
Haley, B. F., 91
Hall, M., 445
Halmos, P. R., 81
Hamberg, D., 102
Handfield-Jones, S. J., 339, 428
Harrel, C. G., 133
Harrod, R., 94, 95, 100, 101, 204
Heady, E. O., 392
Healy, K., 312
Heflebower, R. B., 106
Hennipman, P., 108
Hensley, C., 305, 307, 308
Hicks, J. R., 88, 89, 90, 91, 92, 93, 94, 95, 97, 98, 102, 105
Hirsch, N. D. M., 45
Hirsch, S., 441, 446, 451
Hirschman, A., 229, 420, 444
Hitch, C. J., 165
Hoch, I., 393
Hoffman, W., 338
Hoffmeyer, E., 441
Hogan, W. P., 103, 353
Houthakker, H. S., 358, 400
Hufbauer, G. C., 444, 445
Hunter, L. C., 275, 276, 277, 278, 281
Hutchinson, W. T., 235, 238, 240, 244, 246, 250, 252, 260, 262, 264, 266, 267, 268

Ichimura, S., 456

Subject Index